The Action Plant

THE ACTION PLANT

Movement and nervous behaviour in plants

PAUL SIMONS

BLACKWELL
Oxford UK & Cambridge USA

Copyright © Paul Simons 1992

The right of Paul Simons to be identified as author
of this work has been asserted in accordance with the
Copyright, Designs and Patents Act 1988.

First published 1992

Blackwell Publishers
108 Cowley Road, Oxford, OX4 1JF, UK

3 Cambridge Center
Cambridge, Massachusetts 02142, USA

All rights reserved. Except for the quotation of short passages for the purposes of criticism and review, no part of this publication may be reproduced, stored in a retrieval system, or transmitted, in any form or by any means, electronic, mechanical, photocopying, recording or otherwise, without the prior permission of the publisher.

Except in the United States of America, this book is sold subject to the condition that it shall not, by way of trade or otherwise, be lent, re-sold, hired out, or otherwise circulated without the publisher's prior consent in any form of binding or cover other than that in which it is published and without a similar condition including this condition being imposed on the subsequent purchaser.

British Library Cataloguing in Publication Data

A CIP catalogue record for this book is available from the British Library.

Library of Congress Cataloging-in-Publication Data
Simons, Paul.
The action plant: movement and nervous behaviour in plants / Paul Simons.
p. cm.
Includes bibliographical references and index.
(acid-free paper)
1. Plants—Irritability and movements. I. Title.
QK771.S56 1992
581.1′8—dc20   91–35225   CIP
ISBN 0–631–13899–4

Typeset in 10 on 12 pt Sabon
by Hope Services (Abingdon) Ltd.
Printed in Great Britain by
T.J. Press, Padstow.

This book is printed on acid-free paper

Contents

Preface — viii
Acknowledgements — xi

1 THE SENSITIVE WORLD OF NERVOUS PLANTS — 1
   Classification — 9
   The daisy chain — 10
   Evolution — 11
   Chapter by chapter summary — 12
   The history — 13

2 EXPLODING PLANTS — 15
   Fungal guns — 15
   Violence in the lower plants — 21
   Motorized seeds — 21
   Exploding flowers — 27

3 FLOWER POWER — 34
   Touchy stamens — 35
   Stretchy stamens — 42
   How do moving stamens work? — 43
   Trigger plants — 45
   Sensitive females — 49
   Moving petals — 52

4 HUNTING AND KILLING — 61
   Sex in the lower plants — 62
   The marauding slime — 67
   Sense of taste — 68
   Lassooing fungi — 70

CONTENTS

5	**BLOODY PLANTS**	72
	The most wonderful plant – Venus's flytrap	75
6	**ELECTRIC SELF-DEFENCE**	93
	How *Mimosa* works	96
	Electrical signals	98
	The nerve of *Mimosa*?	100
	Movement	102
7	**SEEING THE LIGHT**	107
	A sense of vision	107
	Swimming under the sun	110
	Green motors	117
8	**SUNBATHING, SLEEPING AND RHYTHM**	121
	Kissing mouths	121
	Keeping track of the sun	123
	Solar-heated flowers	128
	Perception	128
	Sleepy plants	129
	Autonomy in performances	133
	How leaves move	135
	Evolution of leaf movements	145
9	**PLANT MUSCLES**	148
	Power in the cell	148
	Amoeboid muscles	156
	Plant cytoplasm	159
	Chloroplast choreography	161
	Dancing the cell division jive	163
	The chromosome motors	165
	The conductor of the chromosome orchestra	168
	Muscle protein again	169
10	**EXCITABLE CHEMISTRY**	173
	Anaesthetics: waking plants up!	173
	Plant aspirin	177
	Nerve transmitters in plants	182
	Opening and closing gates	184
	Plant and animal hormones	188
	Animals turned on by plants on steroids	191
11	**GOOD BEHAVIOUR**	199
	Plants in training	201
	Plant and animal behaviour	202

CONTENTS

12 THE ORDINARY PLANT — 205
- Touchiness in fungi — 209
- Wounding plants — 213
- Temperature stress — 215
- Touchy sex — 216
- Touchiness in climbing plants — 219
- Touchy plants: freak or fact? — 221
- The evolution of the plant motor — 221

13 THE EVOLUTION OF THE NERVOUS PLANT — 226
- Nervous plants? — 233
- The future of plant electrophysiology — 234
- Academic work — 236

14 EXPERIMENTS — 238
- The *Mimosa pudica* water pump — 238
- *Mimosa pudica* behaviour — 238
- *Mimosa pudica* and anaesthetics — 239
- *Mimosa pudica* and wounding — 239
- Touch-sensitive stamen movements — 240
- Venus's flytrap — 240
- Sundew — 241
- How to record electrical activity from plants — 242
- Apparatus for measuring leaf movements — 244
- Aspirin — 246
- Light-sensitive flowers — 246
- Rhythmic autonomic movements of leaves — 246
- Motorized seeds — 247
- Experiments needing a microscope — 247

How to grow some mobile plants — 253

Glossary — 268

Specialist carnivorous plant nurseries and societies — 271

Mobile plant statistics — 274

Taxonomy of mobile plants — 279

References — 285

Further reading — 308

Addendum to proofs — 309

Index — 311

PREFACE

'Nervous' is a strong word to describe plants, but this is a book about how plants can get excited like animals and move. What makes this such a bizarre story is that, even though plants have no nerves, they have nerve-like behaviour. They use this facility to control a wonderful collection of movements, of all kinds, at speeds sometimes so fast that they cannot be seen with the naked eye, in alpine to tropical plants from microscopic primitives to exotic orchids. The movements alone will astonish and amuse you. But for scientific intrigue they upset many an applecart in the world of botany. For they show how uncannily similar plants and animals can be – the way they both use their senses of sight, touch or taste to switch on movements using electrical signals. Even some of their movements are similar. And intriguing questions are raised about exactly what the differences between plants and animals really are. Ironically, a book that starts out looking at rather quirky plants ends up tracing the origins of our own nervous and muscular system.

I also make no apology for constantly talking about the electrical behaviour of plants. Too often the chemistry of plants has been waved like a magic wand to explain away all manner of plant phenomena, and I feel that it's about time that electrophysiology was given a decent airing for once. For this is an important science with a long history which still receives scant attention.

The botanical side will fascinate anyone even remotely interested in plants – gardeners, naturalists, amateur scientists, secondary school children – and with these readers in mind I have included chapters describing how to grow some examples of performing plants and simple experiments that can be done on them at home or in a modest school laboratory. However, for those with little background in biology, I strongly recommend referring to the glossary of names, especially for plant anatomy. Some of the details are complex, though, particularly as the book progresses, but if you read it in sequence you'll get a flavour of how plant performances – and also animal neuromotors – probably evolved.

This is also an invaluable book for researchers and lecturers interested in the deeper biological implications of plant movements. For these readers I have included important literature references, and some more advanced experiments. Also, do not be put off from using this as a reference book – tables of data and an index of plant names have been included with this in mind.

The one area I have tried to avoid – with a few exceptions – is growth movements. These would have taken the book down a rather different road, into the growth and development of plants, and away from the more rapid plant performances. For an excellent review of plant growth movements I highly recommend James Hart's recently published book *Plant Tropisms and Other Growth Movements*. Otherwise, I have tried to show the immense variety of plant performances thorughout the plant kingdom. And many of them are probably going on in your own home or garden.

Throughout this book there is one scientist whose name appears repeatedly, in almost every chapter. Charles Darwin was the greatest observer ever of mobile plants, and the first to realize that plants and animals behave similarly in so many respects. If all his ideas on plant sensitivity and movement had been accepted as his theory of natural selection was, then we'd be looking at plants as excitable creatures instead of the 'vegetables' we're taught of in school. Hopefully this book sets the record straight.

Acknowledgements

Apart from the many editors at Blackwell Publishers who have been involved in this book at some stage or other, I would particularly like to thank Dr Bernard Dixon who originally commissioned the idea and gave splendid support thereafter.

I

THE SENSITIVE WORLD OF NERVOUS PLANTS

There are living beings in your garden that are not animals, yet they move. They have no proper muscles, yet the contents of their cells churn around using muscle proteins. They have no nerves, yet they have nerve-like electrical signals. They have no brains, but they have senses of taste, vision, touch, gravity, temperature, humidity, pressure, electricity and sometimes magnetism. Some of them even catch and eat meat by their own movements. These organisms are called plants.

'Nervous plants', epitomized by the Sensitive Plant (*Mimosa*) or the Venus's flytrap, behave like animals. Compared with 'ordinary' plants, their tactile leaf movements seem to be bizarre, and probably the closest that the real world has ever come to a triffid. That's why these plants have been castigated throughout history as the oddballs of the plant world – nice to play with, but not worth taking seriously. Yet in this book I am going to show you a vast world of plants which are largely glossed over or forgotten in modern textbooks. From flowers to fungi, to leaves and algae, the 'vegetable' world is alive with sensitivity and movement:

stamens sensitive to touch bend over and coat insect pollinators with pollen;
carnivorous fungi garrotte living worms with an inflatable noose;
the leaves of *Biophytum sensitivum* are so sensitive that they fold up just *before* an insect lands on them;
the single-celled *Chlamydomonas* swims guided by an 'eye' containing a light sensor identical with that in our own retinas;
a waterflea brushing past the carnivorous bladderwort gets sucked into the trap so fast it can barely be seen by the fastest camera – the complete action takes a hundredth of a second.

But why should any plant need to move? After all, plants usually stay rooted to one spot, where they make their own food from sunlight, water and air by the ancient alchemy of photosynthesis. Animals, on the other hand, cannot make their own food, and so they have to move around to find it. That at least is the picture that is presented in most textbooks. As far as most scientists and students are concerned, the animal-like movements of the Sensitive Plant *Mimosa pudica* or the Venus's flytrap are simply freaks of nature. In fact, an enormous variety of plants have special movements for a variety of survival strategies:

catching animals for food,
defending themselves from animals,
enhancing their prospects for cross-pollination,
scattering fruits or spores to new homes,
keeping their photosynthesis mechanism well supplied with sunlight, water and air,
draining rainwater from leaves

and so on. But there's much more than can be seen by the naked eye. The contents of plant cells (figure 1.1) dance in tune with the rest of the plant and the outside world. Apart from the constant churning of the cytoplasm itself, their specialized bodies (so-called organelles) also move:

chloroplasts (which play an important role in photosynthesis) slip, slide, bend or shuffle around for the best light;
when the cell divides into two identical copies, the nucleus and its chromosomes dance with such exquisite precision that exactly the right bits of DNA (deoxyribonucleic acid) are packaged into each new cell;
vesicles carry packets of cell components and substances to wherever they are needed.

In fact you could think of the inside of the cell as a factory, where each organelle is a machine which needs constant supplies of nourishment and energy, the right working conditions and the ability to make replicas of itself. Movement is just one of the many ways in which this can be achieved.

Many of the motors driving the movements are unique to plants. Flower and leaf movements often involve the expulsion of water from special motor cells – you won't see hydraulic power of that type in animals (table 1.1). Yet there are other plant movements, particularly microscopic movements, that are so close to those in animals that I have no hesitation in drawing similarities between them.

The sensitivity of plants also shows many parallels with the animal world – being able to 'taste' the presence of a nearby animal, or 'feel' a mechanical blow, or even react to electric shock are senses not normally considered plant-like (a list of senses that trigger plant movements is given in table 1.2,

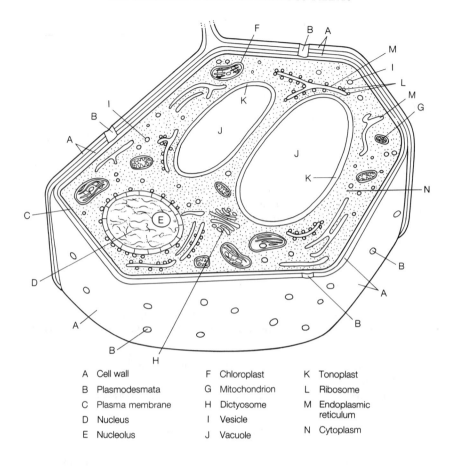

A	Cell wall	F	Chloroplast	K	Tonoplast
B	Plasmodesmata	G	Mitochondrion	L	Ribosome
C	Plasma membrane	H	Dictyosome	M	Endoplasmic reticulum
D	Nucleus	I	Vesicle		
E	Nucleolus	J	Vacuole	N	Cytoplasm

FIGURE 1.1 Contents of the typical higher plant cell. The cell wall A and plasma membrane C envelop the cell except for small channels, plasmodesmata B, linking one cell with its neighbour. The cell contents (the protoplast) consist of cytoplasm (stippled area), cell bodies F (organelles) and vacuoles J (used for storing chemicals). The nucleus D holds most of the cell's DNA, responsible for designing proteins. The nucleolus E inside the nucleus organizes much activity of the nucleus. The organelles carry out specific tasks: ribosomes L manufacture proteins from amino acids; endoplasmic reticulum M helps make and transport proteins; chloroplasts F carry out photosynthesis; mitochondria G carry out respiration; dictyosomes H help process chemicals and package them into vesicles I.

a list of signals that control movement in the living world is given in table 1.3, and a comparison of animal and plant sensitivities is given in 'Mobile plant statistics', pp. 275–7). For this reason, I have borrowed many useful terms from zoology, including the following.

TABLE 1.1 *Motors driving movement in the living world*

Movements inside cells driven by actin/myosin
  Streaming of cytoplasm, driving locomotion
    Streaming backwards and forwards – animal and fungal amoebae
  Streaming of cytoplasm, not leading to locomotion
    Rotation – plant and fungal cells
    Streaming backwards and forwards – animal and fungal amoebae

Movement of organelles
  Nuclei migration in plants, fungi, animals – probably driven by actin/myosin filaments, but not certain
  Chloroplast migration in response to light
  Vacuoles in plants, fungi, animals – probably driven by actin/myosin filaments
  Separation of daughter animal and yeast cells after cell division

Movements inside cells driven by unknown motor
  'Independent' particle movements (role of ground cytoplasm not established)
  Rotation of nuclei – plants, fungi, animals
  Rotation of chloroplasts
  Amoeboid movement of chloroplasts
  Amoeboid movement of mitochondria – plants, fungi, animals
  Amoeboid movement of proplastids (precursor organelle to mitochondria and chloroplasts)

Cellular and subcellular movements driven by actin/myosin
  Gliding of diatoms
  Amoeboid movements – animal and fungi amoebae, fungal plasmodia, mammalian white blood cells

Cellular and subcellular movements driven by kinesin/dynein
  Cilia/flagella
    Whip action of flagella for swimming, e.g. plant, fungi, animal sperm
    Gliding by stationary flagella on solid substratum, e.g. *Chlamydomonas*
    Oar action of cilia for swimming, e.g. some plant and fungi spores and gametes, *Paramecium*, clam gill cells, mammal tracheary tract
  Cell division movements
    Movements of chromosomes away from equator of spindle (possibly in conjunction with tubulin – the 'building blocks' of microtubules) in plant, fungi, animal cells
  Bacterial flagella – different structure and motor from the above cilia and flagella

Subcellular movements driven by tubulin
  Movements of chromosomes during cell division (possibly in conjunction with kinesin/dynein) in plant, fungi, animal cells

Tissue movements driven by turgor (expulsion of fluid) from motor cells
  Touch sensitive, reversible
    Stamen bending, e.g. *Sparmannia africana*
    Stamen tube contraction, e.g. *Centaurea*

Stigma closing, e.g. *Mimulus*
Style bending, e.g. *Arctotis*
Fused style/stamen (trigger) bending, e.g. *Stylidium* – possibly assisted by tension of trigger
Petal bending, e.g. *Pterostylis*, probably turgor driven, but not certain
Leaf movements, e.g. *Biophytum sensitivum*
Carnivorous plant traps, e.g. *Drosera* – probably turgor driven, but not certain
Carnivorous fungi trap, e.g. *Dactylella* – probably turgor driven, but not certain

Not touch sensitive, but reversible
Leaf 'sleep', e.g. *Albizzia*
Leaf 'sun-tracking', e.g. *Dryas octopetala* – probably turgor driven, but not certain
Stomata

Touch-sensitive growth movements
Carnivorous plant trap, e.g. *Dionaea muscipula* – probably assisted by turgor in motor cells
Tendrils, probably assisted by turgor in motor cells
Pollen tubes
Root tips
Twining stems of climbing plants, e.g. *Clematis*
Fungal hypae, e.g. bean rust fungus

Tissue movements driven by actin/myosin
Possibly *Mimosa pudica* leaf movement, assisted by turgor changes
Animal muscles

TABLE 1.2 Stimuli that trigger rapid movements in plants

Touch
Free-swimming single cells, e.g. *Paramecium, Euglena, Noctiluca*
Pollen bags, e.g. *Catasetum*
Stamen bending, e.g. *Sparmannia africana*
Stamen tube contraction, e.g. *Centaurea*
Stigma closing, e.g. *Mimulus*
Style bending, e.g. *Arctotis*
Fused style/stamen (trigger) bending, e.g. *Stylidium*
Petal bending, e.g. *Pterostylis*
Leaf movements, e.g. *Biophytum sensitivum*
Carnivorous plant traps, e.g. *Drosera*
Carnivorous fungi traps, e.g. *Dactylella*
Parasitic fungi traps, e.g. *Haptoglossa*
Fruits, e.g. squirting cucumber (*Ecballium elaterium*)
Seeds, e.g. *Geranium*

TABLE 1.2 *cont.*

Electric shock (direct current)
All touch-sensitive, reversible plant and fungal movements

Electrostatic field
Fungal spore dispersal, e.g. mushroom spores
Leaf movements, e.g. *Biophytum sensitivum*
(Probably many more touch-sensitive reversible plant movements respond to this type of stimulus, although this has not been tested)

Light
All free-living cells
Blue-green waving movements
Chloroplast movements
All leaf movements
Stomata
Flowers

Specific chemicals
Carnivorous plant trap, e.g. *Dionaea muscipula*
Carnivorous fungi trap, e.g. noose of *Dactylella*
Parasitic fungi spores, e.g. *Polymyxa betae*
(The above all respond to chemicals from their animal prey)
Stomata
All free-swimming single cells; these respond to their food, harmful substances and sex pheromones
Amoebae (animal and slime mould) respond to food, harmful substances, sex attractants, 'aggregation' attractants
Pollen tubes – respond to female attractants
All touch-sensitive reversible movements respond to salty solutions

Temperature
All free-living cells
Flower opening/closure, e.g. tulip
Leaf movements, e.g. *Mimosa pudica*
Stomata

Inherent, as yet unknown, rhythmic stimuli from within the organism
All leaf movements
Flower movements
Stomata
Cell organelle movements

Wounds
Touch-sensitive leaf movements
Stomata

Humidity
Spore dispersal, eg. *Dryopteris* fern

Humidity *cont.*
   Seed dispersal, e.g. gorse, *Geranium*
   Stomata, e.g. *Tillandsia*
   Leaf rolling, e.g. marram grass
   Anther sac movement, e.g. *Plantago*
   Leaf bending, e.g. cassava

Carbon dioxide
   Stomata

Atmospheric pressure
   Leaf movements, e.g. *Mimosa pudica*
   Possibly many other movements

TABLE 1.3 Signals that control movement in the living world

Cohesive forces caused by dehydration
   Spore dispersal, e.g. *Dryopteris* fern
   Seed dispersal, e.g. gorse, *Geranium*

Shrinkage caused by dehydration
   Stomata movement of e.g. *Tillandsia*
   Leaf rolling, e.g. marram grass

Turgor forces caused by fluid pressure
   Spore dispersal, possibly mushroom spores
   Fruit dispersal, e.g. squirting cucumber (*Ecballium elaterium*)

Stress caused by growing tissues kept under tension
   Stamen dispersal, artillery plant (*Pilea microphylla*), stinging nettle
   Spore discharge, possibly parasitic fungus *Haptoglossa*
   Carnivorous plant trap, possibly bladderwort (although an electric signal may be involved)

Hormones
   Touch-sensitive growth, e.g. tendrils (although an electric signal may also be involved)
   Reversible 'sleep' movements, e.g. leaf movements of *Albizzia* (although electric signals may also be involved)
   Stomata movements (in response to certain stimuli)

Electric signals
   All animal movements
   All touch-sensitive reversible plant movements (as far as is known)
   Stomata (possibly)

A sense of vision is 'The reception by means of light-sensitive tissue of information conveyed by light rays' (*Encyclopaedia Britannica*, 15th edn, vol. X, p. 461, 1982).

'Taste is the detection and identification by the sensory system of dissolved chemicals in contact with the organism' (*Encyclopaedia Britannica*, 15th edn, vol. XI, p. 837, 1982).

Touch is the ability to perceive mechanical stimulation of the organism.

Sometimes plant sensitivity surpasses human perception. The tickle of a single woollen fibre across the tendril of *Sicyos* makes it instantly start coiling up, yet we humans can't feel this stimulus at all on our skin and the touch sensitivity of many other plants surpasses that of animals (see 'Mobile plant statistics', pp. 274–6). There is something else even more animal-like about plant movements – in the reversible touchy movements detailed in chapter 3 and subsequent chapters. They are controlled by electrical signals, even though plants don't have nerves. Yet in all other respects these impulses behave exactly like animal nerve signals – they carry messages from sensors to motors, telling the plant to take action. This behaviour in itself is a challenge to orthodox botanical thinking, obsessed as it is with the belief that plants only speak a chemical language, usually involving hormones. So it is nothing short of a revelation to discover that *all* plants – both mobile and immobile – carry electrical signals, although for what purpose is not always clear. This is a truly alarming gap in our knowledge, and just one of the reasons why the study of plant movements is so revealing about the vegetable world in general. Many readers may feel that I emphasize electrical signals too much, but at this stage in our knowledge it is a good idea to give the imagination free rein. Fortunately, we now have some new tools – microscopy, chemical analysis, electrodes and chemical markers – which should give a much better picture of just how important electrical signals are to plants, and we can only hope that scientists rise to these challenges.

I have used zoological terms to describe the excitability of plants, with emotive words like 'excitation'. In fact, it is not too extreme to delve into psychology dictionaries to find ways of describing plant behaviour, such as 'memory', and learning processes, such as 'habituation'. But I must add a note of caution here. Plant sensitivity and movements are nothing like as sophisticated as those in the higher animals – amphibians, reptiles, fish, birds and mammals. They cannot even be compared with the complexity of lower animals such as insects, shellfish or worms. We are talking here of something akin to the simplest forms of animal life – jellyfish, sea anemones, *Hydra*, sponges and the single-celled protists. There can be no question of 'conscious' decisions and other thought processes.

Classification

Questions can also be raised about the artificial classifications that we apply to the living world, and what exactly a plant is. No two textbooks seem to give the same classification of the living world, but amongst this taxonomic mess most biologists seem agreed that life can be divided into five kingdoms (table 1.4).

TABLE 1.4 Classification of the living world

Prokaryotes

Monists
  Viruses, bacteria, blue-green algae; their nuclei are not contained in a special nuclear membrane

Eukaryotes

Protists
  Single-celled organisms, embracing single-celled algae and their close relatives, and single-celled animals such as amoebae

Fungi
  Feed from rotting or living matter, instead of photosynthesizing, as well as reproducing using spores formed by sexual or non-sexual means

Plants
  Many-celled organisms which photosynthesize, reproducing by spores or seed

Animals
  Many-celled organisms that feed by breaking down living material; they cannot photosynthesize, contain no cellulose and reproduce without spores

The descriptions in table 1.4 are very simple, but even the most detailed definitions are still not clear cut. The truth is that taxonomists have grave problems sorting living things into neat categories – nature did not evolve to suit our convenience. Animals, plants, fungi, protists and prokaryotes are only the labels that we give them, and, as I will show, mobility makes a mockery of many of these artificial boundaries. To give you a good example consider the fungi known as true slime moulds. At one stage in their life history they form large composite amoeboid (variable shape) sheets that can move about and ingest food – the characteristics of an animal. Yet when they arrive at their reproductive stage, they produce spore-bearing bodies remarkably like those of fungi. Therefore different scientists call them fungi or protists or animals. I have taken the liberty of blurring these distinctions, and treat fungi as one kingdom in the vegetable world, with blue-green algae, photosynthetic protists and 'true' plants making up the rest of the vegetable world.

There is one last group of definitions that needs to be explained here. In this book, the word 'move' refers to *independent* movement: 'to cause to change place or posture; to set in motion; to impel; to excite to action' (*Chamber's Twentieth Century Dictionary*, 1970). So the movement of leaves in the wind certainly doesn't count. Movements can be divided into four broad classes.

Movement inside a cell – e.g. chromosomes moving during the division of a cell, chloroplasts moving out of strong light, cytoplasm streaming round in a cell.

Independent movement of a whole cell (also known as taxis, pronounced 'tack-sis') – e.g. sliding amoebae, gliding blue-green algae filaments, single-celled spores moving by beating minute hairs, slithering slime moulds.

Movement of a whole organ – e.g. folding-up of the Venus's flytrap, collapse of the *Mimosa* leaf, ejection of the fruit of a squirting cucumber, flexing of a whole muscle.

Movement of a whole multicellular organism *en bloc* – e.g. swimming of a *Volvox* algae colony.

This loose classification is only a guide – sometimes one movement involves two of the above categories. For instance, the streaming of cytoplasm inside a slime mould cell also propels the *whole* organism along. A more detailed breakdown of movements throughout the living world is given in table 1.1.

There is another important distinction in plant movements. Some are irreversible – they can only be performed once, e.g. most of the seed and fruit movements described in the next chapter. You could argue that slow growth movements like phototropism should also be included, but that would have taken us down the slightly different path of plant growth and development. Instead, the rest of the book tends more towards reversible actions – those that can be repeated, often countless numbers of times.

The Daisy Chain

So much for the variety of movements – how do they actually work? To answer this you have to strip each movement down into its bare components, all of them linked together like a daisy chain to make a sequence (figure 1.2).

1 The plant receives a stimulus. This can be a cue from the outside world – light, gravity, temperature (see table 1.2) – or a signal generated from inside the plant – hormones, ions, liquid pressure etc. If the stimulus

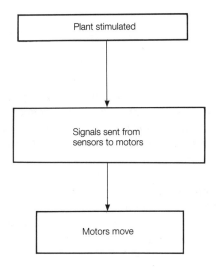

FIGURE 1.2 The events leading to plant movement.

comes from outside, the plant has to extract information from it, e.g. the direction that rays of light are coming from, or the strength or quality of that light. This information is processed, often in a special sensor which becomes excited. The complete process is called perception.

2 The excitation is relayed as a chemical or electrical signal to the motor cell(s) that actually perform the movement (a list of signals in the animal and vegetable worlds is given in 'Mobile plant statistics', p. 278).

3 The motor cell(s) execute the action.

Think of it as a battlefield. The soldiers at the battlefront see that they are under attack (perception), and so they radio to the rear for help (signalling the excitation) and fresh guns are delivered (executing the action).

Evolution

So how did plants evolve their nerve-like behaviour? The animal and plant worlds originally came from a common ancestor: primeval free-living single cells, such as bacteria, which are part of the prokaryote kingdom. These hold the key to the puzzle because, although such primitive organisms are microscopic, many drive themselves to the best feeding places using elegant propellers, guided by sophisticated sensors, nerve-like electrical signals and chemical messengers almost identical with many of our own hormones. And when they need sex, these free-spirited independent cells 'talk' to one another by oozing special chemical signals into the surroundings.

When these prokaryotic cells evolved into larger and more sophisticated single cells called eukaryotes, they brought with them all their sensitivity and mobility. The story of their evolution has been one of the greatest biological controversies of all time. Decades after it was first suggested, we are now fairly certain that the organelles in plant, fungi and animal cells were once free-living prokaryotes that became incorporated into eukaryote cells. Yet the organelles still retain some of their independent primeval past: the chloroplasts and mitochondria still have some of their own individual DNA; many organelles even look like bacteria; and, more importantly from this book's point of view, some of them have kept much of their ancient mobility.

Gradually, the single-celled eukaryotes turned a milestone in evolution – they joined up and lived together as communities of cells. But this feat raised considerable challenges in communication, because one cell had to 'talk' to its neighbours to co-ordinate movement, feeding, sex and all the other features of an organism's lifestyle. The problems were overcome with the equipment already there – cell talked to cell using the electrical and chemical language they had inherited from their free-living ancestors. In time, various cells in these colonies became specialized for particular jobs. Some lost their chlorophyll and became roots for anchoring the plant and absorbing water and nutrients, whilst others kept their chlorophyll and became shoots. These were the organisms that we recognize as plants. Meanwhile, other communities of cells developed their powers of feeding off other organisms, and evolved even greater powers of movement and co-ordination – to become what are now known as animals.

But even though plant evolution and animal evolution went their separate ways, they still have much in common. Both evolved their own motors, guided by sensors and chemical and electrical communication systems. And although it may seem that plants lost their excitability somewhere along their evolutionary path, we can in fact trace a continuous thread of nerve-like behaviour through plant evolution, from the swimming sex cells of the algae, mosses, ferns and other lower plants to the *Mimosa* and Venus's flytrap.

Chapter By Chapter Summary

The book was designed to be read sequentially, but each chapter stands alone if you just want to dip in and out. If you find some of the scientific details heavy going in places, don't worry – you should get the gist of the story in the opening parts of each chapter, and my advice is to skip any sections that are too detailed.

Most of the first plant performances I describe are unique to plants. Spore and seed dispersals are the 'classic' movements that may well bring back memories of old biology classes. Some dispersals are stupendous explosions, often responding to moisture in the air by the wetting or drying of dead cells. Yet not all these movements can be explained simply by mechanical forces alone — we shall also see examples of true animal-like excitability.

In the following four chapters I deal with plant mobility triggered by senses of touch, taste and even electricity. These are examples of truly excitable plant 'nervousness'.

Of all the stimuli from the outside world, plants probably extract most information from light — the time of day, the seasons or even the depth of soil. They also need light for photosynthesis and for keeping warm. Many plants guide themselves to the best light using a sense of vision (chapters 7 and 8).

Chapter 9 reveals again the primeval heritage of plants and animals, with movements powered by actomyosin motors preserved intact for hundreds of millions of years, and now the driving force for creatures as simple as fungi and as complex as ourselves, in our own muscles. The parallels are so close that some fungal cells behave like our own white blood cells!

To understand how movements and electrical signals are controlled, we really need to look at the chemistry of plant and animal excitability (chapter 10). The chemicals that pass messages between and inside cells also have an ancient heritage, and as a result plants and animals share much excitable chemistry, often controlling similar sorts of jobs. And there is increasing evidence that drugs which affect the excitability of animals also affect plants in much the same way. Again, there are lessons here for understanding how 'ordinary' plants work — why, for instance, they contain nerve-transmitter hormones and are affected by anaesthetics and aspirin.

But what else do mobile plants share with 'ordinary' plants? They both show signs of primitive behaviours: memory, training, habituation (chapter 11). All plants are excitable: they respond to touch and wound stimuli in ways that we are only just starting to appreciate (chapter 12). This is the whole crux of *The Action Plant* — that plant movements are simply the manifestation of a deep-seated 'animalness' in plants, using the tools of a nervous system without actually having any nerves (chapter 13).

The History

But, you might well ask, why has little of this been printed in textbooks? In fact much *was* written about performing plants in the last century. Plant 'irritability', as it was known, fascinated naturalists of the time, especially Charles Darwin, a pioneering figure referred to throughout this book. But there was a drawback. Since the days of Aristotle, plants were thought to have a soul in their pith, and hence emotions. Plant movements were driven by a mystical 'vital force'. But during the later decades of the nineteenth century, the emerging science of plant physiology showed by experiment how plants really worked. The idea that plants had anything akin to an animal's nervous system became anathema. Charles Darwin went against the grain of current thinking when he suggested that the Venus's flytrap behaved like an animal. It led to the first ever discovery of nerve-like signals in a plant, in 1873. Unfortunately it was the right discovery at the wrong time.

The work was readily scorned. As Julius von Sachs – a leading German plant physiologist – pointed out, how could plants have nerve-like signals if they had no nerves? And how could the relatively slow electrical impulses in the Venus's flytrap compare with the exceedingly fast electrical signals of an animal nerve? No, he said, it must be the side-effect of a more important messenger, perhaps a rush of water through the trap. Because the German botanists were world leaders in plant physiology for over fifty years from the mid-nineteenth century onwards, plant electrophysiology sank into a backwater.

In fact plants send their messages through tiny pores (plasmodesmata) between neighbouring cells. But plasmodesmata have one significant drawback – unlike nerves, they cannot channel electrical messages along different routes, to select different movements. Therefore one electrical signal tends to swamp the whole plant, restricting it to one or two visible movements. Compare this with the way that animals co-ordinate their behaviour with a wide variety of movements. Even worse, most of the larger plant movements are driven by hydraulics which, although often very fast in one direction, are very slow in reverse.

By the turn of the century, interest in plant sensitivity and movement was waning. As the textbooks of the day reflected this decline, students became ignorant of performing plants. Today little is taught on the subject in schools, colleges or universities. Research is rather patchy, and largely carried out in France, Germany, Japan, the United States and Russia. Each scientist generally studies one type of movement or excitability. If only we had a better appreciation of the vast spectrum of plant performances, we might capitalize on them: exploiting novel hormones, or breeding better types of crops for coping with drought, or improving crop management.

By giving an overview, I hope to renew interest in this intriguing area. Above all, *The Action Plant* should open our eyes to another, rather bizarre, world of plants.

2

Exploding Plants

Warning: plants can explode! Much of the vegetable world is a minefield of violent movements. Spores, pollen, fruits and seeds can be hurled far and wide, powered by water pistols, catapults, harpoons, springs and even static electricity guns.

Explosions are the extreme route that some plants have evolved to overcome one of the greatest problems in their lives – being rooted to one spot. The problem is worst when they need to find a mate for sex or despatch their spores/seeds/fruits home-hunting away from the mother plant. The solution is usually to hitch a ride on wind, water or animals: grass pollen blows away on the wind, coconut fruit floats on water, teasel fruits are caught up in the coats of furry animals and so on. In this way a plant can sometimes travel hundreds of miles before settling down.

But for some species hitch-hiking alone is simply not enough – they power themselves by quite literally blowing up. It's not nearly as peculiar as you might think, since explosions occur throughout the fungal and plant kingdoms, and have evolved in several different ways. But despite their importance for understanding the spread of fungal diseases, the pollination of flowers or the spread of weeds, today few people seem interested in them. Yet the small number of scientific investigations of recent times reveal a plant world bordering on the bizarre and sometimes downright savage.

Fungal Guns

Fungi are surprisingly violent. They reproduce with sexual spores and disperse to find new homes using non-sexual spores. Both sorts of spore are microscopic single cells, with diameters ranging from 0.05 to 0.5 millimetre.

Once these have settled down in a suitable home they germinate into a fresh colony.

Take the ascomycete fungi, so called because their sexual spores are bottled up in a squeezy tube known as an ascus. The ripening ascus grows increasingly turgid and strains itself so much that eventually its tip bursts off. The spores are sent flying for a millimetre or more in many species, but in *Dasyobolus immersus* and *Podospora fimicola* they are flung 30–40 centimetres – some several hundred times more than their own size! The idea of the shot is to launch spores into a passing breeze which will carry them far away. In some ascomycetes, so many asci explode at once that the spores burst out in puffs of 'smoke'. In fact they can be heard hissing for a second or two if held close enough to the ear, and in *Rhizina inflata* the noise can last for several minutes.

Cow dung is the favourite home of a mould fungus called *Pilobolus*, which uses a light-guided gun. It fires capsules packed with thousands of spores, held aloft above the dung on a stalk (figure 2.1). The top of the stalk is swollen into a transparent bulge and serves two purposes: as a launch-pad for the spore capsule, and as a lens to focus the sun's rays onto a light-sensitive 'eye'. The lens and eye guide the spore capsule into the light – a convenient navigation guide for shooting spores up and away. Light striking the lens head on is focused directly onto the eye and tells the stalk to grow straight towards the light. But any deviation makes the stalk bend over to face the light again. This ensures that the stalk tilts over at a good angle for launching the spore capsule towards nearby grass, where animals will eat it and then recycle the spores through their dung.

And so to the ballistics of the gun itself. The launch-pad swells up with so much water that eventually it bursts open at the top and blasts the spore capsule off. The explosion itself is spectacular enough to launch the capsule for a distance of up to 2 metres at an initial launch speed approaching 50 kilometres per hour! Considering that the spore capsule is only about 0.08 millimetres in diameter, this is an enormous distance. Or, to put it relative to its size, this is further than a man can throw a cricket ball. In fact, it is such a powerful explosion that one species of nematode worm joy-rides inside it. The worm usually lives on the dung, but it burrows its way into the spore capsule and eventually is propelled onto nearby blades of grass with the spores. There the worms are eaten and reinfect the cows along with the fungi.

Some fungi, such as *Conidiobolus coronatus*, a parasite of termites and aphids, blow their spores off using special turgid cells held in a state of strain. The tension eventually breaks, the cells suddenly change their shape, and the spores are pinched and rocket off. Some fungi take the art of pinching further. In the bird's nest fungus (*Sphaerobolus*) the spores are carried in a ball which is held within a cup of many layers during its

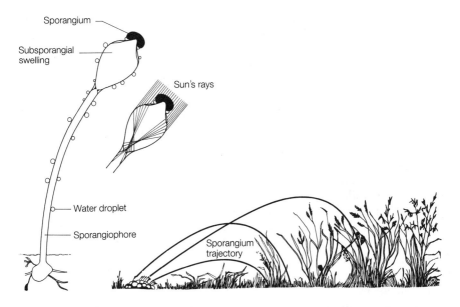

FIGURE 2.1 In the fungus *Pilobolus*, a capsule packed with spores is shot towards the light using a light-sensing gun. The stalk bearing the capsule grows towards the light using a bulging tip which acts as a lens, focusing the light onto a light-sensing 'eye'. Any deviation from the light makes the stalk bend to correct its growth towards the light.
The tip of the stalks also acts as a launch-pad. It swells with so much water that eventually it explodes, blasting the spore capsule off to a distance of 2 metres or more at an initial speed approaching 50 kilometres per hour. Considering that the spore capsule is only about eighty-thousandths of a metre in diameter, this is an enormous distance. The spore capsule sticks wherever it lands, and if it is eaten by an animal it is eventually passed through its dung to begin the cycle all over again.
(After Raven et al., 1986)

development. Tension builds up between these layers as they dry out, so that the cup is finally and suddenly turned completely inside out, hurling the ball 5 or 6 centimetres up into the air. The giant puffballs of Central America are relatives of the bird's nest fungus, and are reputed to choke nearby onlookers when their several billion spores explode all at once.

Yet simple changes in tension cannot explain some strange ballistics observed in other fungi. *Drechslera turcica* is the cause of northern leaf blight, a virulent disease of maize. To give some idea of its importance, in 1970 its cousin *Drechslera maydis* destroyed about 15 per cent of the US corn crop, worth about 1 billion dollars. The fungus breaks open on the surface of the leaves as dark brown lesions, where the non-sexual spores are launched into the air.

The launch of spores generally takes place at noon. The spores are also sensitive to changes in humidity, and so it has always been assumed that they are ejected as in the *Pilobolus* water pistol or by the tension produced as the underlying cells dry out. But Charles Leach of Oregon State University has systematically debunked this orthodoxy over a decade of painstaking work. He noticed something very odd about the trajectory of the spores: instead of flying off in all directions, they always flew perpendicular to the infected leaf. This behaviour is even more puzzling in another ascomycete fungus called downy mildew fungus, as its spores grow in clusters like grapes. According to the conventional turgor/tension theories you would expect these spores to fly off in all directions, and yet they too are ejected in parallel trajectories.

Leach has proposed a startling new theory to explain this phenomenon – electrostatic ballistics. He claims that, under the influence of atmospheric conditions, plants build up electric fields on their leaf surfaces and on any fungi growing there. The static field becomes so intense that eventually it rips the spores off their weak anchorage on the fungus and repels them up into the air in their perpendicular trajectory.

Leach tested this idea by placing electrodes near the leaves. Ejected spores always flew to the positive electrode, showing that they themselves carried a negative charge (opposite charges attract each other). Neutralizing the leaf with an antistatic gun or passing positive ions blocked the discharge of the spores. Changing the surface voltage on the leaf altered their velocities – the larger the voltage, the faster the discharge.

This is all compelling evidence, but how does static electricity collect on plants in the first place? It is not often appreciated that, under normal fair weather conditions, potential gradients of about 150–300 volts per metre commonly occur near the earth's surface, and these will affect plants. The potential gradient varies throughout each day, and is influenced by pollution and the weather – fog, mist, rain, snow and lightning storms all disturb the atmospheric current (Ellis and Turner, 1978).

Leach discovered that there is an electrical potential of up to 120 volts between the ground and the surfaces of leaves. Moreover, the voltage follows a daily cycle, peaking towards positive charges in mid-afternoon and dropping to the most negative values at night. Most notably, claims Leach, the cycle mirrors the daily pattern of spore launches.

There might be some useful spin-offs for the farmer from this theory. Leach suggests that fungicides could be applied more economically according to the weather, the time of day and the electrical strength of their crops (Leach, 1976; Leach and Apple, 1984).

For other fungi, exploding spores need to be launched very precise distances. The spores of the common mushroom are tucked inside the familiar gill-like folds under the 'umbrella' of the mushroom stool. An uncontrolled explosion might well throw a spore from one gill onto another so that it cannot escape into the wind. Instead, spores are popped off just far enough to escape into the gap between the gills (0.1–0.2 millimetres) and into a passing breeze.

The mystery of the mushroom's precision ballistics were solved by a group of biologists and a physicist headed by John Webster at Exeter University. They too discovered that electrostatics were involved. They studied the mushroom *Itersonilia perplexans*, which discharges spores at roughly 5.5 metres per second in the first stages of a 1.5 millimetre launch into the surrounding air. Like Leach's work on the downy mildew fungus, Webster's group found that the mushroom spores are highly attracted to positive electrodes and that an electrostatic charge triggers the spore off its stalk (Webster et al., 1984).

Perhaps the most violent fungus of all fires its spores from a gun aimed at an animal. This fungus is *Haptoglossa*, and it parasitizes small animals by harpooning them with an infection spore. Professor George Barron of Guelph University, Ontario, is an expert on all manner of meat-eating fungi. He and his colleague John Davidson were looking at *Haptoglossa* when they were amazed to see a nematode that was brushing past it writhe in pain before swimming off (Davidson and Barron, 1973). Whatever the attack was, it happened so fast that Barron couldn't make sense of it. Some time later he saw a rotifer impaled by a spike which suddenly shot out from a cannon-shaped fungal cell in *Haptoglossa mirabilis* (figure 2.2) (Barron, 1980). When the rotifer brushed past the gun cell, it suddenly halted, and then wriggled and writhed. A few seconds later it moved off.

The only way to solve what was happening during this mysterious attack was by examining *Haptoglossa* under an electron microscope. Barron and his student Jane Robb found an impressive-looking harpoon only 0.002 millimetres long housed in the muzzle of the gun cell (Robb and Barron, 1982). When a microscopic animal bumps into the muzzle, the fungus explodes: a delicate plug holding the harpoon in place ruptures, the harpoon is shot through the muzzle of the cannon into the animal and a large part of the gun cell rolls inside out like the inverted finger of a glove from the end of the muzzle, forming a tube behind the harpoon. The whole remarkable explosion takes about a tenth of a second (Barron, 1987). The tip of the tube then swells into an infection spore and passes into the punctured animal, where it germinates into a substantial network of cells which feed on the animal. Quite how this gloriously elaborate apparatus actually fires its bullet is still not absolutely certain, although Barron believes that it is all driven by powerful hydraulic pressure in the gun cell, which forces the harpoon and its accessories out like a violent pop gun.

Two other genera of fungi, *Polymyxa betae* and *Plasmodiophora brassicae*, use a slightly gentler hypodermic to infect plants. First, the spores

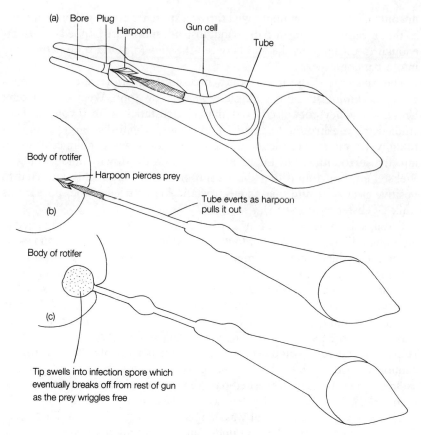

FIGURE 2.2 The parasitic fungus *Haptoglossa mirabilis* uses an explosive harpoon to spear passing animals and protozoa. (a) The harpoon lies poised under tension, held inside a gun cell by a plug. (b) A knock on the gun cell breaks the plug, the harpoon is shot out of the muzzle, pulling the tube behind it, and spears the prey. (c) The tip of the tube swells into an infection spore, which eventually breaks off from the rest of the gun and infects the prey.
(After G.L. Barron)

swim through the soil using flagella, tracking down the tell-tale 'scent' of secretions oozing from cabbage roots. When a swimming spore collides with a root hair, it wraps up its flagella and sits on the outside of the root cell for two hours as it develops an inverted gun and bullet. Suddenly, the bore of the gun slowly turns inside out, and a bullet-like structure inside is fired down the bore and penetrates the root cell. The infection unit then slithers amoeboid-like into the host cell. In *Plasmodiophora*, the whole process takes about a minute (Keskin and Fuchs, 1969; Aist and Williams, 1971).

Violence in the Lower Plants

Amongst the lower plants, ferns such as *Dryopteris* catapult their spores in spectacular fashion (figure 2.3(a)). Sacs containing thousands of microscopic spore capsules are arranged in dark brown clusters which can be seen underneath their fronds. Each capsule is almost entirely ringed by a collar of special cells, in which the side-walls are strong and the outer walls are weak. As the cells dry out, the side-walls collapse inwards and create such tension that the whole collar eventually snaps. The catapult is now set. The spore capsule rips open, first jerking back and then whiplashing forward with such force that the spores inside are thrown out.

In liverworts (figure 2.3(b)), the spores are ejected from their capsules by special four-pronged cells called elaters. These cells swell up in wet weather, and flick out the spores in dry weather like miniature springs.

Mosses are not nearly as violent, mostly relying on the wind or rain to blow or wash away their spores to new homes, but some species can also move of their own accord with small flicks. The capsule containing the spores has a small lid, and when this falls off it reveals two rows of teeth on which many of the spores are carried (figure 2.3(c)). In wet weather the teeth curl inwards, but in dry weather they curl outwards and flick the spores out during the movement. In the aquatic mosses *Wardia* and *Scouleria* the spore capsule lacks teeth, but instead its lid opens and closes according to the weather. In dry weather the lid opens, exposing the spores inside to the wind, whilst in wet weather the lid seals the spores shut inside the capsule (Edwards, 1980). These hygroscopic (moisture-driven) movements can be repeated almost indefinitely, and are not exclusive to the lower plants.

Motorized Seeds

Flowering plants also use hygroscopic movements. This explains why dandelion seeds only fly on dry days. The hairy parachute attached to each seed is water sensitive, and on dry days opens out and catches the wind. When the weather turns humid, the hairs close up, the parachute folds in and the seed falls to the ground or simply does not take off from its launch-pad on the mother plant. Similarly, the stalks holding the cluster of fruits of the wild carrot close over in wet weather, preventing the fruits from escaping. But in dry weather the stalks bend back and allow the fruits to blow off in the wind. All these moisture-sensitive movements are purely mechanical – in other words, they can be performed by dead tissue. In the wild carrot, the movement is driven by a pad of moisture-absorbing cells along one side of each flower stalk. In wet weather these cells swell with water and push the stalk over; in dry conditions they lose the water, contract and pull the stalk back the other way.

Possibly the strangest hygroscopic movements are in seeds that sow themselves! Heronbills, storksbills (both called *Erodium*) and geraniums

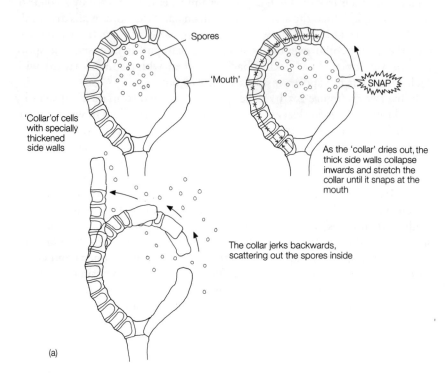

FIGURE 2.3 The lower plants use a variety of explosions or flicks to disperse their spores. (a) In ferns (e.g. *Dryopteris*), a collar of special cells in the spore capsule acts as a catapult. As the collar cells dry out, their side walls collapse inwards, straining the capsule until it bursts open so violently that the collar flings back and catapults the spore out. (b) Liverwort spore capsules, like *Marchantia*, split open and special cells called elaters flick the spores out by coiling and uncoiling like springs, as they dry up and moisten respectively. (c) The moss *Brachythecium* exposes 'teeth' to disperse its spores. The outer row of teeth flick outwards as they dry, spilling the spores out. (d) Specialized mosses like *Scouleria* split open at the top as they dry, exposing the spores to the wind. In wetter weather the lid is forced back on the capsule, protecting the spores inside.
(After Edwards, 1980)

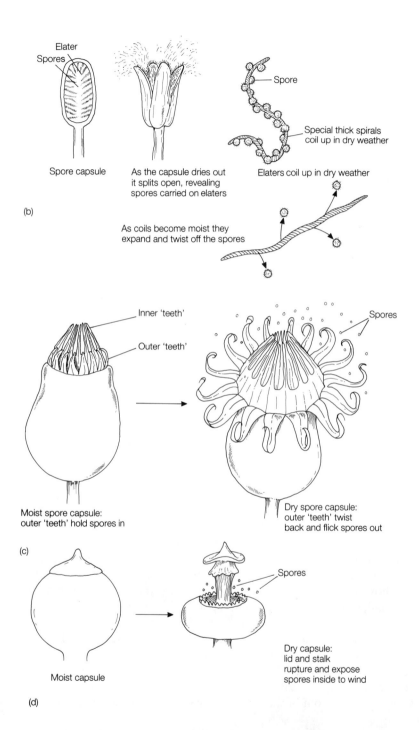

hold several seeds together in a capsule. As the seed capsule ripens it dries out, and the strain eventually rips it apart in an explosion that flings the seeds up to a metre away. But *Erodium* seeds also carry their own motor. Each seed carries a moisture-sensitive spike called an awn which behaves like a spring. When it absorbs water in damp weather it uncoils itself, but in drier conditions it coils up again. So whenever the humidity changes, the coiling or uncoiling drives the awn around in a spiral, and with added traction from barbs on the awn the seed somersaults along the ground (figure 2.4). The movement can be dramatic. In some grass seeds a small prong alongside the awn keeps it coiled up like a spring in a watch, but eventually the strain becomes so great that the awn rapidly uncoils and hurls the seed several millimetres through the air.

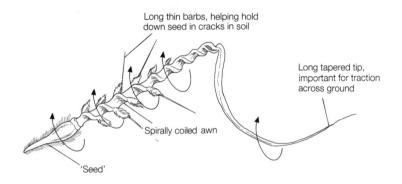

FIGURE 2.4 'Seeds' (really dry fruits) of *Erodium* (storksbill) motor across the ground using spring-like awns. As the awns dry they coil up, and when they get moistened they uncoil. The movement sends the seed scuttling along the ground until it falls into a crack. Then the awn pulls the seed down. When the seed germinates, the bristles on the side of the awn help anchor the seedling down into the ground, counteracting the force of the young root pushing down. (Strictly speaking, the 'seed' is a seed inside a dry fruit.)

The acrobatics of motorized seeds help to find them a good home in which to germinate. Martyn Peart of the Queensland Agricultural College saw how awns drive seeds over barren stones and rocks until they stumble on small, usually damp, cracks in the soil. The awn then pulls the seed down into the crack with the awn facing upwards (Peart, 1979). The seed germinates, but the awn can still perform a useful role. By pointing upwards, the bristles on the sides of the awn anchor the seed in the soil and counteract the upward thrust of the young root as it pushes down into the soil (Peart, 1981). Without the awn the germinating seeds often push themselves back out of the soil, and die of lack of water on the surface. Not surprisingly, self-burying seeds like these produce better seedlings than unburied seeds.

Some motorized seeds can be surprisingly choosy about where they germinate. Nancy Stamp of the State University of New York, Binghamton, found that *Erodium moschatum* seeds preferred to bury themselves amongst cracks in the soil covered in dead leaf litter rather than on bare soil. This makes sense for the future seedling's well-being, as it is wrapped in a nutritious wet blanket of compost rather than being deposited on bare ground which, in the wild semi-arid lands where *Erodium moschatum* grows, is most likely to be around infertile gopher mounds (Stamp, 1989).

Some plant awns perform more dramatically than others, probably as a way of adapting to different environments. Nancy Stamp studied the behaviour of the seeds of two species of storksbill. She found seeds of the common storksbill (*Erodium cicutarium*) in the Sonoran Desert, California, where rainfall is patchy. It has possibly the most sensitive awn known, which uncoils completely with just one spray of water and then leaps into the air. This helps it to locate cracks in the soil quickly, where there is a better chance of finding moisture. *Erodium botrys*, on the other hand, grows in wetter areas, and its awn unwinds and moves much more slowly and steadily. This slower and less spectacular performance enables it to drill into wetter soils, helping it to colonize bare areas and avoid being eaten by predators, or to avoid extreme temperatures on the soil surface during the summer.

Fruits and seeds can also explode. Take this description of the shrub *Dorycnium herbaceum*, by the noted Victorian naturalist Anton Kerner von Marilaun (1904):

> As I sat reading near the table, one of the seeds of the *Dorycnium* was suddenly jerked into my face. Shortly afterwards I saw a second, third, fourth and ultimately about fifty seeds let fly from the small clusters of flowers, and each time I heard a peculiar sound . . . .

The fruits of *Dorycnium* are only one example of explosive 'sling-fruits'. When the fruits are ripe, the tissues holding the seeds become more and more turgid until they eventually rip apart. The tissues split into segments, double back and roll up like window blinds, showering the seeds out far and wide.

In the dwarf mistletoe (*Arceuthobium*) the seeds are discharged by a very high fluid pressure built up in the juicy fruit, which shoots the seeds up to 15 metres sideways at an initial speed of 100 kilometres per hour. Using either their ballistics or an adhesive which sticks to passing birds, the seeds spread from tree to tree. The plant then parasitizes the trees. In the western United States these plants have inflicted serious economic damage on forestry, thanks partly to their remarkable dispersion.

EXPLODING PLANTS

In fact, plants are very often named after their violent behaviour, such as the touch-me-not (*Impatiens noli-me-tangere*) and *Cyclanthera explodens*. And as for the delightfully christened squirting cucumber (*Ecballium elaterium*), the whole fruit actually flies off the plant like a vegetable rocket (figure 2.5). The fruit of the squirting cucumber looks like a hairy gherkin, and the fleshy tissue inside inflates with a slimy juice. Eventually the pressure of the juice becomes so strong that the fruit bursts off its stalk, squirting seeds and slime out through the rupture hole. A successful eruption has been known to launch a cucumber 12.7 metres.

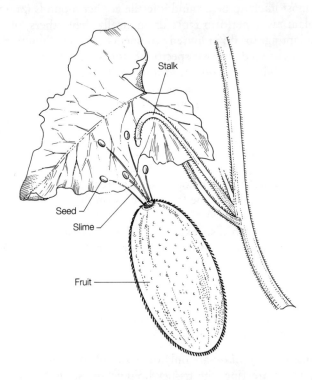

FIGURE 2.5 The squirting cucumber (*Ecballium elaterium*). The fleshy fruit inflates with water and eventually bursts off its stalk, travelling up to 12.7 metres and squirting seeds out in a sticky slime.
(After Kerner von Marilaun, 1904)

But plants don't have to be exotic to display ballistics. As we saw earlier, as the garden geranium fruit ripens and dries out, the uneven stresses in its seed chamber suddenly split its sides apart, catapulting seeds out in all directions (figure 2.6). And on a hot sunny day on heath or moorland you

EXPLODING PLANTS

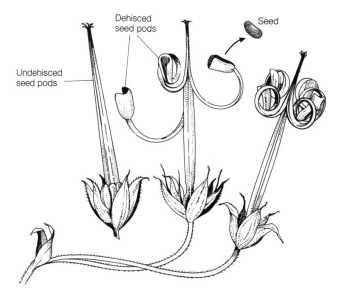

FIGURE 2.6 Fruits like those of *Geranium palustre* split open as they dry, violently rolling up and catapulting their seeds.
(After Kerner von Marilaun, 1904)

can even hear the snap, crackle and pop of splitting fruits on exploding broom and gorse bushes.

Quite possibly the most elaborate exploding seed is the Australian quinine bush (*Petalostigma pubescens*), which embarks on a three-stage journey before it finally settles down to germinate. In the first stage, an emu plucks off and eats the round yellowy green fruit of the bush. The bird digests all but the stone of the fruit, eventually voiding it in its dung. As the dung dries out in the sun the stone explodes, suddenly twisting apart and flinging its thousand or so seeds up to 2.5 metres away. But the story is not finished yet. Each seed carries a small food parcel that specially appeals to ants, which rapidly carry off the seeds to feed off the generous food offering. They take them back to their nests where the seeds have the advantage of protection from fire (a common hazard) and a fertile place to germinate. This Byzantine journey is the only known record of a three-stage seed dispersal (Clifford and Monteith, 1989).

Exploding Flowers

Exploding *flowers* tend to be more muffled than exploding fruits. Broom and gorse keep guns in their flowers, ready to shoot insects on touch. The stamens (male sex organs – see figure 2.7) are pinned down into hood-shaped petals. As the petals develop and grow larger, so the filaments of the stamens inside become stretched and under tension, and the whole device becomes primed like the cock of an old flintlock gun. When an insect alights on the petal it fires the gun. The stamens are released from their petal holsters and spring up, and their anthers punch their pollen onto the underside of the visitor. On visiting another flower of the same species, the insect rubs pollen off onto the receptive region (stigma) of the female sex organ (carpel), and the flower is cross-pollinated.

This flintlock gun technique for blasting pollinators with pollen is a favourite trick of other legume flowers, one of which, lucerne (*Medicago*

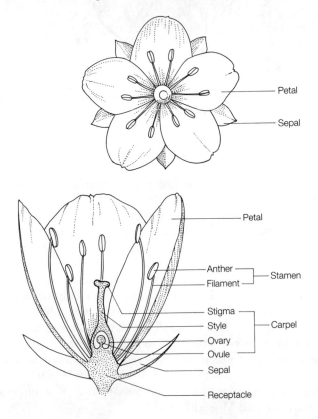

FIGURE 2.7 The basic structure of a flower (based on a geranium) consists of a sepal on the outside petals. The sex organs are carried on a receptacle and consist of stamens and carpels. The stamens carry pollen in anthers carried on a stalk called a filament. The carpels collect pollen on a receptive surface called a stigma where the pollen germinates and grows down through the style to the ovaries where the eggs lie.

sativa), which is also known as alfalfa in the United States, is an important crop plant. In fact it is the world's longest cultivated and most important forage crop, from Norway to New Zealand. When a sufficiently heavy insect such as a bee lands on a lucerne flower, its weight unlatches the flintlock gun mechanism, flicking the stamens upwards and showering the visitor with pollen (figure 2.8). For good measure, the stigma then presses against one of the petals, helping to prevent self-pollination by stopping the insect from pressing home any pollen it collects from the same flower.

But there is a problem. Honeybees are the farmers' favourite pollinator, but they get fed up with lucerne's explosions and soon learn to steal the nectar through a natural slit in the flower. From then on they cheat all the lucerne flowers that they visit, only tripping them by mistake (about once in twenty visits), so that they are terrible pollinators. However, bumblebees and various solitary bees are far better trippers of the flowers, so much so that farmers in North America encourage two important species – the alkali bee (*Nomia melanderi*) and a leafcutter bee (*Megachile rotundata*) – with specially built nest sites. Another problem is that some strains of lucerne become so trigger happy when they are grown in greenhouses that the flowers sometimes fire themselves automatically.

Exploding flowers are not all designed to be triggered by insects. *Hyptis pauliana* is pollinated by hummingbirds, although it belongs to the same family (the legumes) as gorse, broom and lucerne. The flower looks like a deadnettle or mint – the petals are fused into a tube with a crinkly-shaped lip hanging down from the entrance to the tube. Its explosion works on the same flintlock gun principle: the stamens become trapped inside two small pockets inside the lip of the developing flower, and as the lip slowly grows down like a drawbridge the stamens become increasingly strained.

The hummingbird is lured by the flower's red colour, tubular shape and nectar. As it pops its bill into the petal tube to feed on the sweet nectar, the bird inevitably jerks the hanging lip and triggers the explosion – the stamens slip off their restraining holsters, releasing their tension, and fly up and smack the hummingbird's beak with pollen. This works satisfactorily with hummingbirds, but in species of *Hyptis* pollinated by bees the bees have learnt how to manipulate the trigger mechanism by grasping the hidden anthers and defusing the explosion.

The detonator for the exploding tropical artillery plant (*Pilea microphylla*) is even more spectacular. The style lies coiled up like a spring inside the flower bud. When the buds open, the style uncoils in a jack-in-the-box jump, shedding pollen collected from the stamens in small puffs that look like smoke. The two-lipped stigma then opens, ready for cross-pollination.

The common stinging nettle (*Urtica dioica*) also explodes. The flowers – usually separate male and female – hang down like catkins. Before the male flower buds open, the four stamens are curved inwards and fixed between

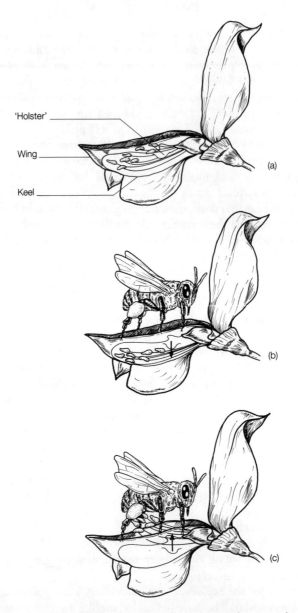

FIGURE 2.8 Legume flowers, like commercially important lucerne (also known as alfalfa, *Medicago sativa*) explode. (a) Unexploded flower (cut in half to show the inside): the stamens are held under tension in the keel by a 'holster' in the keel petals. (b) When an insect lands on the flower its weight forces the keel down and open, releasing the highly sprung stamens and style hidden inside the keel. (c) As they spring up, the stamens cover the underside of the insect with pollen, and the stigma on the style collects pollen from the insect.

the petals and ovary, or against each other. As the flower develops, the filaments of the stamens grow more and more strained. When the bud eventually opens, the filaments of the stamens suddenly straighten out and scatter pollen from the anthers. The movement occurs spontaneously or by a glancing blow. You can frequently see stinging nettle flowers firing themselves in early morning sunshine.

But the sharpest shooter of all plants lives in the tropical forests of Latin America. It is an orchid called *Catasetum*, and recent evidence shows that it deliberately uses violence to *frighten* its bee pollinators – a ploy to stop the bees pollinating more than one flower at a time.

Like the stinging nettle, *Catasetum* goes to such extremes to prevent self-pollination that it segregates its male and female sex organs into separate flowers on entirely separate plants. The flowers of the two sexes are so different to look at that you would hardly guess that they were the same species. However, they both give off the same tell-tale intoxicating perfume. But as more male than female flowers are produced, there is intense competition amongst the males to pollinate the females first.

And so to the ballistics (figure 2.9). The male flower attracts its bee pollinator with the seductive perfume and a gloriously ornate flower. As the animal enters the flower, it inevitably touches one or both of a pair of long fleshy horns, known as antennae, in the centre. The antennae then somehow release from overhead two bags of pollen (pollinia) which are held under tension in a holster. The bags are catapulted out on a sticky disc, tumble through a partial somersault and hit the bee with the sticky disc first.

The impact is so strong that biologists Gustavo Romero and Craig Nelson of Indiana University found that bees hit once seldom ever visited a male *Catasetum* flower again. Instead the insects would often hover over the flowers studying them intently for their maleness to make sure they weren't maltreated again (Romero and Nelson, 1986). In fact the ballistics are staggering: pollen bags weighing fractions of a gram and about a millimetre long can be flung out almost a metre, with a punch of up to a quarter of the bee's own weight, at speeds of around 300 centimetres per second. Small wonder that the insect is sometimes knocked out of the flower. And, as many of Charles Darwin's contemporaries found out when they looked at specimens in Kew Gardens, being hit on the face by an exploding *Catasetum* is quite a painful experience.

The experience with the male flowers was so intimidating that the bees would thereafter only visit female flowers, where their reception was much more civilized and where the pollinia were scraped onto the female sex organ and cross-pollinated – a clear case of aggression paying off.

An interesting evolutionary story is that violence in *Catasetum* may have evolved hand-in-hand with the segregation of its sexes. In species of *Catasetum* with the least visual differences between the male and female

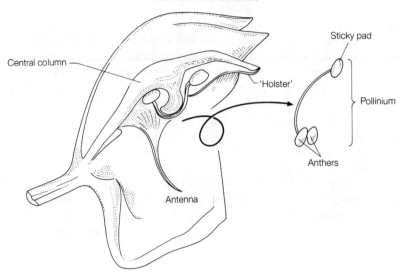

FIGURE 2.9 The exploding flowers of *Catasetum* use violence to persuade pollinators to avoid visiting the flowers! When an insect visits a male flower, it touches one of a pair of antennae and triggers the explosion of the pollen bags from their holster, where they have been held under great tension. The pollen bags stick to the insect, which is now so frightened that it only seeks out female flowers. The reception there is much more gentle as the stigma collects the pollen for cross-pollination. (The figure shows the flower cut in half.)

flowers, the pollinia pop off with the lightest punch and the bees are sometimes undeterred by the assault. But in species where the visual differences are most pronounced, the males pack the heaviest pollinia and bash the bees in a thoroughly provocative manner.

One outstanding question remains, however. How does the *Catasetum* antenna feel the touch of its insect visitor, relay the message to the pollinia and then release them? The message from antenna to pollinia has to travel anything up to a centimetre or so, and touching the pollinia directly cannot fire them. Charles Darwin, in his book *The Various Contrivances by Which Orchids are Fertilised by Insects* (Darwin, 1862), looked at *Catasetum* and two other closely related exploding orchids – *Mormodes* and *Cycnoches* – but could not find out how the trigger worked. Alcohol and chloroform, which, as we shall see later, anaesthetize most other touch-sensitive movements, had no effect. But it is interesting that many other orchids are also touch sensitive. The twayblade (*Listerata ovata*) is a rather dull-looking member of the orchid family. If the central column inside the flower is touched, two drops of sticky fluid are instantly expelled from two depressions on either side of the column. As Darwin remarked: 'a touch from the thinnest human hair suffices to cause the expulsion'.

In fact, touch sensitivity is surprisingly common in many other types of flowers, playing a key role in plant movements that can be performed not just once, like the ones described here, but over and over again.

3

FLOWER POWER

Somewhere in your back garden, perhaps tucked away near the gooseberry bushes, there are plants driven to move over and over again by the need for sex. If only we took more trouble to look, we would see that the flowers of many garden plants have touchy sex organs capable of moving not just once, but almost indefinitely!

To recap the previous chapter, plants need to cross-pollinate – to mate with another plant. We said that most of them rely on outside agencies – wind, water or animals – to deliver their gametes, aided in some cases by explosions. But in most cases these are purely mechanical movements, needing no active chemistry to trigger or power them. Other flowers, though, behave more like animals. They have touchy movements that can be repeated over and over again in an extraordinary spectrum of performances: curtsies, bows, flips, flicks, kisses, pinches, nods, shrugs and many more. Stamens, carpels and petals are all capable of moving (figure 2.7), and over a thousand species of flower perform like this, from cacti to orchids, from thistles to lilies (see table 1.1 and pp. 279–81).

Over the past fifty years the scientific community has shown little interest in finding out how tactile flowers work, but that's not surprising because few people realize that they even exist. Yet botanists of the eighteenth and nineteenth centuries were fascinated with the 'animalness' of flower 'irritability', as they called it. Indeed, by the turn of the century hundreds of scientific papers had been published on flower movements, mostly by German plant physiologists and naturalists. They recognized their sensitive movements in the whole theatre of plant irritability.

Those were the days when the German universities were the international powerhouses of the world, drawing postgraduate students from all over Europe and America. Most of all, the universities of Germany offered freedom of thought and a strong emphasis on philosophy, and encouraged scientific innovation. The professors were allowed to teach whatever subject they chose, and this liberalism inevitably rubbed off onto the students. As Erwin Bünning (1977) wrote: 'We students during those years enjoyed the same freedom as the professors. Nobody checked whether we were going to a lecture. Nobody checked whether we were reading textbooks.' In other words, they had all the makings of a modern student. But there was still hard work, as he continued: 'Those who intended to receive the Ph.D. degree had to submit their thesis . . . we (also) had to pass our oral examination in three or four subjects.' This compares with the modern American Ph.D. equivalent. In fact, the qualification of Ph.D. did not even exist in British universities until the early part of this century.

Although this chapter is about flower movements, it is also a tribute to many of the pioneers of plant science – Wilhelm Pfeffer, Gottlieb Haberlandt and Julius von Sachs – and their students such as Erwin Bünning, Karl Umrath, Frederic Newcombe and Robert Chodat. But the science also rested on a solid bedrock of observations by 'classic' old-fashioned naturalists, including the strong British naturalist tradition – the ubiquitous Charles Darwin and his British associates Joseph Hooker, Robert Brown, James Small and John Knapp – and the Germans such as Anton Kerner von Marilaun and Herman Muller. All these observers and recorders of plant natural history followed in the footsteps of the Ancient Greek grandfather of botany, Theophrastus, the first person ever to record the classification, growth and development of plants. These, and many more, were also the people who drove out from botany the mystical 'vitalism', which had hitherto supposedly explained all living phenomena, and made botany into the science that we know today.

Touchy Stamens

Flowers that entrust their cross-pollination to insects face at least two problems. First, how do they get their insect go-between to carry their pollen? Second, how do they then stop the insect from rubbing the pollen onto the stigmas of the same flower, causing self-pollination?

Touch-sensitive stamen movements can help. They respond to an insect's touch by bending over and dabbing pollen onto it from the pollen bags (anthers). This is a reflex action, in which no 'thought' or decision-making is needed. The motor is located in the stamen filament, but there often isn't any clearly visible site where the sense of touch resides.

Stamen movements are particularly common in the cacti, the sunflower family, the barberry family and many others. Take the stamens of the garden shrub barberry (*Berberis*) and its cousin *Mahonia*. Each tiny bowl-

FIGURE 3.1 *Berberis* is one of many flowers with touch-sensitive stamens. In the 'resting' flower (a) the stamens lie pressed against the petals, but after touching a stamen they respond in forty-five-thousandths of a second by flicking inwards (b). But unlike most other touch-sensitive stamen flowers, the petals of *Berberis* also appear to bend as well. The flower recovers its original position over several minutes.

(Photographs by Alan Mawson)

shaped flower carries six stamens, tightly pressed up against the petals (figure 3.1(a)). As an insect rummages through the flower searching for nectar, it is bound to brush against at least one of the stamens. At the slightest touch the stamen springs up in a flash towards the middle of the flower, striking pollen against the insect (figure 3.1(b)). If the stamen misses its insect target, it bangs into the bottom of the stigma to avoid self-pollination. In *Berberis*, the stamen then slowly bends back and recovers its original position, although it loses some of its excitability, whilst in *Mahonia* it only returns to an erect position. Different species of *Berberis* and *Mahonia* also have different sensitivities (table 3.1).

Consider this description of *Berberis* by the naturalist James Smith (1788):

> I then touched some filaments which had perfectly returned their former situations, and found them contract with as much facility as before. This was repeated three or four times on the same filament.

If this seems little different from the mechanical flower movements covered in the last chapter, he went on to warn:

> We must be careful not to confound them [barberry stamens] with other movements, which, however wonderful to the first sight are to be explained merely on mechanical principles. The stamina of the *Parietaria*, for instance, are held in such a constrained curved position by the leaves of the calyx, that as soon as the latter become fully expanded ... the stamina, being very elastic, fly up, and throw their pollen about with great force.

The famous Austrian plant anatomist and physiologist Gottlieb Haberlandt was particularly interested in trying to find out how plants sensed the touch of an insect, and he thought he had found what appeared to be special sensor cells covering the sensitive side of the *Berberis* stamen. They looked like tiny pimples which were rich in cytoplasm and flexible enough to crumple slightly at a slight touch. Haberlandt grew more convinced of their importance when he discovered similar sorts of pimples on other touchy stamens and, even more interestingly, on some tendrils (the touch-sensitive 'fingers' used by some climbing plants).

Unfortunately we now know that some touchy plants have no such pimples or have them on insensitive parts of their motors. All we can safely say after a hundred years of research is that touching a stamen, a trendil or any other sensitive organ stretches the skin of the cells on the surface of the sensitive organ. That stimulus is then relayed to the motor cell as a message through hundreds of tiny channels known as plasmodesmata, through which one cell 'talks' to its neighbours. The only clue we have as to how the touch stimulus is 'understood' by a sensitive plant is from the Venus's flytrap, which has an electrical signalling system, as we will see later in chapter 4.

TABLE 3.1 Sensitivities of stamen filaments of *Berberis* and *Mahonia*

	Duration of inward movement of filament on first stimulation	Recovery time after first stimulation	Duration of movement after second stimulation	Recovery time after second stimulation	Temperature (°C)	Relative humidity (%)
Mahonia japonica	Instantaneous	11 min		15 min	8	100
Mahonia aquifolium	Instantaneous	13 s	1.8 s	4 min	20.5	57
Berberis antoniana	Instantaneous	12 s			14	75
Berberis stenophylla	Instantaneous	12 s	0.75 s		20.5	53
Berberis darwinii	Instantaneous	10 s			14	75

Source: Percival, 1965

Meanwhile, modern research equipment has helped in collecting more precise data on the stamen movements. We now know, for instance, that *Berberis* stamens react to a stimulus in 0.045 seconds, and take about 0.12 seconds to move through an arc of 30° (Millet and Thibert, 1976). But the recovery of the stamens to their original position is much slower – about 6 seconds. In *Mahonia*, the stamen reacts to a stimulus in 0.05 seconds with a movement lasting 0.265 seconds. It then takes over 30 seconds to recover to just half of its original position. This scale of recovery is typical of sensitive plant movements.

So much for the flower. How does the insect feel about this sort of intimate petting with a stamen? The touchy flowers attracted one of the greatest chroniclers of plant phenomena, the Austrian naturalist Anton Kerner von Marilaun. He looked at the stamens of the cactus *Opuntia*, which bend in towards an insect pollinator. They are probably the largest and most easily visible movements of any flower, and he described the cactus as it opens its cup-shaped golden yellow flowers at 9 a.m. when the sky is clear, revealing its large whorls of stamens:

> When a bee visits the flower, it settles first on the large stigma, which projects above the anthers, and then tries to clamber down to the honey. During this process contact with the irritable portion of the [stamen] filaments is inevitable, and the moment it occurs the stamens that are touched bend over the bee and load it with their pollen which is easily detached from the anthers. It is amusing to watch this phenomenon and observe how quickly one after another the filaments bend over the insect, and administer their blows as it crawls down. The bee is not much alarmed by the inflection of the filaments, or by the taps it receives, but suffers itself to be loaded with pollen without making any fuss.
>
> (Kerner von Marilaun, 1904)

The stamens of *Berberis*, *Mahonia* and *Opuntia* all bend inwards, welcoming their insects with dabs of pollen. But the protruding clusters of stamens of the beautiful white flowers of *Sparmannia africana* (the rumslind tree or African hemp) move outwards, like some sort of animated shaving brush (figure 3.2). Why *Sparmannia* behaves like this is anyone's guess. Perhaps the stamens spread more of their pollen over a larger part of the insects. Or perhaps the pollen is prevented from falling onto the stigma of the same flower during the insect's visit.

All these movements are preprogrammed, however. Once the stamens are touched, they can only move in one predetermined direction, in a kind of 'knee-jerk' reflex. But some flower species go one stage further – they sense the *direction* from which they are touched.

Mordechai Jaffe, of Ohio University, is a leading expert in touch sensitivity in all sorts of plants, and his research group has studied *Portulaca grandiflora*, an attractive herbaceous garden plant with bright flowers (Jaffe

(a)

(b)

FIGURE 3.2 The stamens of *Sparmannia africana* bend from the centre (a) outwards (b) after being touched. Like many other touchy flowers, the movement is triggered by electrical signals that signal the stamens to flush water out of their motor cells.
(Photographs by Paul Simons)

et al., 1977). *Portulaca* also behaves in some very strange ways. Its finger-like stamens react within a second of stimulation, bending through an arc of about 23° *towards* the stimulus in 2–5 seconds. But if you touch a stamen twice in quick succession from different directions, it stops dead in its tracks, as if refusing to move any further (this is typical of many other sensitive plants capable of responding to the direction of a stimulus). This may be because the motor cells on opposite sides of the stamen filament are caught in a tug-of-war, with one side's pull exactly counterbalancing the pull from the other side.

There is also another strange phenomenon. The stamens somehow communicate with each other. For instance, if the stamens on one side of the flower are touched, those on the other side become *inhibited* from moving. There is no ready explanation for this.

After completing its bending, the stamen slowly returns to its original position in about 5 minutes – about a hundred times slower than it first took to respond (a similar time-scale to the touchy movements of other touchy stamens).

Previously it had been assumed that the *Portulaca* stamen movements helped brush pollen onto visiting insects. But Jaffe's group discovered another paradox. The flower's behaviour is also governed by daily rhythms: the petals close over at certain times of the day (which, as we shall see in chapter 8, is nothing unusual), preventing insects from visiting the flower, but to the scientists' amazement these are the times when the stamens are most sensitive! Therefore they had to completely rethink their ideas on why these flowers are touchy, and came to the conclusion that the plant actually stimulates itself! The pressure of the closed petals presses down on the stamens, stimulating their movement away from the stigma, so helping to prevent self-pollination. However, the stamens retain just enough touchiness when the flowers are open – between noon and 1 p.m. – to react to insects. All this goes to show that nothing in nature should be assumed until it is tested.

The strangest of all stamen movements belongs to a totally unrelated species – *Apocynum androsemifolium*, which commonly goes by another tongue-twister name, tutsan-leaved dogbane, although the French give it its most appropriate title: *goube mouche* or the fly-eater. This flower more than just touches small insects – *it traps and kills them*!

The English naturalist John Knapp, in his *Journal of a Naturalist* (a book 'which gives the pleasing idea of the pursuits by which a country gentleman imbued with a taste for natural history may amuse his leisure' said the *Proceedings of the Linnaean Society*, 24 May 1845), painted the dogbane in its most lurid colours, describing what happens as a fly settles on the flower:

> The [stamen] filaments close, and catching the fly by the extremity of its proboscis, detain the prisoner writhing in protracted struggles till released by

death, a death apparently occasioned by exhaustion alone; the filaments then relax, and the body falls to the ground. The plant will at times be dusky from the numbers of imprisoned wretches.

(Knapp, 1838)

Why does the flower go to such extremes? Maybe the death trap excludes all but the most desirable strong insects for pollination. But Knapp gives us his thoughts: '. . . our ignorance favours the idea of a wanton cruelty in the herb'. Sadly, the dogbane remains as obscure now as it did in Knapp's days.

Stretchy Stamens

The stamens that we have discussed so far display sideways movements, either inwards or outwards. But there is one group of stamens with an extraordinary ability to contract downwards – they curtsy! These flowers belong to the Asteraceae (which used to be known as the Compositae – the dandelion and sunflower family) and include the globe artichoke (*Cynara scolymus*) and the cornflower (*Centaurea cyanus*). Their graceful dexterity is due partly to the stamens' astonishing elasticity. In fact they are so elastic that you can stretch them to twice their usual length and they still snap back to their original length.

The stamens in these flowers are fused together by their anthers into a tube, but their filaments are separate and free to move (figure 3.3). This whole arrangement looks like a cage, inside which the carpel is hidden. When the stamens are touched, the filaments contract downwards by up to 30 per cent of their original length, exposing the stigma to cross-pollination by an insect. The contraction also squeezes pollen out of the anthers like toothpaste from a tube. Consider this observation on the touchiness of *Centaurea cyanus* by the great nineteenth-century pollination expert, Hermann Muller:

> The anther-cylinder was drawn down 2 to 3 millimetres very quickly, and then more slowly to an extent of 5 to 6 millimetres; in a few seconds a mass of pollen emerged . . . and then gradually the style protruded 3 to 4 millimetres above the orifice [of the anther tube].
>
> (Muller, 1883)

The list of plant species with touchy stamen movements is truly astonishing (see 'Taxonomy of mobile plants', pp. 279–81). The intensity or sensitivity of their stamen contractions alone is quite variable; in *Perezia multiflora* the contraction is so vigorous that all the pollen is squeezed out at once. In the daisy-like flower head of *Layia elegans*, the stamen filaments of the freshest florets are so sensitive that just blowing on them is enough to make them shorten.

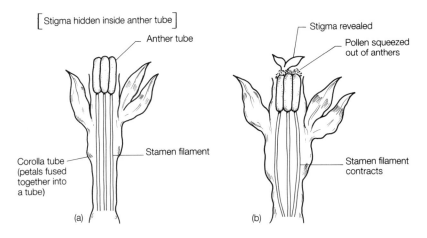

FIGURE 3.3 The stamens of *Centaurea* shorten after being touched. (a) The anthers are fused into a tube, leaving their filaments free (the flower is shown cut in half). (b) When the stamens are touched by visiting insects the filaments contract, pulling the anthers down and squeezing their pollen out like toothpaste. Meanwhile, the stigma lies exposed, ready to receive pollen brought by the insect from another flower. Evidence from electric shock treatment suggests that the movement involves electrical signals. After an interval, the filaments recover their original length and the process can be repeated. Percival (1965) found that in a freshly opened floret of *Centaurea montana* about a quarter of the anthers' pollen was squeezed out by the contraction.

Muller's work was followed up by the English botanist James Small (1917) who, on returning to his lectureship at Bedford College, London University, at the end of the First World War, made a very extensive search of these stamen movements in the Asteraceae family. He found that 95 of the 149 species in the family were touch-sensitive movers.

How Do Moving Stamens Work?

Unravelling the mysteries of how flowers sense touch and then move has been frustrating. For one thing, the sensitive parts of the flower are often small and fiddly to work with. And yet the first experiments were extremely promising. By the end of the nineteenth century the English medical scientist, Sir John Burdon-Sanderson, had discovered electrical signals in plants.

One simple test of electrical excitability in both animals and plants is to mimic the touch stimulus using an electric shock, although it isn't a foolproof test because it can produce unnatural side-reactions. But when Burdon-Sanderson applied a mild electric shock to the stamen tube of

Centaurea cyanus it contracted as if it had been touched. It was the first indication that electricity could be involved in sensitive flower movements, and yet inexplicably Burdon-Sanderson didn't follow the discovery up by searching for the stamen movement's own electrical patterns. Instead, his pioneering work went unnoticed until the 1920s, when a later generation of plant physiologists renewed the search for electrical signals in plants. One of these botanists was Erwin Bünning, a student at Berlin University.

In those days the German universities were still revelling in the prestige that they had inherited from the previous century, and Berlin University was the home of Nobel Prizewinners Otto Warburg, Otto Meyerhof and Hans Spemann. Also working there was Carl Correns, one of the rediscoverers of Mendel's genetic work, Gottlieb Haberlandt (emeritus professor) and Max Hartmann. In the second year of Bünning's undergraduate studies, his professor suggested that he look at the flowers of *Sparmannia africana*, and after having his curiosity aroused Bünning turned his interest into a doctoral thesis.

> I became especially fascinated by the similarities between the processes going on in these stamens and those occurring in animal nerves. There are the periods of latency, the 'all or none' type of reaction, the absolute and relative refractory [rest] period, the action potentials, etc. This brought me to a study of general physiology. It brought me also to the books and other publications of Wilhelm Pfeffer.
>
> (Bünning, 1977)

Bünning discovered action potentials not only in the bending of *Sparmannia*, but also in the *Berberis* stamen (Bünning, 1934). The beginning of the action potential always precedes the movement, usually by 1–2 seconds in *Sparmannia* but only by 0.003–0.007 seconds in *Berberis*. Thus, the electricity appeared to signal the start of movement. But the mechanism that translates the touch stimulus into an electrical signal is still not known.

So how do the stamens themselves move? The microscope reveals few secrets of the stamen motor. According to Haberlandt (1884), the motor tissue of *Portulaca* consists of ordinary-looking packing (parenchyma) cells sandwiched between the skin (epithelium) and central plumbing strands (vascular tissue).

Similarly, the stamens of *Berberis* and *Centaurea* look very little different from stamens that cannot move. Yet Wilhelm Pfeffer suspected the presence of a hydraulic motor, which pumped water in or out of the motor cells. So he cut off the anthers of a *Centaurea* floret, and saw a drop of liquid ooze out of the cut end during the movement (the stamen was otherwise surprisingly unimpaired by the surgery). When the movement is complete the water is slowly reabsorbed, and the motor cells inflate and push the stamen back to its original position.

How does the water get jettisoned in the first instance? Again according to Pfeffer, the hydraulics are driven by osmosis. To put it simply, water is sucked from dilute to more concentrated salt or sugar solutions through the cell's plasma membrane. In the case of the stamens, their motor cells start off with a concentrated salt solution that retains water, keeping the cell full and turgid. But if the salts are lost, the water is drawn out and the cells collapse. Pfeffer even tried to determine the strength of these osmotic forces by stretching the contracted stamens with small weights strapped onto them. This led him to achieve a landmark in biological science: the first ever construction of an artificial membrane with which he made the first measurements of osmotic pressure (Pfeffer, 1877).

When Bünning took a close look at the cells of the *Berberis* stamens he too suspected osmotic forces at work in the movements. He found that they were in a state of considerable turmoil after reacting: their protoplasmic contents were rolled up into clumps, similar to the agitation that Charles Darwin had noticed in stimulated tentacles in the trap of the carnivorous sundew. Bünning found tiny droplets of liquid extruded from the walls of the stamen after the movement had occurred. Clearly, liquid had been lost from the cell, as would be expected from Pfeffer's osmosis theory (Bünning, 1930).

So what triggers the osmosis in the motor tissues of the stamen filaments? The motor certainly relies heavily on energy generated from respiration, as is the case in all touch-sensitive movements. When the chemical energy carrier ATP (adenosine triphosphate) is added to *Portulaca* stamens, they respond even more vigorously than usual (Jaffe et al., 1977). The same fuel drives the movements of *Mimosa* and Venus's flytrap. But this still does not explain how water is kept under such great pressure and is then suddenly jettisoned out of the stamen filament. For a probable explanation we must turn to an exceedingly bizarre flower.

Trigger Plants

The flower of *Stylidium*, the trigger plant, at first sight looks almost like an orchid, but it is quite different and quite extraordinary. It has no stamens and carpels as such – they are both fused together into a magnificent 'trigger', which is a crooked column hanging out of the flower like a beckoning finger. When the column is touched, it unleashes a sweep through more than 180° in 0.01 seconds. The whole movement takes a staggering 0.02 seconds – one of the fastest and largest touchy movements in the plant kingdom. Afterwards it resets to its original position in about 3 minutes.

In nature, an unwitting insect setting foot on the flower is bound to brush the column and make it flip. Then the insect is walloped with fresh pollen and at the same time any pollen already on its coat, brought over from another flower, is rubbed off onto the receptive stigma.

What makes this display even more enticing is that we now know more about how *Stylidium* works than about any other flower movement. Thanks to a recent intensive research programme in Australia, we may have the key not just to *Stylidium* but also to most other touchy flowers. Geoff Findlay and his colleagues at Flinders University, South Australia, found that the ingredients of the motor were potassium and chloride ions driving water by osmosis, although they still remain baffled by the sheer speed of the trigger (Findlay, 1975; Findlay and Pallaghy, 1978). Their results also show that Pfeffer was on the right track over a hundred years ago when he speculated that, not only do plasma membranes let water pass through by osmosis, but they also actively select what passes through them, in this case potassium and chloride ions.

When the flower is primed to fire, the motor cells in the crook of the column are full of potassium and chloride ions. This concentrated solution makes the inside of the cells very salty, and they swell up with water drawn in by osmosis from surrounding reservoirs between the cells. Because they are swollen and stiff with water, these cells keep the column cocked outside the flower. In contrast, the cells stretched opposite, on the outer bow of the crook of the column, are relatively elastic. But all this changes when the column is triggered. During the movement, the potassium and chloride are flushed out into the reservoirs on either side of the bend of the trigger, dragging water along with them by osmosis. The motor cells lose 99.6 per cent of their stiffness and deflate, and the trigger flips over. The whole process is shown in figure 3.4.

But are ions and hydraulics fast enough to explain the astonishing speed of the trigger movement entirely? Unfortunately, it is so fast that there is simply no way of tracking the rush of ions during the movement. The same problem prevents measurements of any changes in electricity, although there is a hint of some electrical activity because changes in voltage accompany the slow return of the trigger to its original starting position. Given these drawbacks, Findlay and his colleagues feel that ion movements alone are hardly likely to explain the whole movement, simply because the ions and water are unlikely to move fast enough. Instead, they suggest that

FIGURE 3.4 Schematic illustration of the rapid column movement of a *Stylidium* flower. (a) Touching the column triggers it to move through over 180° in 30 milliseconds (0.03 seconds). (b) (i) On touch stimulation, the motor cells in the top half of the bend of the column rapidly lose K^+ ions and water. (ii) The cells in the top half become flaccid whilst those in the bottom half remain turgid, pushing the column over.

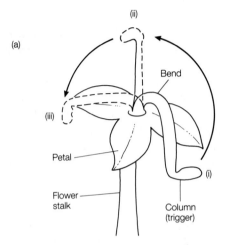

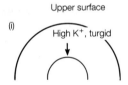
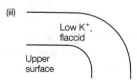

the osmotic engine fires a 'gun'. In the 'resting' flower the column is held under great strain. The inflated motor cells keep the natural spring of the column's joint in check, rather like a gun kept at half-cock. But when the motor cells deflate, the spring is released and the 'gun' is triggered. Afterwards, the motor cells reflate and pull the column back to its starting position, and the 'gun' is ready to fire again. You could argue that *Stylidium* combines some of the explosive power of the mechanical flower movements we discussed earlier in chapter 2 with the excitable movements discussed in this chapter.

Stylidium species flourish throughout Australia, Tasmania and Southeast Asia. They have homes in coastal plains and swamps, and on the bushlands and uncleared farmland, even scattering onto the desert fringes. In Eastern Australia many kinds grow in the mountains, while others live in the grasslands of Northern Australia. In all there are 136 species, ranging from tiny ephemerals to tall woody shrubs, and the bizarre column movement has undoubtedly helped in their success.

There are some even odder plants in the same family as *Stylidium*, sharing many of the same habitats. However, they are so small that they are named *Levenhookia*, after the pioneering Dutch microscopist Antonie van Leeuwenhoek who discovered hundreds of tiny creatures with his microscope. The common name of *Levenhookia* is the styleworts, and they are hardly taller than a thumbnail with leaves only a few millimetres long. In fact they only stand out when growing in their thousands, carpeting the ground in pink or white blooms. The flowers are not easy to study because, apart from their small size, they have a marked habit of closing their petals indoors, so that most investigators have had to work outside in the plant's natural habitat.

Like *Stylidium*, the styleworts have a column, but its tip is capped by a special hood-shaped petal (the labellum). When a knob or spot at the tip of this labellum is touched, the petal instantly recoils, releasing the column with a jerk that sprays pollen over the visiting insect. The flower can only fire once because the trigger's stalk grows slightly longer during the movement, and the labellum 'hood' fails to re-cover the trigger.

Insects are not always needed for catapulting the trigger and its pollen. According to Australian naturalist Rica Erickson, some styleworts, such as *Levenhookia pusilla*, indulge in group pollination. Their blooms are arranged in circles and mature at about the same time. When their labella are jerked away (perhaps by a gust of wind or pressure of the trigger inside), the pollen is sprayed across to blooms on the other side of the circle.

> It is a floral game of catch ball, in which every flower sprays out a shower of golden grains and has its fielders [stigmas] ready to catch what comes their way.
>
> (Erickson, 1958)

Sensitive Females

So much for the movements of touchy male organs. The female's stigmas and styles can also exhibit touch-sensitive movement.

In the early part of this century, the American botanist Frederic Newcombe studied as a postgraduate at Leipzig, where he came under the influence of the German physiologists, particularly Wilhelm Pfeffer, and their fascination with plant movements. Newcombe was the bridgehead for the later American entry into this subject, and he took a particular interest in sensitive stigmas.

The flowers of the monkey flower (*Mimulus*) are a good example. From the throat of these attractive bright yellow, often speckly, flowers protrudes a style bearing a stigma shaped like a pair of pouting lips waiting for a kiss. When either of the lips is touched, they slowly and elegantly close together (figure 3.5). If pollination is successfully performed, the stigma lips stay shut, but if not they reopen (Newcombe, 1922).

The closing stigma movement occurs in four other plant families: Bignoniaceae, Scrophulariaceae, Martyniaceae and Lentibulariaceae. Perhaps it is just coincidence, but *Martynia lutea* of the Martyniaceae and the *Pinguicula* species (butterworts) and *Utricularia* species (bladderworts) of the Lentibulariaceae are also carnivorous. What significance this has, if any, I have no idea.

As early as 1861, a Silesian botanist, Wilhelm Kabsch, found that the stigmas of *Mimulus guttatus* were sensitive to an electric shock (Kabsch, 1861). A century later, the Soviet plant physiologists Sinyukhin and Britikov (1967a, b) measured localized action potentials after touching the stigma lips of the *Incarvillea delavayi* flower just before they closed together. That, though, is the only electrical work on female movements, and it is by no means definitive.

How does the movement work? Wolfgang Christalle of Konigsberg University recorded liquid seeping out of the motor cells of *Mimulus luteus* into the spaces surrounding them during the movement of the stigma lips (Christalle, 1931). So perhaps these stigma movements are driven by the same electric signalling, ion pumping and hydraulic motor that we have seen in other tactile flower movements. No one has taken much interest in the stigmas for many years, and so this question remains unanswered.

The speeds of reaction vary throughout the different genera (table 3.2).

The intensity of the touch sensitivity also varies. Newcombe found that the common foxglove (*Digitalis purpurea*) only had a feeble stigma movement, and that many other members of the same figwort family were completely insensitive, even though they had the typical pouting stigma lips. This sort of range of sensitivity is typical of most touchy plant movements,

(a)

(b)

FIGURE 3.5 The two-lobed stigma (the part of the female sex organ that receives pollen) of the *Mimulus* (monkey flower) (a) before and (b) after being touched. The stigma reopens in a few hours if cross-pollination is not achieved, but stays closed if pollination is successful.
(Photograph by Alan Mawson)

TABLE 3.2 Reactions of touchy stigmas

Species	Time to complete closing after stimulation (seconds)
Martynia	3
Catalpa bignoniodes	7
Torenia fourenieri	2
Tecoma radicans	3
Mimulus glabratus var Jamesii	10

Source: Newcombe, 1922

like the stamen movements we've just considered and the touchy leaf movements in the legume and Oxalidaceae families (chapter 6). These gradations may reflect the evolution of touchy responses in groups of similar plant species.

But why do stigmas need to close up? After all, most plant species manage perfectly well without any sort of moving female parts. Newcombe (1920) suggested two reasons. First, a closed stigma prevents a retreating insect from coating it with pollen collected from the same flower, i.e. it helps prevent self-pollination. However, if pollination is successful, the stigmas behave rather oddly: they reopen and then close permanently some two to ten hours afterwards. Newcombe suggested that this helped the pollen to germinate better, because moist air is essential for the germination. Certainly the pollen itself is an important trigger to movement, because inorganic powders of the same size or sterile pollen killed in hot water cannot mimic it.

The styles of a few species can also move, although only three genera, in the Asteraceae family, perform these types of behaviour. For instance, when the styles of the garden plant *Arctotis* are gently stroked, they bend towards the stimulus. The 'flower' that we usually refer to is actually a collection of dozens of small flowers (florets) arranged in whorls. As earlier shown for *Centaurea*, each floret comprises a tube of stamens through which the style pokes. But in *Arctotis* the styles grow above their male counterparts, and in a highly active specimen the style only rests for about half a minute before it is ready to respond again.

In 1839 the French botanist Charles Morren carefully described and beautifully drew another flower with a touch-sensitive style, *Strobilanthes anisophylla* (which used to be called *Goldfussia anisophylla*) (Morren, 1839). Even though Morren saw that insects were needed to effect fertilization, he thought only of self-fertilization, and so he misinterpreted the mechanism of the flower. The stigma is curved upwards so that an insect entering the flower immediately touches it and coats it with pollen. When it is touched (or even breathed on!), it immediately straightens out and then curves towards the right, pressing tight to the wall of the petals. As the insect passes further down into the flower, it is dusted with fresh pollen. And when it carries the pollen out, it cannot touch the stigma, which is tightly pressed up against the corolla, thus preventing self-pollination.

Moving Petals

So far we have only looked at the movements of excited sex organs in a flower, but to the naked eye these generally tend to be rather small scale. Much more dramatic, although not so common, are some outlandish petal movements in a few species of orchids.

One of the first people to discover the secret of the moving orchid was Mr W. Bean, the foreman of the orchid house at Kew Gardens (Anon., 1887). He was looking at *Masdevallia muscosa*, a flamboyant orchid from the West Indies. To explain how the movement works, you need to know something of the incredible complexity of any orchid flower (see box 3.1).

Box 3.1: General structure of an orchid flower

Orchid flowers generally consist of very showy petals arranged around a central column which holds the all-important sex organs. The idea is to advertise the flower to passing insects, which land on a special platform (the labellum) made from the lower petal. At the centre of the flower an insect is often loaded with a pair of special bags of pollen equipped with exceptionally sticky pads (in fact these have one of the strongest known natural adhesives), which they will carry off to another flower of the same species. There the pollen bags will be knocked and break open onto the receptive female carpels. Sadly, the insect is often saddled afterwards with the sticky discs.

The important feature from our point of view is the labellum. It hangs seductively out of the flower, like a small hand on a slender wrist. The merest touch of a hair or an insect's foot on the crest-like 'palm' of the hand sends a message along to the 'wrist', which flips up in a second or two and imprisons the insect in a cage made from the labellum, the side petals and the centre of the flower (figure 3.6). Its only means of escape is one small opening, but as the animal desperately clambers free it has to run a gauntlet of sticky slime. The flower uses this to peel off any pollen that its prisoner may already be carrying, and also for planting fresh new pollen

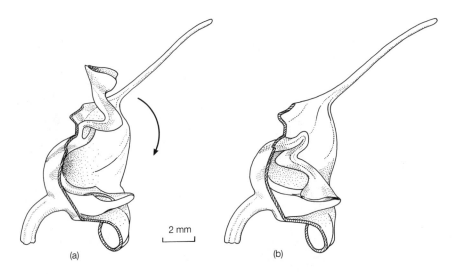

FIGURE 3.6 The touch-sensitive labellum (petal) movement of the orchid *Poroglossum*. Side view of the flower cut in half, (a) before movement and (b) after movement. When an insect touches the labellum it closes down rapidly like a trapdoor, pinning the insect against the flower's sex organs. A short time later the labellum lifts up, releasing the insect after it has hopefully executed cross-pollination.

sacs onto it. If the animal repeats the same wretched performance in another *Masdevallia muscosa* flower, the pollen will be carried to another carpel. This may seem like some sort of plant sadism, but it all enhances cross-pollination (Oliver, 1888).

A similar trap works in the touchy Australian orchids, the greenhoods (*Pterostylis*), *Poroglossum* and *Acostaea* (figure 3.7). These orchids often grow alongside trigger plants and styleworts in Western Australia, an area particularly rich in touch-sensitive flower species. The greenhoods have probably the most elaborate prisons to trap their pollinators. Two of the petals and one of the sepals from a hood (hence the name greenhood) enclosing the column of the flower.

Once again, we owe some of our best descriptions of these orchid movements to Charles Darwin. A quarter of a mile from Down House, his home in Kent, was a rich field of orchids. In the summer of 1838 or 1839, he had begun studying the ways in which flowers were cross-fertilized by insects. No better examples exist than the orchids – petals, stamens and carpels are all developed to promote cross-pollination. His book *The Various Contrivances by Which Orchids are Fertilised by Insects* (Darwin,

FLOWER POWER

(a)

(b)

FIGURE 3.7 The touch-sensitive labellum (petal) movement of *Pterostylis* (a) before and (b) after being touched. Insects landing on the labellum are tossed back into the depths of the flower by the sudden flip of the labellum, encouraging the insect to cross-pollinate the flower's sex organs. After a while, the labellum draws down and releases its visitors ready to visit another *Pterostylis* flower.
(Photographs by Alan Mawson)

1862) includes this passage about the Australian greenhood orchids, and *Pterostylis longifolia* in particular:

> The labellum affords a landing-place for insects . . . when the organ is touched it rapidly springs up, carrying with it the touching insect, which is thus temporarily imprisoned within the otherwise almost completely closed flower. The labellum remains shut from half an hour to one and a half, and on reopening is again sensitive to a touch.

Violence forces insects into cross-pollinating another Australian orchid, *Caleana* (the flying duck orchid) (figure 3.8). The great nineteenth-century botanist and explorer Joseph Hooker, who was a friend of Darwin, described the labellum in his book *Flora of Tasmania*:

(a)

(b)

FIGURE 3.8 The touch-sensitive labellum of *Caleana* (flying duck orchid) (a) before and (b) after movement. The labellum's movement catches insects in the flower, helping cross-pollination.
(Photographs by Alan Mawson)

(a)

(b)

FIGURE 3.9 The labellum of *Drakaea* (hammer orchid) looks, smells and even feels sufficiently like a female wasp to fool the male of the species into landing on the labellum and attempting to mate with it. Instead, the orchid's labellum flings the wasp back on a hinge onto the flower's sex organs. (a) Before labellum movement; (b) after movement; (c) the hinge of the labellum; (d) sex organs: stamens at the top, stigma in the hollow.

(Photographs by Alan Mawson)

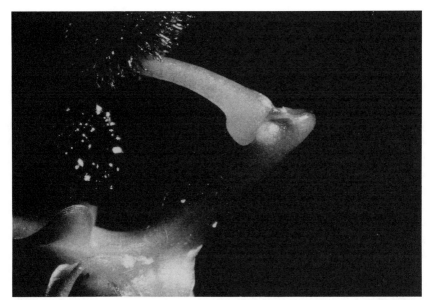

(c)

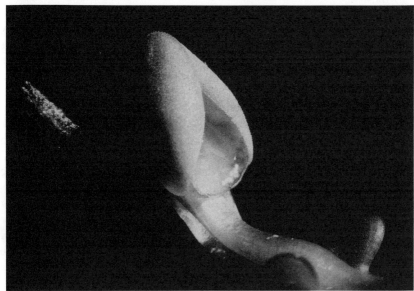

(d)

It stands erect, as it were, over the flower, but when the plant is shaken or rain comes on, it suddenly shuts down, like a lid, on the flower, and its lamina becomes closely applied to the large petaloid column. A Swan River species is said thus to catch insects, whose struggles appear to disengage the pollen from the anther, and apply it to the stigma.

(Hooker, 1860)

Having said that, Robert Fitzgerald in his masterly book *Australian Orchids* published in 1882 claimed that *Caleana* was not irritable. When an insect landed on the flower, it dragged the labellum down under its own weight (Fitzgerald, 1882). The movement has remained a mystery ever since.

The Australian hammer orchid *Drakaea* (figure 3.9) combines a violent labellum with simulated sex with its pollinator. The plant flowers at the same time that one particular species of male wasp is on the wing, looking for a female to mate with. Unfortunately for him he has to wait a couple of weeks before the female emerges from the ground to join him in a nuptial flight. During this crucial fortnight the orchid exploits the male wasp mercilessly. The flower carries a thoroughly convincing likeness of the female wasp at the end of its slender labellum, down to her size, colour and texture, and even her seductive scent. Having flown around in frustration looking for a mate, the male embraces the dummy female with gusto. In return the labellum flips him into the centre of the flower, dusts off any pollen that he might be carrying and plants two fresh bags of pollen on his back. A most frustrating scenario for the wasp, but when the real females eventually take to the air their seductive powers prove far stronger than the flower's. By that time, though, the flower should have been cross-pollinated.

Darwin also described another sort of pollinium movement that most orchids perform. After the pollinia have attached themselves to an animal, each of the club-shaped pollen bags stands erect on tiny stalks. About 30 seconds later, as the insect flies about, the pollinia gently bend forwards, anticipating meeting the next flower and ready to collide head on with the receptive stigmas.

Possibly the most elusive of all the flower movements is the contraction of an *entire* flower. If a cloud passes over tulip, crocus and a few other flower species they close up (figure 3.10). The German plant physiologist Julius von Sachs found that these flowers are acutely sensitive to temperature. The lower parts of the tepals (fused petals–sepals) grow best at lower temperatures ($3-7°C$), while the upper parts grow best at $10-17°C$. So when the weather is warm the top grows faster than the bottom, and the flower opens out. The reverse happens under cooler conditions, when the lower parts grow faster and close the flower over. A sudden change of only $0.2-1°C$ can open or close the tulip or crocus flower.

The same applies to gentian flowers, but with a difference. Gentians have

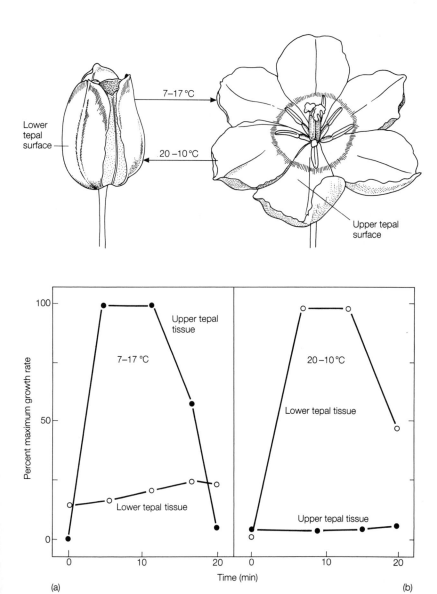

FIGURE 3.10 The heat-sensitive tulip flower folds up in cooler temperatures and opens up when it gets warmer. The upper and lower surfaces of each tepal (fused sepal and petal) drive the movement by growing at different rates: (a) as the temperature rises, the growth of the upper surface speeds up and the flower opens; (b) a cooler temperature slows the growth of the lower surface and the flower closes. (After Salisbury and Ross, 1985; courtesy of Wadsworth Publishing Company)

trumpet-shaped flowers with a range of stunning colours which appeal to both gardeners and bees. On open sites in the higher parts of the Javanese mountains, the 6–8 millimetre long flowers of *Gentiana quadrifaria* stand out with their intense blue colour, a rare colour in tropical flowers. During the course of a day the flowers open and close several times, often with a change in temperature or as a result of rhythms in the plant itself. In 1915, the Dutch botanist Cornelius Bremekamp, who was working at Java's Sugar Experiment Station at the time, took to tickling the flowers all over (Bremekamp, 1915). To his great surprise he found that the corolla closed up – it seemed to be touchy!

This extraordinary phenomenon caused a few ripples of excitement in the botany world. A few years later, the German scientist Karl Goebel, who was an assistant of Julius von Sachs before being made a professor, actually published photographs of the gentian movement (Goebel, 1920). Later, in 1940, the noted Dutch pollination expert Leendert Van der Pijl made possibly the most telling discovery. He was visiting the Observational Post of the Volcanological Survey at Mount Papandajan, Java, when he tried touching flowers of *Gentiana quadrifaria*. The flowers reacted to a single touch, folding up first slowly in 10–20 seconds and then more rapidly in another 1 or 2 seconds. However, they very often refused to budge at all. It seemed that a change in the surrounding air temperature was needed before the flowers would close. Van der Pijl concluded that they were in fact largely responding to temperature changes, and that the touch stimulus simply enhanced the excitement (Van der Pijl, 1940). Or perhaps the heat of an experimenter's hand is enough on its own to trigger the movement?

Quite why these flowers should want to close and shrink at the slightest provocation is not very clear. It was thought that by trapping insects they would coat them with a fair dose of pollen. On the other hand, the movements are so slow that any insect would surely have escaped a long time before.

4

HUNTING AND KILLING

Imagine that you are hungry. The smell of cooking coming from the kitchen intensifies your hunger – your mouth waters, your gastric juices flow and your legs carry you towards the source of the delicious smell.

Now imagine that you are a bacterium in a drop of pond water or sea water. You are always hungry, and you do not know where the next meal is coming from. A few molecules of possible food arrive in that drop of water – say a few molecules of amino acids or sugars from a dead alga or a bit of plankton. You sense the presence of that potential meal; your flagella, the long whip-like threads that sprout from your cylindrical body, rotate furiously anticlockwise and come together as one propeller, and you move toward your meal.

The sense of taste is probably the most ancient sense of all living beings, and has evolved throughout the prokaryote, plant and animal kingdoms. Yet although human beings and bacteria are obviously different in size and complexity, this sensing of the chemical stimulus and the response to it show striking similarities: in both cases, molecules of food (whether released into the air during cooking or dissolved in water) are perceived by receptors (in your nose, or on the surface of the bacterial membrane) and generate a signal. The perception of that signal ultimately causes a directed movement to take place.

Perhaps most remarkable of all, bacteria even hunt. Ninety years ago the great German plant physiologist Wilhelm Pfeffer discovered that bacteria tracked down meat extracts like hunting sharks, picking up the scent even in very dilute solutions and chasing the source by sensing the strength of the meat scent: the stronger the scent, the closer to the source. He also showed how fussy the bacteria were, preferring meat over potassium chloride or alcohol (Pfeffer, 1888). Eventually he came to the astonishing conclusion that single-celled organisms had senses as good as those of vertebrate animals.

We have noses, nerves, a brain and muscles – a bacterium lacks all of these. So how does a bacterium sense and find its meal? It was a long time before the food habits of bacteria were looked at again. Around 1960 Julius Adler of the University of Wisconsin discovered how bacteria tasted the difference between chemicals using special 'spots' on their cell surface. These spots, each of which is made of a protein, are sensors, and when the right sort of chemical locks onto one of them the bacterium becomes excited. Adler and his group have since found over twenty different receptors for different chemicals in the bacterium *Escherichia coli*, a favourite specimen for studying in the laboratory. Of the twenty receptors, twelve are for attracting the bacterium to glucose, maltose and other sugars, and the other eight receptors are for repelling it from chemicals such as fatty acids (Adler, 1969).

How do the sensors work? The protein receptor is made up of two parts. On the outside of the cell, its 'antenna' lies fishing in the water ready to lock onto a target chemical – that which has just the right shape to fit the antenna. The rest of the protein lies embedded inside the cytoplasm, ready to tell the rest of the cell if a catch is made. When something is caught, the protein fires off a signal into the cell telling it in which direction to swim.

Adler and his colleagues are now trying to find out what this signal is, and, as in the case of the cilia of *Paramecium* described later, electricity is already suspected. Szmelcman and Adler (1976) detected changes in voltage on the surface of bacteria when they were excited with tasty chemicals. The electrical activity was definitely part of the response to the chemicals – for instance, substances that the bacteria do not recognize had no effect. This electrical activity may be a signal to the flagella telling them which way to beat: clockwise or counterclockwise. It is certainly good evidence for a sudden influx of ions (electrically charged atoms or molecules) across the membrane surface of the bacteria, which is something common to animal and plant cell movements as well.

Sex in the Lower Plants

Like bacteria, many plant cells are swimmers and come equipped with tails called flagella that whip them through their watery environment, or they have hairlike cilia that beat rhythmically to propel them along. These flagella and cilia are particularly common in the sex or spore cells of algae, ferns and other lower plants; in fact, apart from the gymnosperms (conifers, gingkos and gnetophytes) and the flowering plants, almost all living things possess cilia or flagella at some stage of their lives, including our own bodies: cilia in the respiratory tract, the Fallopian tubes and the ventricles of the brain, whilst sperm cells swim with flagella like single-celled animals and plants. The remarkable thing is that although they are different from the bacteria flagella just discussed, all plant, fungi and animal flagella and cilia are essentially built the same. As we'll see repeatedly in this book, throughout evolution nature often holds onto successful engineering.

Cilia and flagella are supported along their length by hollow tubes, rather like flexible scaffolding, called microtubules, arranged in pairs. Each pair is joined to its neighbour by tiny arms, and when these arms relax their grip one pair of microtubules slips over its neighbour, bending the cilium over. And with several pairs of microtubules all slipping and sliding over each other the cilium bends backwards and forwards, beating through the water in breast-stroke fashion.

The carefully synchronized waves of beating cilia are a stunning piece of choreography, worthy of the best Busby Berkeley dance formation. The difficult question is what masterminds their beating? Part of the problem in studying them is that they are exceedingly small, only hundredths of a millimetre long. But two physiologists, Roger Eckert and Yutaka Naitoh of the University of California at Los Angeles, have pieced together an astonishing picture of nerve-like activity in the relatively large (about 0.2 millimetres long) protist *Paramecium* (figure 4.1).

Because the cilia of *Paramecium* are virtually identical with those in plants, Eckert and Naitoh's discoveries have made an enormous impact on plant research. This protozoan swims using thousands of cilia embedded in its plasma membrane. The cilia beat in a carefully synchronized wave, but when *Paramecium* is subjected to an electric shock the beating goes wild. This gave Eckert and Naitoh the idea that the cell's own electricity could have something to do with the beating, and so they inserted electrodes into a single *Paramecium* and recorded remarkably nerve-like electrical signals bursting along the plasma membrane (Eckert, 1972; Eckert and Machemer, 1975). The effects of these impulses were dramatic. Both the direction and frequency of the cilia beats were controlled by the intensity and frequency of the signals. But more than that, the cilia themselves behaved as sensors, tasting any changes in chemicals around them. They then triggered the electrical signals that fired across the cell.

As we will see later with the light-sensitive movements of single-celled algae, the cilia or flagella are excellent sensors; their electrical activity allows them to react rapidly to various stimuli. It probably arose in evolution in the mating of different gametes, with each sex having its own particular patch of membrane on the cilium or flagellum which recognized the opposite sex and fused with it cilium to cilium or flagellum to flagellum. Animals evolved the sensory aspect of the cilium or flagellum in highly sophisticated organs, so that the cilia inside our own ears, for example, have evolved to decode

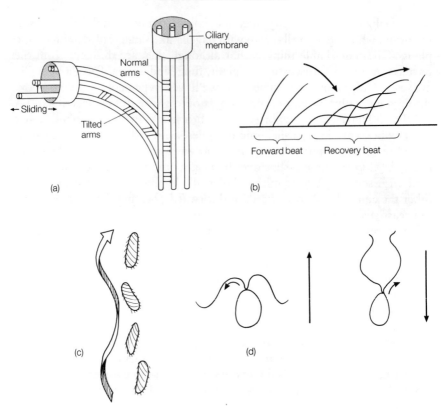

FIGURE 4.1 Cilia and flagella are tiny hair-like propellers on the surface of many different cells, such as sperm and spores. (a) The cilia or flagella of eukaryotes beat backwards and forwards using tiny arms holding pairs of microtubules together. When these arms relax their grip the microtubules slip, bending over the cilium or flagellum. (After Satir and Ojakian, 1979; courtesy of Springer-Verlag.) (b) Cilia beating. The arrow shows the stroke of the cilia, beating forward and then bending back to their original position. Nerve-like electrical signals are now known to control the cilia beats of *Paramecium*, a protozoan that has given many insights into the workings of plant cilia and flagella. (c) *Paramecium* spirals along using rows of beating cilia, arranged in a helix on its body. (d) Beat of *Chlamydomonas* flagella shown by small arrows; the direction of the organism is shown by large arrows.

various sound frequencies into electrical impulses understood by the brain.

Eckert and his colleagues even measured the chemistry driving the electrical signals. As the action potentials sweep across the membrane, the change in voltage opens up special pores to calcium ions (Eckert et al., 1976). Calcium and potassium ions then carry the electricity across the plasma membrane, and the calcium closes the pore behind it (Saimi and Kung, 1982). Indeed, the leakage of calcium ions seems to spark off the action potential in the first place.

The engine driving this sliding is a special enzyme called dynein, which works by gripping one pair of microtubules onto its neighbour. The dynein itself is an enzyme fuelled by the energy-carrying chemical adenosine triphosphate (ATP), and when the energy is released the dynein changes shape, relaxing its grip on the microtubules so that they slide past one another (Satir and Ojakian, 1979).

Here is another remarkable feat. The calcium ions that send messages from one cilium to another also turn on the dynein motors. As we shall see later, electricity and calcium drive muscle contraction in a similar way. Thus, electricity controls the swimming of the little *Paramecium* by setting off a cascade of events, from perception to movement (figure 4.2).

So much for *Paramecium*. The cilia and flagella of plant and fungi cells almost certainly behave in the same way, because their structure is much the same and because circumstantial evidence also increasingly indicates that electricity and calcium ions are involved in the control of the movement. We will look at this in more detail in our discussion of the light-sensitive movements of plant cilia and flagella in chapter 7.

What is important to realize here is that cilia and flagella are widespread throughout sex cells and spores in the fungus and plant kingdoms, except for the conifers and flowering plants. Sperm or egg cells, or sometimes cells of both sexes, swim to one another to achieve fertilization.

But fertilization can be a risky business. Millions of sperm are usually despatched in search of a minute egg cell, and it's often like looking for a needle in a haystack. To increase the chances of successful mating, the female egg cells release powerful scents called pheromones. These chemicals are so irresistible that they attract sperm at concentrations as low as 0.000,000,02 grams in 1 litre of water. Concentrations as low as this match the astonishing sensitivity of insect pheromones, and the sense of taste drives the sperm closer to the female on the trail of her pheromone.

Pheromones have long been known in animals, where they help one sex find another, and because each species has its own characteristic scent it only attracts its own kind of sex cell. Thus when the pheromone sirenin is produced by the egg of the water mould fungus *Allomyces*, it only excites the sperm of that species. The sensitivity of the sperm to that one chemical is so strong that closely related compounds (even mirror images of sirenin molecules) have far less effect. Another strange thing is that when the sperm encounters the sirenin, the molecule simply vanishes. Its disappearance may help to put other sperm off the scent of the egg cell, and so improve the first sperm's chances of successfully fertilizing her.

In fact the taste for pheromones is so ancient that even bacteria use them.

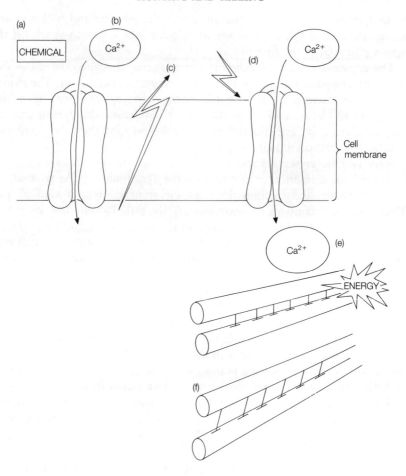

FIGURE 4.2 Sequence of events leading to cilium movement: (a) a chemical stimulates the membrane of a cilium; (b) excitement opens up a pore to calcium ions; (c) the flux of calcium ions upsets the voltage across the membrane, triggering an action potetial across the entire cilium; (d) as the action potential spreads across the cilium it opens more calcium pores; (e) the calcium ions lock onto the microtubules, releasing energy from ATP molecules; (f) the dynein arms lose their grip, letting the microtubules slide past each other, and so the cilium bends over.

Sex is also a strong force in the lives of bacteria. They do not have males and females as such, just plus and minus strains, but they perform the basics of sex by exchanging deoxyribonucleic acid (DNA).

Apart from sex cells, many of the fungi and lower plants also shed swimming spores into their surroundings to find new habitats to grow in. Many of these spores can swim, and they also use an acute sense of taste for finding food and avoiding harmful substances. However, we know little of how they sense these various chemicals, transmit their excitement to the motors or move. The sense of taste also drives the swimming of single-celled algae such as *Euglena* and *Chlamydomonas*, but these too are poorly studied.

We know far more about how the sense of taste drives the animal-like slitherings of some fungal creatures. These movements have much in common with animal embryos and even our own blood.

The Marauding Slime

If you go down to the woods today you might see slimy bright orange pancakes slithering along the ground. They are fungi (although some scientists say they are more animal than fungus), and because of their rather disgusting consistency these endearing creatures are called slime moulds – strictly speaking, acellular slime moulds. They have a strong sense of taste and enjoy a good picnic, usually chasing bacteria and other tiny morsels. They also have rather a sweet tooth and are partial to sugars such as maltose and galactose, but find other substances such as chloride salts repulsive.

A closer look shows that an acellular slime mould is a blob of protoplasm ranging in size from microscopic to 0.2 metres square. An even closer look shows that the whole body is seething backwards and forwards along stringy veins. When a vein pushes it usually squeezes a finger of protoplasm out of the body of the mould. If the surroundings are agreeable – and a slime mould likes shady warm places with plenty of moisture, as well as something tasty to eat – the rest of the protoplasm squeezes into and expands the exploratory finger, and so the whole beast slithers along.

How is the protoplasm, and with it the rest of the organism, pushed along? The answer lies in the veins. They consist of a stiff tube of jelly wrapped around a more liquid core of protoplasm, and when the jelly tube contracts it spurts the protoplasm out in a particular direction. 'Spurt' is no exaggeration: the protoplasm flows at speeds of up to 1.3 millimetres per second, equivalent to 76 metres per minute.

There is one other sort of slime mould, called the cellular slime mould. It is distinguished from its large acellular cousins by its preference for crawling around as tiny cells, each with its own nucleus, whereas acellular moulds have thousands of nuclei in each cell. Although it is too small to be seen with the naked eye, it has an astonishing sense of taste.

Each cell feeds on bacteria living in the mulch on a woodland floor, and moves around by pushing out fingers of protoplasm. Because of this movement and their shape they look very much like an animal amoeba. Like their animal counterparts, the fungal amoebae track down their prey by tasting tell-tale trails of folic acid scent that many bacteria leave behind in the soil. Once on the scent, an amoeba follows the track as it gets stronger and stronger, leading eventually to the microbe itself. Then the amoeba literally wraps itself around the bug and swallows it. This is typical amoeboid behaviour.

At some stage in their lives, the amoebae usually become starved of food or threatened by some other hazard. When this occurs, something truly remarkable happens. Hundreds and thousands of amoebae gather together and make a new creature: a slug. The metamorphosis starts when the amoebae rest for a few hours and then suddenly start huddling together in concentric rings, rather like an archery target. Time-lapse cine film shows that the aggregating cells pulsate in these rings (or streams) for several hours, in a determined effort to congregate together at one central bull's-eye. Once all the amoebae have gathered into a ball, they join up with one another to form a large slug. This then grows a tall stalk holding a bag of spores which are liberated into the air to find new homes elsewhere – with a better supply of food.

It is quite one of the most astonishing pieces of metamorphoses in the living world, involving precise co-ordinated behaviour between cells that had previously led a life of isolation. It is a phenomenon that greatly excites biologists because it tells us something about one of the great mysteries of life – how living cells organize themselves into complex new tissues, organs or even whole beings. It is particularly important in developing embryos, in which layers of cells flow and fold to form limbs and organs, with each cell behaving like an amoeba.

Sense of Taste

The secret signal that transforms slime mould amoebae into a communal slug is a chemical that the fungi taste. Although the amoebae usually huddle together on a solid surface, they can also be persuaded to make slugs on a very thin film of water, and this has an added benefit for the investigating scientist. Chemical messages sent by the amoebae have been collected in the water film, and when added to fresh amoebae made them also turn into slugs. Some thirty years after aggregation was first observed, Professor John Bonner's group of researchers at Princeton University discovered in 1967 that the signalling chemical was cAMP (cyclic adenosine $3',5'$-monophosphate), a substance common in many other plants and animals (Bonner, 1971).

When the amoebae were given drips of cAMP they began shuttling around in great excitement, dancing a conga routine. They then spewed out their own cAMP, amplifying the chemical signal and agitating their

HUNTING AND KILLING

neighbours even more. But how do the cells taste and then respond to the cAMP?

Experiments tagging cAMP molecules with a tell-tale radioactive label have shown that the cAMP does not actually enter the innards of the amoebae. Rather, it briefly sticks onto special receptors on the cell surface which recognize cAMP, and these send a message to the motor inside the cell to start squeezing a finger of protoplasm out towards the cAMP. The contraction eventually steers the whole amoeba cell.

The signal to the amoeba's motor may well be calcium ions, and the evidence for this is persuasive. As the cell tastes the cAMP, the level of calcium in the amoeba rises; chemicals that block this rise also block the movement. In animal amoebae, current flows in the area of the animal where an exploratory finger will later be squeezed out. In fact experiments on the amoeba are surprisingly easy: the animal thinks that the measuring electrode resembles its usual protozoan prey, and immediately sends out a feeler in the appropriate direction. The movement is by no means exclusive to these primitive organisms. Our own white blood cells slither around in amoeboid fashion, and they also have well-defined electrical signals – action potentials – triggered by an electric shock (McCann et al., 1983). In all these movements the cytoplasm contracts. Interestingly, a squirt of calcium and an electric current also herald the contraction of cytoplasm in the division of animal cells.

Electric currents also presage an influx of calcium and movement in the acellular slime mould. A group of Japanese biologists headed by Tetsuo Ueda detected electrical disturbances in the acellular slime mould *Physarum polycephalum* as it tasted new chemicals (Ueda et al., 1985). Another thing which animal amoebae, animal cell division and fungal movements have in common is that they are all driven by the same motor, which is itself extraordinarily sensitive to a slight rise in calcium levels. This motor will be discussed in depth in chapter 9. So the slime mould tastes with its cell surface, is electrically excited, and sends calcium messengers rushing into the cell interior to switch on the motors which eventually drive protoplasm into the exploratory finger that investigates the new taste sensation.

The action is even more dramatic in some fungi that parasitize nematodes (eelworms) living in the soil. Once they have a whiff of their prey's secretions, they swim off to catch it using their flagella or cilia. As soon as they have encountered the animal at close quarters, they burrow through its skin and live off the juicy interior – board, lodging and transport courtesy of the worm. After gorging themselves, the fungi then break out through the worm's skin, shedding spores into the soil. At this point the worm probably collapses. The fungal spores then lie in wait in the soil for the next prey.

There is even intriguing evidence that spores can track by electricity. Scientists at Aberdeen working with the veteran electrophysiologist Marvin

HUNTING AND KILLING

Weisenseel of Karlsruhe measured an electric current flowing in the surface of healthy growing roots, around wounds on injured roots. It may help draw nutrients into the root and direct its growth (Miller et al., 1986). Even though 3,000 roots would be needed to power a light bulb, the currents might also attract fungal disease spores in the soil. Pfeffer showed long ago that spores are attracted by electric currents of the right polarity and strength, and the scientists at Aberdeen have now detected spores congregating at the very spots on the root where current flows into it. The electricity could be a navigation guide, directing the spores to the best bit of the root to infect. It might also help tell the fungus whether the root is alive or dead, as dead roots lose their electricity but continue to leak attractive chemicals into the soil which might confuse the fungi (Miller et al., 1988).

Lassooing Fungi

Some fungi are carnivorous, and trap, kill and digest nematodes in special snares (figure 4.3). Some of the traps, like that of *Dactylella brochopoga*, move on a truly explosive scale. The snare is a three-celled ring, and a nematode passing through it inevitably rubs the sensitive inner surface. Within 0.1 seconds the cells swell inwards and garrotte the animal. It is not clear whether the nematode is then poisoned, but we know that the fungus grows a network of cells into the animal's body after its death and sucks it dry.

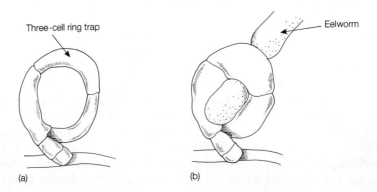

FIGURE 4.3 The lassoo fungus trap of *Dactylella bembicodes* relies on touch and speed to snare its prey. When an eelworm brushes the inside of a three-celled noose (a) they inflate within a tenth of a second and garrotte the animal (b). The fungus then slowly penetrates the animal's body with hyphal cells, and extracts nutrients from it.
(After Couch, 1937, and Barron, 1977; courtesy of Canadian Biological Publications)

In some cases these fungi can recognize their prey by identifying the taste (probably lectin molecules) of the worm's skin, and some fungi are so fussy that they will only ensnare certain species of nematodes. In *Arthrobtrys dactyloides*, the trap also exudes a nematode-attracting pheromone to lure the animal (Balan and Gerber, 1972). In all these ring-trap fungi, the chemicals given off by the nematodes stimulate the fungi to form more traps in anticipation of the food to come. The eminent mycologist Charles Duddington described the nematode's problems as follows:

> It must be remembered that eelworms have no eyes, nor is there reason to believe that Nature has given them any extra sense to warn them when a deadly ring stands just ahead. An eelworm passing through a dense patch of *Dactylella doedycoides* is in a similar situation to a tank crossing an unsuspected minefield, and its peril is hardly less great.
>
> (Duddington, 1957)

Sadly, little is known about how the trap works. Recent work on a mutant strain of *Dactylella* with 'megatraps' eight times larger than normal has shown that the three-ring cells pump themselves up with water. Once the trap is triggered it cannot recover its original position, unlike the flowering plant carnivorous traps. Also, the rings do not appear to communicate with each other.

Yet the ring cells are extraordinary. Each is wrapped in two coats: the outer one is thick, and the inner one appears to be folded, concertina fashion, towards the inside of the trap. When the trap is touched the outer coat rips open along a small break in the wall, letting the inner coat burst through like a jack-in-the-box. The pressure of water inside the cell is probably the driving force behind the explosion (Insell and Zacharaiah, 1978).

5

BLOODY PLANTS

When most people think of carnivorous plants they conjure up pictures of a rather oddball collection of flowering plants such as the sticky-leaved sundew, pitcher plants, the Venus's flytrap and the bladderwort. There are some more, and even a few new species which have only recently been classified as carnivorous, but here we are concerned only with those carnivores that move. And none can equal the breathtaking speed of the bladderwort (*Utricularia*).

The bladderwort lives largely in water, although some species are adapted to living on wet soil. Its bladder-shaped traps are only a few millimetres or less in length, but are highly effective deathtraps (figure 5.1). What makes the bladderwort such an exciting plant is that, more than most other carnivorous plants, it relies entirely on its own speed to surprise and trap its victims. And when you consider that its diet consists largely of fast-swimming *Daphnia* waterfleas, this is no mean feat.

The bladders steer their prey with long whiskery hairs towards smaller touch-sensitive trigger hairs at the mouth of a trapdoor. When a passing

FIGURE 5.1 (a) The carnivorous bladderwort trap. The long slender guide hairs help deflect passing animals into the doorway of the trap, where they bump into small trigger hairs. Once touched, these spring the trapdoor open, suck the animal in, and slam shut again almost immediately. The whole response takes a hundredth of a second, one of the fastest reversible movements in the plant kingdom. (b) (1) Water is pumped out from the trap by special hairs on the bladder wall, creating a vacuum inside. (2) A passing animal knocks into a trigger hair, the trap door is sprung open, and water and prey are sucked in. (3) Almost immediately, the trapdoor shuts and the prey is trapped. (4) The prey is digested by the plant's enzymes and its remains are absorbed.

(After Fineran, 1985; courtesy of the *Israel Journal of Botany*)

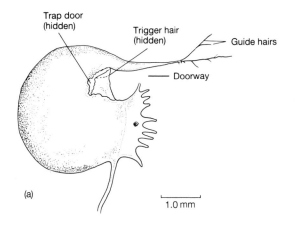

(a)

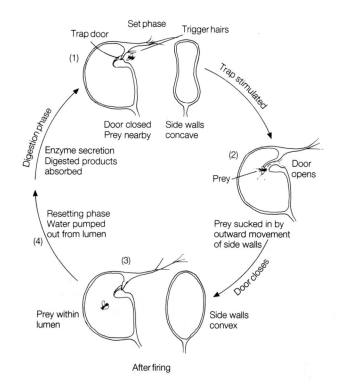

(b)

insect or crustacean even glances against a trigger hair, the trapdoor bursts open, sucking in both water and creature. As soon as the suction is over, the door slams shut again. The whole action is over in a few thousandths of a second, making this one of the most explosive movements known in the plant kingdom. The impressive numbers of waterfleas caught by each plant testify to its success.

Once the animal is trapped, it can still be seen through the transparent wall of the bladder, struggling to escape for minutes or hours before succumbing to the digestive juices of the plant or perhaps just to its own exhaustion. The bladderwort then absorbs the digested remains and uses it for its own nutrition – a classic carnivore.

The secret of the trap's suction is the microscopic hairs dotted along the walls of the bladder. These pump water out of the trap's interior by osmosis, creating a vacuum inside which sucks in water from outside as soon as the trapdoor opens (Sydenham and Findlay, 1975). The trap resets its original vacuum by pumping out the water inside, ready to catch another prey. Quite what makes the trigger hairs touch sensitive is not known. The sheer speed of response suggests that electrical signals might be involved. However, all attempts to record electrical signals between the trigger hairs and the trapdoor have failed simply because the trap is sprung as soon as an electrode is placed near it (Sydenham and Findlay, 1973).

The butterwort (*Pinguicula*) belongs to the same family (Lentibulariaceae) as the bladderwort, yet its trap is so utterly different that there is no comparison between them whatsoever. Its leaves are covered in sticky glands which catch insects like a flypaper. The leaf margins also have a sloth-like rolling movement as they curl over any insects caught there, but this only improves the digestion and has no effect on catching the prey. It is a slow growth movement, and Darwin (1875) found that the curling depended on a sense of taste for chemicals found in abundance in insects (proteins, nitrogen-rich salts and many others).

The sundew (*Drosera*) also tastes its prey (figure 5.2). The leaves are peppered with finger-like tentacles, each holding a dab of very sticky slime – one of the stickiest substances known in the living world – on its tip. A glancing blow against the slime is often enough to trap a passing insect, and as it tries to wriggle free it often collides with more tentacles and their slime, and sinks deeper into a sticky morass. As the insect becomes increasingly desperate and thrashes around, digestive enzymes in the slime start to eat into it and it dies. A very sticky end!

Some species of sundew go one step further. Their tentacles are also touch sensitive; as an insect brushes against them, they bend in from the bases of their stalks towards the kill, like fingers of a grasping hand, daubing the beast with even more slime. In fact one bending tentacle very often rubs against and stimulates another, until eventually so many tentacles crane over that the leaf itself becomes doubled up, like the palm of a hand folded over. Once the insect has been digested, usually after several days, the tentacles straighten out. Even in sundews without touchy tentacles, the whole leaf blade can also perform a slower rolling-over movement.

Charles Darwin took a fancy to the animal-like features of the sundew. He found that proteins, ammonium salts, phosphates and many other substances that mimic an insect activated tentacle movement. Ammonium phosphate was the most active substance he tested, and a drop of water containing a trifling 0.000,000,423 grams of the salt induced curvature when placed on the head of a tentacle. He also tested the touchiness of the tentacles: a human hair weighing 0.000,000,822 grams caused a tentacle to move, and yet the tentacles could still differentiate between this and a raindrop, which had no effect.

The fast movements of the tentacles are driven by electrical signals – more of that later after we've looked at the Venus's flytrap. Little is known of how the sundew's taste sense works, but two German plant physiologists have shown that it involves the plant hormone IAA (indoleacetic acid), which regulates ordinary growth in plants.

Martin Bopp and Inge Weber of Heidelberg University discovered that drops of IAA mimicked the sundew's bending towards cheese, and that IAA blockers prevented the movement (Bopp and Weber, 1981). They also found another interesting feature of the trap: it could tell the difference between small pieces of cheese and glass – it ate the cheese but virtually spat out the glass. But how the sundew tells the difference is a mystery.

There are over a hundred species of sundew spread throughout the temperate and tropical regions of the world, making this a very successful carnivorous plant. And yet in the same family as the sundew is a species so lonely in both geography and evolution that it is only found in one part of the world, and it is the sole member of its genus.

The Most Wonderful Plant – Venus's Flytrap

In the sandy bogs of North and South Carolina lives a plant so bizarre that it has inspired more science fiction writers than scientists. It is the Venus's flytrap, and of all its devotees, none was more besotted than Charles Darwin. 'The most wonderful plant in the world,' he called it, and certainly it has one of the most incredible movements in the entire plant kingdom.

The Venus's flytrap (*Dionaea muscipula*) has a round leaf divided into two cushions fringed with a fearsome row of 'teeth' used for imprisoning, not biting, its prey. Each cushion also carries three slender and highly tactile needles called trigger hairs (figure 5.3). Any reasonably sized insect, crustacean or spider strolling across the trap easily bumps into these hairs

(a)

(b)

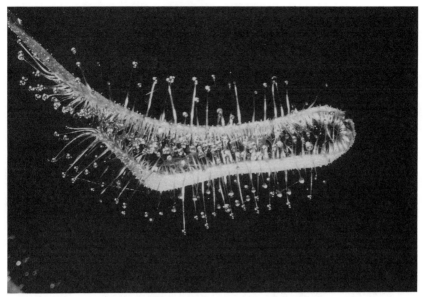

(c)

FIGURE 5.2 (a) An insect landing on a sundew tentacle simulates the plant's sense of touch and taste. (b) Electrical signals fired along the sundew's tentacles provoke bending movements. The insect becomes engulfed in sticky digestive slime and dies. (c) In *Drosera capensis* (the Cape sundew) and some other sundews, the sense of taste also triggers the leaf blade to curl over as well, smothering the prey in even more slime.
(Photographs by Alan Mawson)

and springs the trap. The two cushions snap together, the 'teeth' lock together like the grills on a prison cell and the bug is imprisoned. As it thrashes around seeking escape, the prisoner continually knocks into the trigger hairs, exciting the trap to grip even more tightly. In fact, it squeezes so hard that the outline of the animal actually bulges out through the walls of the trap. Once the trap takes a firm grip, a potent concoction of digestive enzymes contained in acid juice oozes out from special glands into the pouches of the trap. We can only guess that at this point the animal perishes.

Very little was known about how the Venus's flytrap moved, or even if it killed animals for food, until Charles Darwin began studying it. He established that it caught insects for nourishment, and saw how the trap had senses of both touch and taste, finding that meat, cheese and many other animal products could activate it into a slow movement (Darwin, 1875).

Unlike the sundew, the digestive glands of the Venus's flytrap don't start

FIGURE 5.3 The Venus's flytrap (*Dionaea muscipula*) ready to fire. The two lobes each carry three trigger hairs. When any one of these is bent over twice within 35 seconds, an electrical signal fires across the trap telling cells on the outer surface to expand rapidly. In fact, this is the fastest known rate of plant growth. The trap snaps shut, imprisoning the prey inside. As the animal thrashes around it stimulates the two lobes to squeeze tight together and ooze digestive juices from glands (seen as pimples on the lobe surfaces).
(Photograph by Alan Mawson)

secreting their juices until a suitable animal is caught. Some sort of signal is needed to start the secretions, and Philip Rea of Oxford University discovered that small molecules containing nitrogen, or potassium and sodium salts, activated the glands. The secretions themselves are rich in hydrochloric acid and play a vital role in the plant's stomach (Rea, 1982). Once the gland is activated it draws in chloride from the surrounding tissue, gathering up hydrogen ions on the way to make hydrochloric acid. The acid then sucks up water, creating such a powerful hydraulic pressure that it erupts with all the enzymes through the gland, through the skin covering the surface of the gland and into the pouch of the closed trap (Joel et al., 1983; Rea et al., 1983). In fact, it behaves somewhat like an animal stomach.

But the touchiness of the Venus's flytrap is even more dramatic. It responds to touch in 0.3 seconds, using its sheer speed and strength to pounce on prey that would escape if the movement was too slow. It seems obvious now, but in the latter half of the nineteenth century no one guessed that the trap was behaving like an animal until Charles Darwin realized in a flash of inspiration that it had all the hallmarks of an animal nerve reflex. In other words, electrical signals might control the trap's movement. However, in those days such an idea was inconceivable.

Not having the necessary equipment to test his theory, Darwin turned instead to one of the great Victorian medical physiologists, Sir John Burdon-Sanderson, who was then working at University College London on the nerve signals that make muscles contract. On 9 September 1873 Darwin informed Burdon-Sanderson of the first delivery of the experimental plants:

> I will send up early tomorrow two plants with five goodish leaves, which you will know by their being tied to sticks. Please remember that the slightest touch, even by a hair, of the three filaments on each lobe makes the leaf close and it will not open for twenty four hours.

Burdon-Sanderson was immediately intrigued by the behaviour of the Venus's flytrap. Working at the Royal Botanic Gardens, Kew, he used virtually the same apparatus as in his muscle work, probing for any sign of electrical excitement in the plant with a pair of electrodes resting on the surface of the trap. To his delight, just three days after Darwin had sent him the specimens he recorded electrical signals from the surface of the trap after nudging over a single trigger hair. He immediately telegraphed Darwin at his country home in Kent with the news, but we can only guess what he said because the telegram has been lost. However, we can get some idea of Burdon-Sanderson's reactions from a lecture he gave many years later:

> The touching of the sensitive hairs was immediately followed by an electrical disturbance, which preceded the visible motion of the leaf. As the electrical phenomena strikingly resembled those which presented themselves under similar conditions in animals there seemed no room for doubt that the analogy which had suggested the discovery was a true one.
> (Burdon-Sanderson, 1899)

Darwin's response to the news, though, *is* recorded:

> How very kind of you to telegraph me. I am quite delighted that you have got a decided result. Is it not a remarkable fact? It seems [so] to me, in my ignorance. I wish I could remember more distinctly what I formerly read of Du Bois Reymond's results. My poor memory never serves me for more than a vague guide.

(The German physiologist Emil Du Bois Reymond discovered in 1843 that electrical impulses carry messages in nerves, so laying one of the cornerstones of modern electrophysiology. Interestingly, his discoveries were inspired by studying another obscure group of organisms, the electric fish, which shows the scientific value of apparently freakish creatures.)

This was the first detection of nerve-like activity in any plant, showing that high speed electrical signals could stimulate plant and animal cells. We now call these signals 'action potentials', because the electrical potential (i.e. the voltage) causes an action at its final destination. On a chart recording, the typical action potential looks like a wave, with a sharp upward spike followed by a slower recovery to the original 'resting potential' of the cell (figure 5.4). The wave sweeps through excitable cells until it hits its target and triggers movement (or activates secretion in glands). If it was not for electrical signals, animal life as we know it could not exist. From single-celled creatures upwards, animals rely on their action potentials to react quickly both to the outside world and to events inside their own bodies. An animal without action potentials would be like a city without telephones or any other rapid communication system.

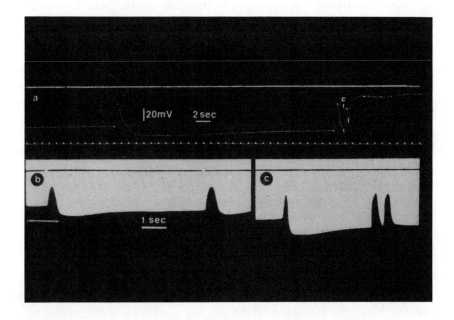

FIGURE 5.4 The top trace (a) is a modern recording of electrical signals in Venus's flytrap (*Dionaea muscipula*). The first and second peaks on the graph are action potentials, but the disturbance thereafter is where the recording electrode was thrown off by the trap snapping shut. The lower traces (b, c) are Sir John Burdon-Sanderson's remarkable recordings of the Venus's flytrap action potentials, recorded in 1873 using a beam of light shone through a column of mercury onto a photographic plate. As the voltage changed, the mercury column rose or fell.
(From Williams, 1976; courtesy of the American Philosophical Society)

Burdon-Sanderson quickly published his results in a brief report (Burdon-Sanderson, 1873), and he followed this up over the following fifteen years with a series of meticulous experiments proving beyond any doubt that this was a very real phenomenon. His recordings are remarkably accurate by today's standards, even though the signals from the electrodes could only be recorded by passing them through a column of mercury and photographing the shadow of the rise or fall of the column as it responded to the current. The recordings showed that flicking one of the trigger hairs on the trap released an electrical signal travelling at 20 centimetres per second – a far cry from the thousands of centimetres per second that signals travel in our own nerves, yet with the same characteristic rise and fall of voltage of an authentic action potential.

But although Burdon-Sanderson's results were soon confirmed by other researchers, he was bitterly criticized for the importance that he attached to the electrical signals. His critics argued that the electrical signal is the *result* of the movement, not its *cause*. Perhaps the most caustic attack was made by the leading German botanist Julius von Sachs (1887), who claimed that the slow speed of the electrical signal and the lack of any nervous tissue in the plant proved that the electricity was unimportant.

With hindsight, von Sachs's shortsightedness seems incredible, but we should bear in mind that the science of plant physiology was then in its infancy and battling against popular myths. As with modern claims that plants have emotions, there was a widespread belief in those days that plants had a soul in their pith (for which, incidentally, there wasn't a shred of evidence). In their pioneering scientific experiments on plants, the German botanists such as von Sachs were attacking anything that they thought smacked of animalness in plants. They emphasized the differences between plants and animals, and as a result they established the foundations of plant science that we learn at school: how plants transport water or sugar through their roots and shoots, reproduction in flowers, the nutritional needs of plants and so on.

So what might have been the glorious birth of plant electrophysiology was simply stamped out. The Germans were then the world leaders in science, and so the textbooks reflected their views, and Burdon-Sanderson's discovery lay virtually unknown for almost a hundred years.

To gain a real appreciation of the power of the action potential, imagine the electrical signals as bullets fired from a gun. When the trigger is squeezed sufficiently hard, the bullet is fired at very special targets: motor cells or gland cells. Once the bullet hits its target, it unleashes a response: movement in a motor cell or secretion in a gland. If, on the other hand, the trigger is not squeezed sufficiently hard, the action potentials are not fired at all. It is a response known as 'all-or-nothing', and as a result the Venus's flytrap ignores stimuli from insects which are too small to matter or gusts of wind or any other trivial event that might waste energy in trap movement.

Let us take the analogy further. Just like a gun, action potentials can only be fired one at a time, with a rest between shots to reload bullets. This is because the excitable cells have to recover from each action potential. As we shall see shortly, they do this by drawing ions through their cell membranes. Therefore if the trigger is squeezed twice in quick succession without a few seconds' rest, the trap cannot fire a second action potential and the second squeeze of the trigger is ignored.

That is about as far as we can take the gun analogy. The excitable movements of animals and plants are infinitely more sophisticated. For instance, the trigger that releases an action potential is made of special sensor organs. In the Venus's flytrap these are the trigger hairs, and what excites both animal and plant scientists today is that these hairs are one of the few obvious tactile sensors in the entire plant kingdom. They are also extremely amenable to study, standing proud of the trap like stiff little antennae and mounted on a springy notch which bounces the hair back after it has been bent over. In 1906 the great Austrian plant anatomist Gottlieb Haberlandt pinpointed the source of the hairs' touchiness as the cells lying in the notch. However, he was not proved right for several decades, and then by an animal physiologist more used to working with insect antennae.

Stuart Jacobson, a physiologist at the University of Minnesota, turned his scalpel and electrodes from insects' antennae to locate the sensory cells in the trigger hair. He carefully sliced slivers from a hair, starting at the tip and working downwards, until he eventually triggered and recorded an action potential (Jacobson, 1965). To his surprise he recorded more than just an action potential — it was immediately preceded by another, quite distinct, electrical signal. On the chart recordings this had none of the distinct curve of a typical action potential. Instead it was a wobbly signal confined to the sensor cell he had stimulated. In animal nerve parlance he had discovered a 'receptor potential'. These signals translate the character of the stimulus — in this case the speed and strength of the hair being hit — into electrical code. The code tells the stimulated cells whether or not to fire off an action potential into the surrounding tissues (figure 5.5). Thus, a strong stimulus generally activates a strong receptor potential and triggers an action potential. Weak stimuli activate weak receptor potentials and don't usually trigger an action potential, although given enough weak stimuli in quick succession the receptor potentials add up into sufficient excitation to do so. So the trigger hair has a simple 'memory', something we shall be returning to in chapter 11. Generally, however, when a decent-sized insect brushes a trigger hair, it produces a sufficiently strong receptor potential to fire an action potential immediately.

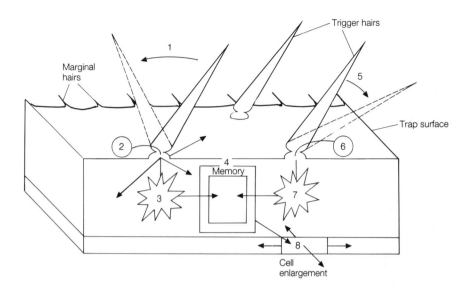

FIGURE 5.5 A chunk of the Venus's flytrap showing some of the sequence of events leading to the trap closing: (1) a trigger hair is bent over; (2) the touch sensation is translated into electrical code, known as a receptor potential, which is confined to just the sensor cells inside the trigger hair; (3) a sufficiently large receptor potential fires a fast-moving electrical wave, known as an action potential, which spreads across the trap lobes; (4) the trap doesn't move, but it somehow remembers being touched; (5) another hair (or the same one as before) is bent over; (6) a second receptor potential is fired in the sensor cells; (7) a second action potential is fired across the trap; (8) if the second action potential comes roughly within 35 seconds of the first one, the trap is sprung shut. The cells on the outer epidermis expand rapidly, folding the trap lobes over. If no prey is caught, the trap reopens about 12 hours later. But if an animal is trapped, the trap slowly tightens around its prey, secretes digestive juices, and absorbs the remains of the animal over about 1–2 weeks.

Confined to its sandy bogs, the Venus's flytrap and its animal-like behaviour are written off by most scientists as a bizarre amusement, a sort of evolutionary cul-de-sac. But although it is the only species of its genus, it does have a less well known underwater cousin. In still ponds scattered across Europe, Asia and Australia, the waterwheel plant (*Aldrovanda vesiculosa*) looks at first sight like a clump of tiny leaves floating in the water (figure 5.6). Indeed, it is so inconspicuous that it remained unrecorded in Europe until the middle of the eighteenth century. But take a closer look with a small magnifying lens and you'll see whorls of leaves looking very much like tiny Venus's flytraps, so delicate that captured prey can clearly be seen inside them.

FIGURE 5.6 *Aldrovanda vesiculosa* is the underwater cousin of the Venus's flytrap, with many similar features. Its two-lobed trap snaps shut when its trigger hairs are touched, fired by receptor and action potentials (electrical signals). However, the signals and movement are faster than those of the Venus's flytrap: the trap closes in about six-hundredths of a second, compared with roughly a second in Venus's flytrap. (a) Traps at 3 and 9 o'clock on the whorl are ready to fire. (b) Traps closed.
(Photographs by Alan Mawson)

Probably because of their size, these traps have not been studied nearly as much as those of the Venus's flytrap, but they behave in very much the same way. Each pair of trap lobes has several relatively long trigger hairs, all capable of sensing touch stimuli using receptor and action potentials which trigger the trap to snap shut. Its reactions are explosive compared with those of the Venus's flytrap: a healthy trap responds in about 0.02 seconds, and the action potentials burst through the trap at between 40 and 120 millimetres per second. It probably needs the extra speed to catch the darting waterfleas that it feeds on.

With these two authentic cases of electrical signals in plants, we can now ask whether any other plant carnivores work by electricity. The indefatigable Charles Darwin suspected that the sundew (*Drosera*) might also have nervous tendencies, yet this time he failed to persuade Burdon-Sanderson to search for electrical signals, perhaps because the sundew's tentacles are much smaller to work with than are those of the Venus's flytrap.

So the story turns instead to Washington University, St Louis, Missouri, in the mid-1970s, where Barbara Pickard was launching a revival into research on electrical signals in plants. Twenty years previously she had been a student of the great plant hormone expert Kenneth Thimann and had become interested in the electrical gradients that accompany the transport of auxins inside the bending growth movements of plants. This in turn led her to look at electrical signals in plants, and so she set one of her postgraduate students, Stephen Williams, the challenge of finding out whether the sundew behaves electrically. With a battery of convincing recordings, he discovered receptor and action potentials in the sundew's tentacles, although with a slight difference. One stroke of the tentacle activated not one action potential but a whole series down from the tip of the tentacle stalk to the motor cells towards the base of the stalk (figure 5.7). The more that the tentacle was stimulated, the wilder the frenzy of electrical signals and the more the stalk bent over (Williams and Pickard, 1972). What makes this particularly intriguing is that sea anemones (these are animals, despite their misleading name) give the same sort of response when their tentacles are touched. The reason in their case is that more stimulation excites more nerves in their simple nerve network. Even though plants don't have nerves, it's possible that each action potential in a sundew's tentacles stimulates more and more motor cells in a similar way. Unfortunately, this is only guesswork, particularly as we don't even know *how* the tentacles actually move.

When Williams dipped an electrode into the sticky slime at the tip of the tentacle, he also recorded a receptor potential. So, again, we see that a plant movement – and the degree of its movement – is controlled first by receptor potentials and then by action potentials. However, one puzzle in the sundew is where the receptor potentials are created. Unlike the trigger hairs of the

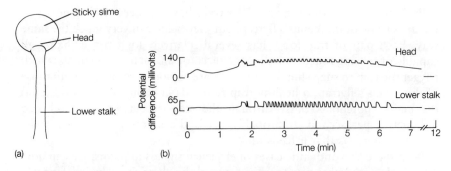

FIGURE 5.7 (a) Like Venus's flytrap and *Aldrovanda*, each tentacle of the sundew translates touch stimulation into electrical signals. (b) Top: (from zero time) the impulses start as a slowly developing receptor potential on the head of the tentacle, which sparks a train of spikey action potentials. Bottom: the action potentials pass down from the head to the base of the tentacle, where they trigger movement in the motor cells.
(From Williams and Pickard, 1972; courtesy of *Planta (Berlin)*)

Venus's flytrap and the waterwheel plant, there is no special touch-sensing organ. Instead, the whole tentacle is touch sensitive, making it somewhat less sophisticated than the Venus's flytrap and its waterwheel cousin. But the similarity between all three carnivorous plants shows how closely related their excitability is.

How do these plants create their electrical activity? There has been a flurry of activity on this question over the past few years and some encouraging results are emerging, although we understand the Sensitive Plant (*Mimosa*) far better. But before continuing with our carnivorous plant story, we need to understand how electrical signals travel in excitable animal tissues (figure 5.8 and box 5.1).

So how do plants create their electrical signals? First, we need to find out how the touch-sensitive cells in the Venus's flytrap trigger hair create their electrical signals.

Electron microscope pictures of the sensor cells reveal whorls of endoplasmic reticulum – a network of membranes in the cytoplasm usually associated with making, packaging and transporting proteins – a veritable protein factory. Buchen et al. (1983) found whorls of endoplasmic reticulum at the top and bottom of the sensor cells. Such whorls are common in the sensory cells of animal ears and eyes, and recent work has shown that the endoplasmic reticulum accumulates calcium, particularly near the plasmalemma in the light-sensitive cells of the eye. Calcium plays an important role in the sensitivity of the eye cells, by controlling their electrical signals. (In muscle cells, the endoplasmic reticulum is developed into a network which permeates through the muscle fibres and is known as a

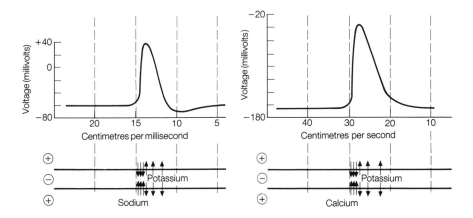

FIGURE 5.8 How animal and plant cells pass action potentials. In both animal and plant cells, stimulation makes the inside of the cell more positive. Given sufficient stimulation, the localized electrical disturbance spikes into an action potential wave that moves along the cell. Note the marked difference in wave velocities in the cells. The spike is triggered by ion gates in the membrane opening to sodium (in animals) or calcium (in plants). Potassium flows out to balance the voltage. Both sodium/ calcium and potassium ions then return to their original 'resting' levels.

sarcoplasmic reticulum. When a nerve signals a muscle to contract, the sarcoplasmic reticulum releases its store of calcium, and this precipitates a cascade of biochemistry that eventually ends in muscle contraction.) Whorls of endoplasmic reticulum may also be involved with perceiving gravity in root tips, where granules of starch press down on them. It is possible that in all these cases the endoplasmic reticulum is strained whenever the cells are squeezed or crushed by a touch stimulus, causing it to release a chemical message of some sort.

Buchen et al. also found tannin vacuoles (bubbles surrounded by a membrane). Tannins are polyphenols, more familiar in tea, that are very effective binders of ions. Therefore they can store ions, and the presence of tannin vacuoles in many other performing plant cells is very significant. The movements of the Sensitive Plant (*Mimosa pudica*) and the chloroplasts of

---

Box 5.1: How animal and plant cells carry electrical signals
The key to understanding electrical signals is ions. These are molecules or atoms carrying either negative or positive electric charges. Wherever the ions go, they carry their charges with them, creating electric currents in a similar manner to the generation of currents in the acid in a car battery. All living cells contain ions, but they

are very fussy over the amounts of each ion that they contain. So, for example, sodium ions are kept at low levels (this is a throwback to primeval days when all life lived in the sea and needed to exclude the sodium chloride from the sea water). Like a car battery, cells have to keep their ions contained and electrically insulated from the outside. The answer to that problem is fat, which is an excellent electrical insulator. The plasma membrane enveloping a cell is made up of a thin sheet of fat sandwiched between two layers of protein. When an ion needs to pass across the fat it has to go through special gates provided by particular types of protein. Because there is an imbalance between ions on the inside and outside of a cell, the electric charges on each side of the membrane do not quite match. So, generally speaking, the inside of the membrane becomes slightly negative, usually of the order of 0.1 volt.

The protein gates open and close when they receive the correct signal. In nerves, the key used to open a protein ion gate is an electric current. As the electrical signal sweeps along a nerve fibre, it opens a line of gates along the nerve's plasma membrane, allowing sodium ions to rush in and potassium to rush out. The shuttling of these two ions unbalances the voltage across the membrane, and so the electrical potential in the cell first becomes more negative and then slowly recovers to its original voltage. This is what gives the action potential its distinct wave pattern. As we see throughout this book, the shuttling of ions is also the key to how plants carry electrical signals.

Protein gates also open up under the command of hormones, ions and even stretching. The sequence of events that leads to the gates' opening can be very complex. For instance, a hormone first attaches itself to a special receptor protein on the plasma membrane which triggers the release of a chemical; this activates an enzyme that releases energy to open the protein ion-carrier through which a certain type of ion will escape or enter. The questions now being asked are exactly how the protein gates work and how the intricate web of chemical and physical commands that controls their behaviour is constructed. One important thing to recognize is that, although the protein gate research bandwagon is now enormous, it is still very new and there is much more to learn about this subject. Yet animal physiologists and, more recently, plant physiologists are becoming highly excited about them, because they are the key to understanding how cells interact with the world around them: how cancer cells 'talk' to each other, what influences nerve cells in the brain, how hormones affect the tissues on which they are targeted and so on.

Mougeotia are affected by particularly high levels of calcium ions stored in tannin vacuoles (see box 5.2). Perhaps the tannin and endoplasmic reticulum act together as an ion store. We await more research for the answer.

German scientists have recently found that action potentials cannot be fired in the Venus's flytrap unless endoplasmic reticulum is present in the sensor cells (Casser et al., 1985). High levels of calcium and potassium have been found in the sensor cells, but so far it has not been possible to locate the ions exactly inside the cells. So we do not know yet whether the ions are locked up in the endoplasmic reticulum, but if animal excitable cells are anything to go by, they should be (Buchen and Schroder, 1986). The sensor cells are shown diagrammatically in figure 5.9.

The electrical state of the sensor cells is probably maintained by the calcium and potassium ions. Stuart Jacobson, now working at Carleton University, Ottawa, perfused Venus's flytrap in various ions, and found that potassium was probably needed to fire the action potential in the trigger hair (Jacobson, 1974).

The second step in the performance of the Venus's flytrap is sending the action potential out of the trigger hair through the trap and so to the motor cells which actually move. Unfortunately, we are less sure of the events taking place here. There are no nerves. Under the microscope the cells which conduct the signal look fairly ordinary, except again for those little tannin vacuoles. The action potentials of both the Venus's flytrap and the waterwheel plant are very sensitive to calcium ions (Iijima and Sibaoka, 1985; Hodick and Sievers, 1986), and these ions may carry the action potential. As potassium ions are involved in keeping the sensor cells electrically 'alert', they may also drive the action potential through the rest of the trap.

The next step is the arrival of the electrical signal at the motor cells, the tissue that actually performs the movement. Dieter Hodick and Andreas Sievers of the University of Bonn have suggested that calcium ions are the link between the electrical signal and the movement. As action potentials pass through the trap, the level of calcium – or some chemical action controlled by calcium – in thè motor cells rises until a critical threshold is reached. So far their hypothesis is a long shot, based on the sensitivity of the trap to calcium and the sudden brake on the streaming cytoplasmic movements in some algal cells observed when calcium rushes in during the passage of action potentials.

As to the mechanics of movements, the textbooks say that the cells on the inner face of the trap jettison their cargo of water, shrink and allow the trap lobe to fold over. In support of this, Toshio Iijima and the veteran plant movement expert Takao Sibaoka of Tokyo University detected large movements of potassium ions during the shutting and reopening of the

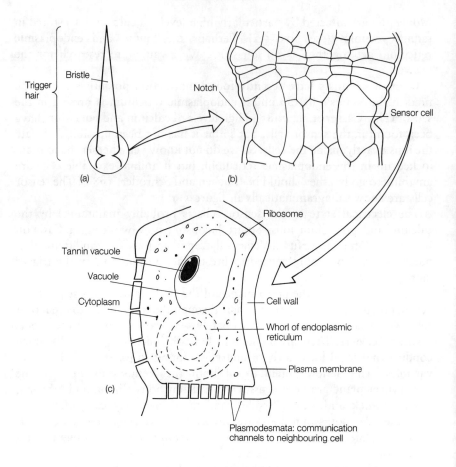

FIGURE 5.9 Sensor cells of Venus's flytrap.

waterwheel plant's trap lobes (Iijima and Sibaoka, 1982). The concentration of this ion was so strong that it pulled water out of the cells by osmosis – just as we saw for the trigger movement of *Stylidium*. So the natural assumption was made that the trap folded up because it lost turgor on one side of each trap lobe. And however the waterwheel plant worked, the Venus's flytrap was assumed to behave in the same way.

But just as things were looking cut and dried, Stephen Williams turned the applecart upside down. Having graduated from Washington University, he moved to a lectureship in biology at Lebanon Valley College, Pennsylvania. There he measured the surface area of the Venus's flytrap lobes before and after movement, and saw that the traps had enlarged after movement. Earlier botanists, including Darwin in the nineteenth century, had noticed

Box 5.2: Calcium
The importance of calcium is worth stressing. Although we generally think of it as something necessary for healthy teeth and bones, it is also vital for carrying messages inside and between cells. As electric current carried by calcium passes through a seaweed egg, for example, it waves a 'magic wand' in a way that no other ion can – it turns on all sorts of biochemical machinery, switching special nutrient channels on or off in the plasma membrane for example. The reason why living cells rely so heavily on calcium probably dates back to primeval times when living creatures in the sea had to cope with high concentrations of calcium, sodium and other ions in the salt water. They shut sodium ions out, and bound up calcium ions on special molecules inside the cell out of harm's way. So the levels of calcium and sodium inside the cells became much lower – sometimes a thousand times lower – than in the sea water outside. But calcium was unusual. If it leaked into the cell it was soon absorbed by 'scavenger' molecules. And this sudden temporary leakage of calcium would provide an ideal way of triggering rapid reactions in the cell. The table lists calcium-regulated processes in plants.

Calcium-regulated physiological processes in plants

Plant growth hormone action	Cell division
Auxin-induced elongation	Osmoregulation
Auxin binding	*Mimosa* leaf movement
Auxin transport	Pollen tip growth
Auxin-mediated H^+ secretion	Tuberization
Cytokinin-induced bud formation	Membrane structure and function
Abscission	Ultrastructural membrane alteration
Senescence and ripening	Membrane damage and leakiness
Cotyledon enlargement	Betacyanin synthesis
Secretory processes	GA-induced hypocotyl growth
Cell wall material secretion	Cation transport
α-Amylase secretion	Transmembrane Ca^{2+} fluxes
Polysaccharide secretion	Wound-induced nuclear migration
Peroxidase secretion	Circadian leaf movement
Tropism	Guard cell swelling
Gravitropism	Protoplast fusion
Spore germination	Freezing injury
Chloroplast movement	Cell polarity
Cytoplasmic streaming in *Nitella*	Ripening
Membrane depolarization of *Nitella* cells	Photosynthesis
Phototactic behaviour	Physiological disorders

something similar: when a trap reopens the epidermis (skin) on the inside of the lobes peels away, as if stretched to breaking point. Indeed, traps sometimes overstretch themselves and lose their mobility altogether.

Together with Alan Bennett of Cornell University, Williams went on to mimic a touch stimulus by gently infusing liquids of various pHs into the traps (Williams and Bennett, 1982). When the infusions were acidic, the trap shut spontaneously, but as the pH was made progressively more neutral the traps became sluggish, until at pH 5 (mildly alkaline) they were virtually paralysed.

What particularly excited the two botanists is that a similar phenomenon occurs during the *growth* of ordinary plant cells. As a plant cell enlarges it needs to loosen the tough corset of cellulose fibres in the cell wall enveloping it. Therefore the cell attacks the wall with acid, which activates enzymes, and these in turn snip away at the cellulose fibres. Because of the high water pressure inside, the cell then balloons out. The inescapable but incredible conclusion is that when the Venus's flytrap moves we are actually witnessing it *grow* at a phenomenal rate. This must be the fastest growth in the living world.

Hardly surprising for such a controversial theory, Williams and Bennett have come under attack from other botanists. Dieter Hodick and Andreas Sievers of Bonn University recently tried to repeat their pH-soaking experiments on the Venus's flytrap and failed, at least when the trap snaps shut (Hodick and Sievers, 1989). However, they did find that the much slower tightening of the trap around its prey (which follows the snap-shut movement) probably relied on pH and growth. But this is a far cry from the high speed of the snapping movement. Instead, they felt that the motor cells in the trap are kept squashed, somewhat like coiled springs, and then spring open. Quite what makes them spring open and force the trap to snap shut is not clear, but it may be that the motor cells leak potassium, water is lost and the cell pressure drops enough to let the cells spring open – on the face of it rather reminiscent of the rapid trigger bending of *Stylidium* (chapter 3).

Frank Litchner and Stephen Williams made another spectacular discovery (Litchner and Williams, 1977). They made a trap close without any food caught inside, and then investigated what would happen if they continued to stroke the digestive glands even after the trap lobes had shut. They found that the stroking provoked the glands into ejecting their juices. Litchner and Williams claim that it is quite natural for the plant glands to respond to touch, because insects caught inside the trap continue to thrash around and rub against the glands. They did not investigate whether rubbing stimulation triggers the secretion by electrical signalling through the gland, but this is a strong possibility. If action potentials are involved, this is the nearest that we have yet come to discovering a plant version of a stomach.

6

Electric Self-defence

For hundreds of millions of years the plant and animal kingdoms have been waging war against each other. Herbivores need to eat plants, and have developed teeth, digestive systems and all the other paraphernalia needed for vegetarianism. But the plants fought back with noxious chemicals, thorns, spikes, sticky hairs and even ants hired as mercenaries. Plants also developed camouflage, and few plants have mastered it more beautifully than *Lithops*, with a stony appearance that blends in with their rocky background. But some plant species have evolved camouflage to perfection: they disappear! The foliage is there one second and gone the next. And there are few better exponents of the disappearing leaf trick than the famous feather-like leaves of the Sensitive Plant (*Mimosa pudica*). Touching any of the pairs of leaflets sends them folding up successively like a row of collapsing deckchairs. Given a sufficiently hard whack, the main leaf stalk droops down. And when a leaf is wounded by cuts or burns, all the leaves on a plant slump.

The collapse of mimosas in their native tropics and semi-tropics is quite extraordinary. Anecdotes about the plant could fill another book, but consider this account by the Scottish botanist Robert Brown in 1874:

> The concussion caused by a horse galloping along the road on the sides of which the plant grows, will often have the effect of causing the leaves to fold up. We have often noticed this effect produced by the passing of a train along the Panama railroad in Grenada, on the sides of which the plant grows abundantly. So sensitive are they, that on one plant folding its leaflet, the contact will irritate its neighbour, and so on – the irritability travelling along the patch almost as fast as the traveller can keep up with it in walking.
>
> (Brown, 1874)

ELECTRIC SELF-DEFENCE

The reaction is reversible: the leaflets and leaf stalk slowly unfold again over a period of minutes following stimulation. The plant lives up to its name admirably – not only is it touch sensitive, but it is also excited by changes in temperature, changes in the direction and brightness of light, electric shocks, vibrations, cuts, burns, various chemicals and even changes in atmospheric pressure. It can be anaesthetized using chloroform or ether (further details are given in chapter 10) or paralysed with caffeine. In fact, rumour has it that a *Mimosa* at Kew Gardens was touched so much that it had a breakdown and shed all its leaves!

The sensitivity of *Mimosa* is so great that flying insects can trigger the movement even *before* they land on the leaves. Extrasensory perception? No, it works by static electricity, as Christine Jones and J.M. Wilson of the University College of North Wales found when they looked at touch-sensitive leaf movements in another plant, *Biophytum sensitivum* (Jones and Wilson, 1982). As insects almost touched them the leaves folded up. It is possible that as an insect beats its wings it builds up a considerable electrostatic charge on its body. Just before it lands on the plant the charge is transferred to the leaves and triggers the movement. You might ask what the insects think of this ploy. We can only guess that they feel something akin to astonishment.

The question of why sensitive leaves should fold at all has been more difficult to answer. It has often been said that it is a self-defence against animals, either by throwing off insects or by fooling larger animals that the leaves have vanished. Neither explanation is very convincing, particularly as no one has actually tested these ideas scientifically. But the world-renowned American biologist Tom Eisner has come up with an intriguing suggestion based on experiments, which lends it added credibility (Eisner, 1981). He looked closely at the touchy leaf of *Schrankia microphylla*. This species is a sprawling vine-like shrub, with leaves and leaflets as feathery as *Mimosa*, which is indigenous to the southern United States. The entire shoot is also covered in extremely sharp thorns which are densest along the stems and branches, like those in *Mimosa*. The art of any self-defence is to deploy weapons first and foremost as deterrents, and only secondly to use them in actual combat. We tend to think of thorns as clear warnings to feeding animals, yet in *Schrankia*, *Mimosa* and many others the thorns are hidden by the leaves, and so their deterrent value seems questionable. But after touching the *Schrankia* Eisner saw over twice as many thorns revealed by the folded leaves. The plant had truly bared its 'teeth', suggesting that a browsing animal – insect, mollusc, mammal or whatever – would get a thoroughly hostile reception if it tried grazing the plant. It is a canny strategy by the plant, holding back its ultimate weapon to the last minute, but otherwise allowing its leaves to carry on photosynthesizing as normal for the rest of the time.

Like all other legumes, *Mimosa* and *Schrankia* also fold their leaves up at night, again exposing their arsenal of thorns. Eisner speculates that, because the plant has no need to hold its leaves out at night, it has nothing to lose by folding them down and keeping the thorns continually exposed, warding off nocturnal predators. Perhaps the touchy leaf movements even evolved from the sleepy movements for just this reason, a theme I'll explore in more detail later on in this chapter.

According to Kerner von Marilaun (1904) the benefits do not stop there. It occurred to him that touchy leaves were good at fending off violent rainstorms, by diverting the falling rainwater down the stem. However, other scientists have found that, although *Mimosa* folds up at the start of heavy rain, it *learns* to ignore the rain after some time and reopens, although it still remains receptive to other touch stimuli (we explore this and other intriguing learning behaviour in chapter 11). Yet Kerner von Marilaun's idea received support from an unexpected quarter several years ago. John Dean and Alan Smith at the Smithsonian Institution's tropical research station in Panama discovered that the leaves of the rainforest liana *Machaerium arboreum* folded down when bombarded with rain (Dean and Smith, 1978). The drooping probably helps water to drain from the leaves' surfaces; otherwise it might suffocate them, reflect useful sunlight, leach valuable nutrients, weigh the leaves down and even encourage the unwelcome growth of epiphytic plants.

The leaf movements of *Mimosa* are so sensitive that they probably have several purposes. Kerner von Marilaun claimed that noonday heat also triggers movement, and may help conserve water in the leaves. That may indeed be true, but the problem with such a sensitive plant is that it's often difficult to disentangle which of many stimuli the plant is responding to. In the case of noonday movements, Kerner von Marilaun may not have realized that the leaves can carry on a daily self-driven rhythm that all leaf-moving plants perform, regardless of their environment. The purpose of these rhythmic movements is a complete mystery.

What is more certain is that *Mimosa* has successfully spread throughout large parts of the tropics. One species, *Mimosa pigra*, has become a weed in the wetlands of northern Australia since it was introduced from its natural home in Central America. It would be going too far to suggest that its touchiness alone contributed to its rampant invasion, but it was probably introduced into a botanial garden in the first place because of its leaf movements (Lonsdale and Braithwaite, 1988).

Before delving further into the mysteries of the *Mimosa*, one common myth must be destroyed. *Mimosa pudica* is no isolated freak of nature: it has several touchy cousins – *M. sensitiva*, *M. casta*, *M. dormiens*, *M. humilis*, *M. speggazinii*, *M. invisa* – as well as slightly further removed relatives in the same legume family, such as *Schrankia microphylla*, *Machaerium arboreum*, *Neptunia plena*, *Cassia fassiculata* and *Aeschynomene indica* (the pith plant, whose stem pith forms the basis of the sun helmet industry in Calcutta). More interesting still, there are tactile leaves in completely unrelated families: the Indian *Smithia sensitiva* of the Gesneriaceae family, and *Biophytum sensitivum* from Brazil in the Oxalidaceae family (the relationships between all these species is shown in 'Taxonomy of mobile plants' at the end of the book).

What is fascinating is how such a wide selection of species all came to evolve touch-sensitive leaf movements. Did they arise spontaneously in several families at various times? Or do all plants have a residual touchiness left over from some sort of ancestral past? One clue is that the leaf movements don't all have the same intensity – some plants are more sensitive and vigorous movers than others. For instance, *Biophytum sensitivum* is a vigorous and sensitive leaf mover, but *Oxalis hedesera* in the same family has a very sluggish and extremely restricted movement, whilst another relative that lives in temperate woodlands, the wood sorrel (*Oxalis acetosella*), is so insensitive that it needs repeated stroking or a heavy rainfall before it even starts responding. Thus some species have evolved touchy movements further than others. But what did they all evolve from? As we'll constantly be seeing in this book, touchiness and mobility are widespread features of the plant kingdom; it's just that some species are better at displaying their excitability than others.

How *Mimosa* Works

How does a touchy leaf work? The answer falls into three parts:

the touch stimulus is perceived by the plant;
once perceived, the excitement is signalled to the rest of the plant;
this signal then turns on the motor cells to make the leaf actually move.

Given its animal-like behaviour, you might think that *Mimosa* works exactly like the carnivorous plant traps, but this is only partly true. For one thing, the *Mimosa* leaf has nothing like a trigger hair for picking up touch stimulation, and indeed has no obvious touch sensor at all. For such a sophisticated plant this has both shocked and frustrated all the scientists who have tried tracking down where and how *Mimosa* detects its tactile stimuli. What they found instead were leaves and cells that look no different from those in 'ordinary' plants like its cousin *Mimosa dealbata*, which has no touchy leaf movements. The only clue as to how *Mimosa* converts a blow to its leaves into a message that excites the rest of the plant came by accident, because it wasn't recognized by its discoverer. Takao Sibaoka of Tokyo University recorded electrical disturbances in *Mimosa* leaves which had the classic hallmarks of a receptor potential (Sibaoka, 1954). He also obtained similar recordings from the unrelated touch-sensitive species *Biophytum dendroides* (Sibaoka, 1973). As we saw earlier for Venus's flytrap, the receptor potentials of the *Mimosa* and *Biophytum* translate the strength of a physical blow into electrical code, and when this is sufficiently intense an action potential is triggered. And that is all we know about touch perception in touchy leaves.

So let us pass to the next stage of movement: how messages are signalled to the surrounding leaves. The history of research into the *Mimosa*'s signalling is long and tortuous, and it is a classic example of how scientists can become flummoxed when they all jump onto one bandwagon.

Seventeenth-century scientists realized that the co-ordination of the *Mimosa*'s leaf movements involved some sort of signalling. But a plausible theory was not put forward until 1837. The great French plant physiologist Henri Dutrochet, who amongst other things discovered the phenomenon of osmosis, tried to find the route along which the excitation passed. He cut into the stem until he cut off the signals. The phloem (sugar-conducting channels) was not involved, because the leaves carried on folding when these tissues were cut. But severing the xylem – the channels which conduct water and salts – did block the signal, and so Dutrochet proposed that touching the plant squeezed water inside the xylem, sending a pressure wave to the other leaves rather like an old-fashioned hydraulic lift, which is perhaps where he got his inspiration from.

It was an elegant theory, but it was wrong. We now know that the signal *can* travel across wounded phloem as well as xylem tissue, but this error misled others for over fifty years. Yet Dutrochet's reputation was such that his theory seduced a succession of German plant physiologists. Their convictions were strengthened when they all saw water oozing out of cuts in the stem as the signal travelled from one leaf to another. The only difference of opinion was whether, as argued by Gottlieb Haberlandt and others, the phloem instead of the xylem conducted the pressure (Haberlandt, 1914). As we will see, they were all barking up the wrong tree.

The old water-pressure idea was finally shattered by a sensational theory in 1916 – which ultimately created even more confusion! An Italian botanist, Ubaldo Ricca of the University of Genoa, had taken an interest in a wide range of plant movements for several years (Ricca, 1908). He was unconvinced by the pressure theory. For one thing, a cut leaf held in the air could be stimulated over and over again, even though its hydraulic pressure had collapsed.

If it was not hydraulic pressure, what else travelled through the stem that could spark off leaf movement? One way to find out was to set a trap for any chemicals that might be involved. Ricca severed a *Mimosa* stem and reconnected the cut ends with a glass tube filled with water. Any chemical signal would have to pass through the water trap to stimulate the leaves on the other side. And that was what he found: when he wounded the bottom of the stem beneath the trap, the leaves above closed as normal. Simultaneous pressure measurements showed that there was no change in water pressure, and so the pressure theory of signalling had finally been put to rest.

The chemical signalling theory was finally clinched with an elegant *coup de grâce*. Ricca found that he could spark off movement in unstimulated leaves simply by using the sap collected from the cut ends of stimulated leaves. He proposed that *Mimosa* movements were controlled by a chemical signal which was released by stimulation into the transpiration stream, where it was swept along and excited all the leaves into movement.

Ricca's factor, as it became known, made a great impression in botanical circles, and was soon accepted as the signalling system in the *Mimosa*. But there is also another signal at work which was completely overlooked, even though it was discovered earlier on and even acknowledged by Ricca, and has a great bearing on how Ricca's factor works.

Electrical Signals

The rapidly developing field of d.c. electricity in the nineteenth and early twentieth centuries led to a craze for electric shock treatment. From the simplest living cells to humans, electrocution was the vogue, and it was inevitable that plants were treated as well. Probably the first electrocution given to *Mimosa* was by the great German physicist and pharmacist Johann Ritter, whose interest in electricity had also led him to invent the dry electric battery. In 1811 he succeeded in stimulating *Mimosa* with an electric shock, but as these were the days before Du Bois Reymond's discovery of electrical signals in animal nerves, the significance of his experiment was lost.

However, there should have been no excuse after the discovery of action potentials in the Venus's flytrap by Burdon-Sanderson in 1873. Yet although electrical signals *were* discovered in *Mimosa* a few years later by the German botanist Karl Kunkel, a professor at Heidelberg University, their significance was still not recognized (Kunkel, 1878). Kunkel dismissed the signal as a by-product of the 'real' signal, the wave of 'water pressure'. Even though the German physiologist Biedermann of Jena University recognized Kunkel's signals as electrical excitability (Biedermann, 1898), it was not until 1907 that the Indian botanist Sir Jagadis Bose established the existence of action potentials and their importance in triggering the movement in *Mimosa* (Bose, 1907). However, the story now becomes rather murky. The German botanists were unimpressed with Bose's work. With their water-pressure theory and then the conversion to Ricca's chemical explanation, they had no need for an electrical theory as well, and they dismissed the electrical signals as an irrelevance.

The attitude of Burdon-Sanderson, the pioneer of plant electrophysiology, was altogether more baffling. He too refused to believe Bose's results. Why he was so antagonistic amazes me. Was it professional jealousy because he himself had not investigated the *Mimosa*? Or perhaps Bose had failed to acknowledge Burdon-Sanderson's pioneering discovery of electrical signals in the Venus's flytrap? More probably there was an edge of doubt about Bose's professional competence. In a meeting of the Royal Society in London where Bose presented his discovery, Burdon-Sanderson went out of his way to criticize the Indian. As he commented in his referee's report on the first paper that Bose submitted on plant excitability in 1903:

> The defect of the research is that the methods of observing the order of succession and duration of the phenomena are crude ... it is to be regretted that the author [Bose] has not used the methods by which alone these can be accurately investigated.
>
> (Burdon-Sanderson, 1903)

Therefore Burdon-Sanderson recommended that the paper should be rejected for publication in the *Philosophical Transactions of the Royal Society of London*.

There is a case that Bose was much to blame himself. He was a controversial figure to say the least, and the Victorian science establishment simply could not tolerate mavericks. On the one hand, he carried out some impressive experiments first on radio waves and then on the electrical signalling of plants such as *Mimosa*. However, he also had some outlandish concepts. He was convinced that lifeless matter such as metals behaved like organic life, with electrical messages susceptible to poisoning. And his interpretations of the electrical signals he recorded in plants were sometimes disastrously wrong – he also persisted with Kunkel's idea that electrical signals were produced by waves of water suddenly travelling through leaves. He failed to appreciate how living things conduct electrical impulses using ions, a phenomenon that was beginning to be appreciated even at the turn of this century. In addition, his descriptions of his experiments and their results were sketchy, often making it very difficult to reproduce them. Nevertheless, Bose made some impressive discoveries on *Mimosa*. Further analysis of Bose's contributions to plant physiology is given in an excellent summary by Galston and Slayman (1979).

So what exactly did Bose discover in *Mimosa*? He was the first person to appreciate fully that electrical signals control leaf movements in this plant (Bose, 1906). By studying the relationship between electrical signals and movements in *Mimosa* and *Biophytum sensitivum*, he showed that plant excitability has many features in common with animal nerves:

the plant becomes fatigued if exercised too frequently;
stimuli too weak to cause movement on their own can build up into a sufficiently strong signal that eventually triggers movement;
each movement has a waiting (latent) period before a response is apparent.

These plants also show the same features that Burdon-Sanderson had already demonstrated in Venus's flytrap:

the all-or-nothing response;
resting (refractory) periods;
successive stimuli of equal strength excite the plant more and more, facilitating its movements.

Therefore we have a striking parallel to the plant carnivore movements: action potentials orchestrating a movement. Bose, incidentally, also discovered the same phenomenon in another species of *Mimosa* (*Mimosa speggazinii*), and later scientists discovered the signals in another legume, *Neptunia plena*; thus *Mimosa pudica* is no exceptional case.

But, because of the demise of Bose's research, confusion has reigned ever since over whether *Mimosa* uses a chemical or electrical signalling system. Many of today's textbooks *still* do not appreciate the importance of the electrical signals.

The confusion between the two theories was finally ironed out in 1935 by the Dutch botanist A.L. Houwink of the Botanical Institute, Utrecht. In a painstakingly thorough study, Houwink (1935) showed that the chemical and electrical theories worked hand-in-hand. Ricca's factor was only released from injured tissues, was sucked into the transpiration stream and triggered a distinctive type of electrical impulse – dubbed the 'variation potential'. This rather erratic wave of electricity in turn triggered an action potential, which swept through the leaves to the motor tissues, telling them to move (figure 6.1(b)).

Later, in chapter 10, I will be looking more closely at what exactly Ricca's factor consists of, the variety of other plants it has been discovered in and how it could be exploited. For the time being, however, we need to find out what creates the electricity and exactly how it travels through the plant.

The Nerve of *Mimosa*?

How are electrical signals carried in *Mimosa* and other plants – surely they must have nerves? The answer is that there are no nerves as such. Ricca's factor moves through the dead cells of the xylem, sparking off the variation potential in the surrounding living xylem parenchyma (packing cells). So Dutrochet's old idea that the signal moved through the xylem *was* partly

ELECTRIC SELF-DEFENCE

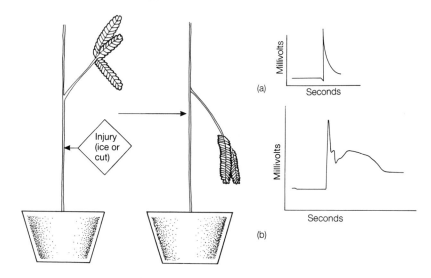

FIGURE 6.1 (a) Cooling the stem of the Sensitive Plant (*Mimosa pudica*) by putting ice onto it triggers the pulvini to move. The sharp spike on the electrical recording is an action potential, passing up the stem from the iced zone, and signalling the leaves to collapse. (b) Injuring *Mimosa* triggers the release of special hormones called Ricca's factor. These substances fire a different type of electrical signal, the variation potential, which in turn triggers an action potential. The electrical impulses travel fast and far throughout the plant's shoot. Similar responses to injury also occur in 'ordinary' plants such as tomatoes and cotton.
(After Houwink, 1935)

correct. But in 1966 Takao Sibaoka found that the action potential also passes through the phloem and *its* surrounding tissue – proving Haberlandt correct as well (Sibaoka, 1966)!

Without having to evolve nerves as such, the *Mimosa* has neatly adapted its existing food and water conduits and their neighbouring tissues for carrying electrical signals. The only specialization of these so-called vascular tissues is a sheath of dead cells around them, rather like the insulation around a cable (Fleurat-Lessard and Roblin, 1982). This merely helps channel the signal to its target motor tissues in the pulvinus instead of letting it fade away into other tissues.

Haberlandt's meticulous microscope work also uncovered what he thought were special 'express' routes: elongated phloem cells, much longer than normal, which he called *Schlauchzellen*. At first sight these look suspiciously like nerve cells, and it was believed that they carried high speed signals, but their significance is now questionable. Another veteran plant electrophysiologist, Karl Umrath, working with Gerald Kastberger at Graz

University, recently found high speed signals in the stem of Neptunia plena (Umrath and Kastberger, 1983). However, Neptunia has no Schlauchzellen, so they are obviously not needed for high speed signals, at least in this plant.

What Haberlandt may have seen is simply 'ordinary' phloem cells called sieve tubes, particularly long tubular cells which carry most of a plant's sugar supplies. The electrical behaviour of sieve tubes is notoriously difficult to measure with ordinary electrodes because these cells are delicate and prone to rupture, but recently an ingenious technique has been developed to solve this problem.

Greenfly and other aphids have an exclusive taste for the sweet phloem sap which they suck out using long slender mouth parts. If the feeding insects are decapitated, their mouth parts continue feeding, providing a perfect conduit for reaching the phloem with electrodes. Using this decapitation technique Jörg Fromm and Walter Eschrich of Göttingen University measured strong action potentials passing along phloem sieve tubes. Once an action potential arrived in the pulvinus, sucrose was suddenly dumped from the sieve tube into the rest of the pulvinus, and this was closely followed by leaf movement (Fromm and Eschrich, 1988). Quite what significance sucrose may have in the movement is not entirely clear, but 'soaking' the motor tissues in sugar may help collapse the motor cells by drawing out their water by osmosis.

How are the electrical signals carried through the cells? Once again they involve the shuttling of electrically charged chloride, potassium and calcium ions in and out of the excitable cells. These ions have all been implicated because altering their concentrations affects the signals. Also, Hideo Toriyama (1962) found that potassium salts moved from the inside of vascular tissue into adjoining spaces following the passage of an action potential. This is analogous to how an animal nerve works (figures 5.8 and 10.2 and pages 184–8).

Movement

How does the motor organ, known as a pulvinus, respond to an electrical signal? Again, there is nothing spectacular in its anatomy that gives the game away. From the outside, a pulvinus looks like a hairy rugby football carried at the base of the leaflets or leaf stalk (figure 6.2). The inside is a spongy mass of thousands of fairly ordinary-looking parenchyma cells surrounding a thin core of phloem and xylem tissue. Normally, the vascular strand is fairly strong, helping to support a leaf, but in the Mimosa pulvinus it is so thin and feeble that it gives the pulvinus great flexibility. The living phloem and xylem cells also have abundant membranes, which would help

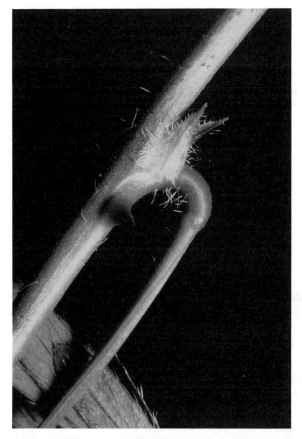

FIGURE 6.2 The motor of the Sensitive Plant (*Mimosa pudica*) is located in the swollen base (pulvinus) of each leaf stalk. After receiving electrical signals (action potentials), the motor cells on one side of the pulvinus flush out potassium ions and then jettison water by osmosis. With the massive loss of water, the motor cells collapse, the pulvinus bends over, and the rest of the leaf folds. There is also evidence that muscle-like fibres may also help to collapse the motor cells. If so, these are probably the closest things to muscles in the higher plant world.
(Photograph by Paul Simons)

with the intense movements of the ions. The vascular tissue is connected to the other tissues through an extensive network of plasmodesmata, which help carry the action potential out to the motor cells. Similar features are also seen in other plant parts concerned with unusually heavy traffic of chemicals, such as nectaries, root nodules which manufacture nitrogen compounds from the air and many others (Fleurat-Lessard and Bonnemain, 1978).

So how does the *Mimosa* motor work? The German physiologist Ernest Brucke, working in Berlin in 1848, made a quite surprising discovery. He removed the top half of the pulvinus, and the pulvinus still moved! Later German plant physiologists, such as Wilhelm Pfeffer, found that the pulvinus moved by pumping water out of its lower half so that it collapsed. In an exciting application of a new technology, the water pumping theory has just been confirmed using NMR (nuclear magnetic resonance), which is now finding applications in botany after its introduction into medical body scanners. Water is lost from cells on the underside of the pulvinus and almost immediately taken up by cells on the top half of the pulvinus (Tamiya et al., 1988). This technique allows us to see, for the first time, the movement of water in living tissues without having to cut them open, and hopefully will herald a new phase in the study of water-driven plant movements.

Blackman and Paine (1918) noted that the escape of water was also linked to an escape of ions. Only in 1955 did Hideo Toriyama, using a special stain on thin wafers of *Mimosa* tissue, observe that potassium escaped from the motor cells on the underside of a pulvinus (Toriyama, 1955). Later work showed that sufficient potassium and chloride was escaping from these cells to draw water out of the motor cells by osmosis, leading to their collapse and the folding down of the pulvinus. These cells – at least those on the 'underside' of the leaf stalk pulvinus – then collapse by jettisoning most of their water. When the ions return to their original starting positions, the motor cells soak up water again and re-inflate the pulvinus, ready to start moving all over again. This is a classic example of ions and hydraulics driving a plant movement.

Is there anything special about the motor cells themselves that would help the engine work? At first sight apparently not. They look like fairly ordinary parenchyma cells, although their cell walls are unusual because they crumple easily like a concertina, and this obviously helps the pulvinus collapse. Similarly, the contents of the cell crumple up, in much the same fashion as the contents of sundew tentacles or stomata (Fleurat-Lessard, 1988). Toriyama (1953) also discovered that some of the vacuoles contain something better known for its taste in tea leaves – tannin. This substance usually deters insect pests, but in the case of *Mimosa* the tannin vacuoles hold unusually large deposits of potassium and, in some motor cells, calcium (Campbell et al., 1979). So the tannin vacuoles are important reserves of ions for the pulvinus to draw on during its movement.

But this is far from the whole story. There may be one, or possibly two, other motors at work. In 1951 Marvin Weintraub of Toronto University mounted living motor tissue onto a microscope and saw tiny vacuoles squeezing shut, rather like the so-called contractile vacuoles common in single-celled protozoans (Weintraub, 1951). These regulate the water balances of the cells, like buckets bailing out a boat with too much water on board – it is a very effective way of preventing cells from blowing themselves up with too much water. And it is interesting that electrical activity has been associated with the emptying of contractile vacuoles in *Amoeba proteus* (Josephson, 1966). Whether anything similar occurs in *Mimosa* is guesswork. Indeed, this theory has received little attention in recent years.

However, there is more than a suspicion that an even more incredible motor is also involved: muscular contraction. The evidence so far is rather fragmented, but when added together it is quite tantalizing. Under the electron microscope, small fibres can be seen straddling the large cell vacuole, or across the cytoplasm, and these shorten after the movement (figure 6.3) (Toriyama and Satô, 1968). A protein capable of contracting has been extracted from the motor cells (Biswas and Bose, 1972); motor cell extracts recognize and bind to actin from animal muscles (Fleurat-Lessard et

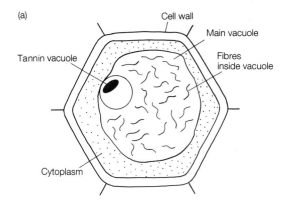

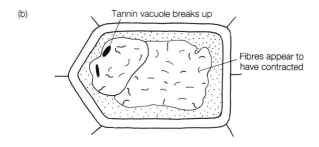

FIGURE 6.3 Schematic diagram of the motor cell of *Mimosa pudica*: (a) before movement; (b) after movement.
(After Toriyama and Satô, 1968)

al., cited by Fleurat-Lessard et al., 1988); and the motor cell movement is blocked by drugs that affect the actomyosin contractions involved in cytoplasmic streaming (Fleurat-Lessard et al., 1988). Toriyama and Jaffe (1972) discovered that calcium held in the tannin vacuoles before the pulvinus moves appear to migrate to the central vacuole, where the fibres may then contract. All this is reminiscent of the contraction of animal muscle fibres, in which calcium plays a key role in the contractions of the protein actomyosin.

Yet there are many puzzles left to solve. What makes the motor cells so different from their neighbours that show no movement at all? How does the action potential trigger such a colossal loss of ions from the motor cells? What is the significance, if any, of the tiny 'contracting' fibres straddling the interior of motor cells? And are these fibres anything like those found in other plant leaves which move more slowly (p. 142)? As the techniques for detecting water, ions and actomyosin improve, we should get a better answer to these questions and find out whether the *Mimosa* motor is a freak or the evolutionary refinement of something much more common in the plant kingdom – something we'll consider in the next chapter.

7

SEEING THE LIGHT

A Sense of Vision

Place a tray of mustard and cress seedlings by a window and a day later the little shoots have craned themselves towards the daylight – they have recognized the direction of light. Plants have a primitive sense of 'vision', albeit very different from ours because they cannot see or understand images, but it is a powerful sense none the less. For convenience, however, I'll refer to it as plant vision.

What plants see is a heavenly world of reds and blues, tinged with all the other colours of the spectrum apart from green (this is the one wavelength of light that plants reject and reflect, which is why leaves look green). They can separate colour tones far too subtle for us to distinguish, using pigments of which some are astonishingly like those in our own eyes. Indeed, plants and animals probably evolved their sense of vision from the same primeval creatures which existed hundreds of millions of years ago.

So why do plants have a sense of vision? A plant tries to get just the right light intensity and spectrum for photosynthesis. Plants also use light for metering their overall growth and development: they can estimate their depth under the soil or under water, how heavily they are shaded by surrounding vegetation, and the season for germinating seeds or for flowering or for setting buds. They even use light for telling the time of day.

The best known plant response to light is probably phototropism, which is admirably demonstrated by the mustard and cress example. When these plants grow towards light they are making a commitment. Their cells have elongated more on one side than the other, and literally push the shoot over. If, at a later date, they need to change direction for any reason, they must grow and bend again. That's the drawback to phototropism – once the plant has finished growing there is no more flexibility left to respond and the phototropism stops.

But there are other light-sensing plant movements that are unfettered by growth and are performed repeatedly throughout a plant's life. Single-celled plants can quite simply swim to the light, or if the light is too bright they can turn tail and head back into the shade.

Swimming under the influence of light originally evolved in bacteria, some of which developed the power of photosynthesis. They show two sorts of movement controlled by light: photophobia and phototaxis. Photophobia is *not* a fear of light! It is movement caused by sudden changes in light intensity, irrespective of the direction from which the light is shining. Thus, photosynthesizing bacteria swim faster as the light grows stronger and slower as it grows weaker. They quite literally follow the bright lights. On the other hand phototaxis is a movement governed by the direction of light – bacteria usually swim towards light.

This rudimentary sense of vision became much more refined in a group of organisms that bacteriologists call the cyanobacteria but botanists claim as the blue-green algae, and which henceforth I'll simply call the blue-greens.

Whether or not they are bacteria or algae, most of these creatures are blue-green in colour. A few are red, and under the right conditions bloom on the surface of oceans into a red tide – hence the name Red Sea. Other blue-greens bloom on fresh water, a worrying phenomenon that is becoming more frequent as polluted rivers collect excess fertilizers, drained from farmlands, which the blue-greens thrive on. Other blue-greens tolerate the harshest conditions that any living organism can survive: ice, hot springs and the surfaces of desert rocks. They even cohabit with fungi as lichens.

The blue-greens are microscopic: some are single cells, but most are clumps or filaments of cells, or are organized in patterns rather like the radiating spokes of a bicycle wheel, and are often covered in jelly (figure 7.1). They can execute the most striking movements. The ends of the filaments can often wave gently; in others the whole thread glides along, or stretches itself and contracts again, coiling up and straightening out like a caterpillar. If there is a sudden drop in the brightness of light, blue-greens (like the photophobic bacteria) will suddenly stop dead and then back-track exactly along their previous path. In nature, this movement helps the algae to find good light for photosynthesis. This phenomenon is also neatly exploited by researchers to trap the algae using pools of strong light.

The blue-greens react to light in only 10 seconds, and the speed of this response prompted the German plant physiologist Donat Häder of Phillips University, Marburg, to look for signs of an electrical signal controlling the movement. After all, what other physical or chemical message could react this fast? Häder tried interfering with the photophobic gliding movement of the blue-green *Phormidium uncinatum* by passing electric currents through

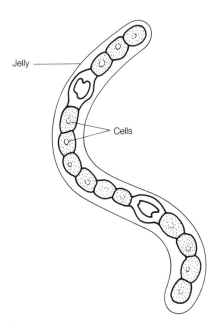

FIGURE 7.1 Blue-green algae sense the direction of light and respond by waving, gliding or caterpillar-like wiggling. They also sense the brightness of light. Their movements seem to rely on electrical activity, but exactly how their signals or movements work is not known.

the liquid in which it was bathed. Those filaments aligned parallel to the electric field stopped gliding (Häder, 1977). When Häder recorded the electrical voltage on the surface of the cell, it responded to the change in light after about 10 seconds – the same time as that elapsing before the start of movement (Häder, 1978). The electrical activity may come from an electrical 'spillage' created by the alga's photosynthesis as it responds to the light, but these results are taken as a strong hint that *Phormidium* uses electricity to trigger its movements. However, we have no idea of how it does this, or how the movement itself works.

Blue-greens also exhibit phototaxis, although not much is known about how they do this. *Phormidium* moves towards the light with a shuffling waltz – moving backwards and then forwards, each time taking larger strides towards the light. The blue-green *Anabaena* has a more relaxed attitude, and in true sunbathing fashion merely twists its body towards the direction of light.

Most other algae exhibit phototaxis. The desmids are a good example – they are a large family of fresh-water algae with exquisitely sculptured bodies that are quite beautiful to look at under the microscope. They are typically divided into two symmetrical halves, each of which is divided into deep constrictions, rather like a walnut. They slither along on a slimy secretion in slow creeping movements, gradually drawing the cell along the ground. Some desmids, like *Closterium* and *Penium*, have a curious rotating movement, with one end fixed to a support whilst the other swings up to the source of light. We know very little about these fascinating displays.

Another class of algae, the diatoms, also rely on slime for their movement. These are probably the most stunning of all algae, best imagined as a rigid pill box with the lid and box heavily embroidered with silica patterns so delicate that microscope makers used to use them to calibrate their lenses. Many diatoms live in mud or at the bottom of pools where they move phototactically towards the light during the day and then retreat back into the mud at night. Other diatom species glide over their substratum, or push forward in fits and starts. They double round objects or push them out of their way with one of their hard points. They too move on a slippery trail of slime, driven by 'muscles', and we'll return to them later (chapter 9).

Light-sensitive movements are also performed by fungi. Häder and Poff (1979) observed that the amoebae of the slime mould *Dictyostelium discoideum* first gathered into clumps under strong light, but then dispersed again. These too are driven by 'muscles'.

Swimming Under the Sun

The slitherings of the diatoms look pedestrian compared with the odysseys of the flagellate algae. We've already seen that taste-sensitive flagella or cilia are ideal motors for sex cells in fungi, algae, mosses, ferns and the other lower plants. It is highly likely that they operate using electrical signals. What I'm going to explain now is how swimming algae also use a sense of light to steer their movements.

This is an exciting time in the investigation of vision in both plants and animals. Many years ago physiologist William A.H. Rushton of Cambridge University wrote:

> Molecules respond to light as people do to music. Some absorb nothing. Others respond by the degraded vibration of foot or finger. But some there are who rise and dance and change partners.

And so it is with light-sensitive molecules, the so-called photoreceptors. Only minute amounts of photoreceptors are needed for sensing light, and once excited they fire off an explosive chain reaction, eventually triggering nerve impulses. Thus the photoreceptors are the guts of any sense of vision.

The power of photoreceptors in plants was discovered by accident, when one day over 150 years ago a pitcher of red wine stood between a plant and light coming from a window. Ordinarily you would expect the plant to bend towards the light, but the wine filtered out the blue light in the sunshine and the plant no longer 'saw' the direction from which the light was shining. So it grew straight instead of bent.

This tasteful experiment inspired the search for a photoreceptor sensitive to blue light. All sorts of plant pigments were tested by comparing their reactions to various wavelengths (colours) of light. The results threw up two prime candidates – carotenoids and flavenoids – and scientists have been squabbling over their significance ever since.

Carotenoids give carrots their orange colour, and also provide the yellow of buttercups, the pink of flamingos and the red of lobsters. The carotenoids were first nominated as light-sensitive receptors in plants as early as the mid-1930s, and were found to govern the phototropism of oat coleoptiles (the sheath covering a young shoot) and the spore-bearing stalks of the two fungi *Phycomyces* and *Pilobolus*.

And there is a strange twist to this story – animals also need carotenoids for their own vision. When you walk from bright light into a dark room everything at first looks completely black, but after a short while your vision returns, and black and white images can be picked out. The transition is caused by a switch-over from our normal daylight sensor cells to dim light sensor cells in our retinas. These dim light sensors have caused great excitement amongst animal physiologists over the past couple of decades. They contain a photoreceptor called rhodopsin, also known as visual purple, which is made up of two partners: retinal, derived from vitamin A and closely related to the carotenoids, combined with a protein called opsin. (Incidentally, the idea that carotenoids from carrots improve your night sight is not true – it was a ploy in the last war to make people eat more carrots!)

The partnership of retinal and opsin works well, and physiologists have now mapped out a large part of how they work together. When the retinal absorbs light it changes shape, causing the opsin to turn on a cascade of reactions which eventually trigger an impulse into the optic nerve. From the simplest animal to our own retinas, rhodopsin helps give the power of sight throughout the animal kingdom. But its evolutionary roots are intriguing.

In hot salty ponds, like those around the Dead Sea, live bacteria that have such an addiction for salt that they can't grow and reproduce without copious supplies of it. They are known as the halobacteria or purple bacteria because of the purple spots on their plasma membrane. Biologists became fascinated in understanding how anything could survive under these conditions, and so they looked for special characteristics in the bacteria. One obvious thing is that they can photosynthesize, not with chlorophyll as you'd expect in a plant, but with their purple spots, turning sunlight into chemical energy. Halobacteria and methane-producing bacteria are the only living beings other than plants that can photosynthesize. Solar energy enthusiasts have even suggested that the system might be usefully harnessed to generate electricity. The halobacteria are so extraordinary that some people have suggested that they and the methane-producing bacteria should belong to a sixth kingdom of life, the Archaebacteria.

As part of their adaptation to life in hot shallow salty pools, halobacteria also use flagella to swim towards bright light but avoid dangerously high levels of ultraviolet radiation by reversing their swimming. That in itself is remarkable, but in 1971 the biochemists Dieter Oesterhelt and Walther Stoeckenius of the Cardiovascular Research Institute, San Francisco, shocked the world of zoology. They identified the purple pigment on the halobacterium's membrane as rhodopsin, the visual pigment in animal eyes. As they commented in their paper: 'The similarity of the purple membrane [of the bacterium] to the visual pigment [in higher animals] is striking' (Oesterheit and Stoeckenius, 1971). In fact *Halobacterium* uses one type of rhodopsin for photosynthesis and another type for some of its photophobic swimming.

Previously, zoologists had believed that rhodopsin was unique to the retinas of higher animals. The halobacteria shattered their illusions, but botanists were in for a rude shock as well.

The single-cell alga *Chlamydomonas* lives in fresh-water ponds, is about 0.01 millimetres in diameter and is powered by two flagella at its front (figure 7.2). It swims in a spiral path with a breast-stroke, flagella first, beating backwards over opposite sides of its body by phototaxis and detecting light with an 'eye' on one side which is known as an eyespot. By swimming towards or away from the light, it locks the axis of its spiral path to the sun's rays.

Carotenoids were long thought to be the photoreceptor in *Chlamydomonas* phototaxis. But in 1980 Kenneth Foster and biochemist Robert Smyth of New York University stunned the world of botany. They noticed that the ideal wavelengths of light for phototaxis in *Chlamydomonas* coincided exactly with the peak activity of rhodopsin from cow eyes, as they reported in a provocatively entitled paper: 'The visual system of green algae: a forerunner of human vision' (Foster and Smyth, 1980a). However, because carotenoids are related to retinal, there was always a faint possibility that the two substances might have been confused (Foster and Smyth, 1980b). So Foster put the animal rhodopsin to the test. He assembled a large team of pharmacologists and chemists, and they performed a minor miracle: they restored the sight of a mutant blind strain of *Chlamydomonas*. This strain cannot make carotenoids and cannot perform phototaxis, but by using various types of retinal usually found in cow eyes the alga's sight was

SEEING THE LIGHT

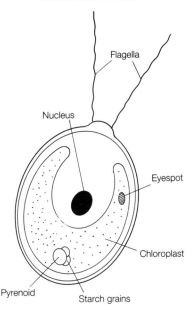

FIGURE 7.2 *Chlamydomonas* swims with two beating flagella, guided by the direction and brightness of light. Recent research reveals that this sense of vision relies on a light-sensitive pigment, rhodopsin – the same as in our own retinas. The analogy with human eyes may carry further, because there's good evidence that the swimming is controlled by electrical signals, although this has not been related to rhodopsin's activity yet.

miraculously restored (Foster et al., 1984). The results were unmistakable: the spectrum of activity of the newly sighted alga coincided precisely with that of cow rhodopsin. In other words, the alga 'saw' with a cow's light sensor, combining the animal retinal with its own opsin already present in its eyespot. Such a combination of animal retinal with an algal opsin is powerful testimony to how well rhodopsin has survived during the course of evolution from bacteria, algae and primitive animals up to the highest animals. The inescapable conclusion is that, because the alga's rhodopsin is so close to our own, human eyesight may well have evolved from the 'eye' of a bacterium or algal-like creature (figure 7.3).

Foster's team also suspect that a variety of other algae also use rhodopsin in their phototactic behaviour. What needs to be worked out now is the pathway that leads from rhodopsin to movement. The signal presumably involves changes in membrane voltage, because the appropriate wavelengths of light trigger electrical signals in a related alga, *Haematococcus* (Litvin et al., 1978). *Chlamydomonas* flagella are also sensitive to calcium ions, and so

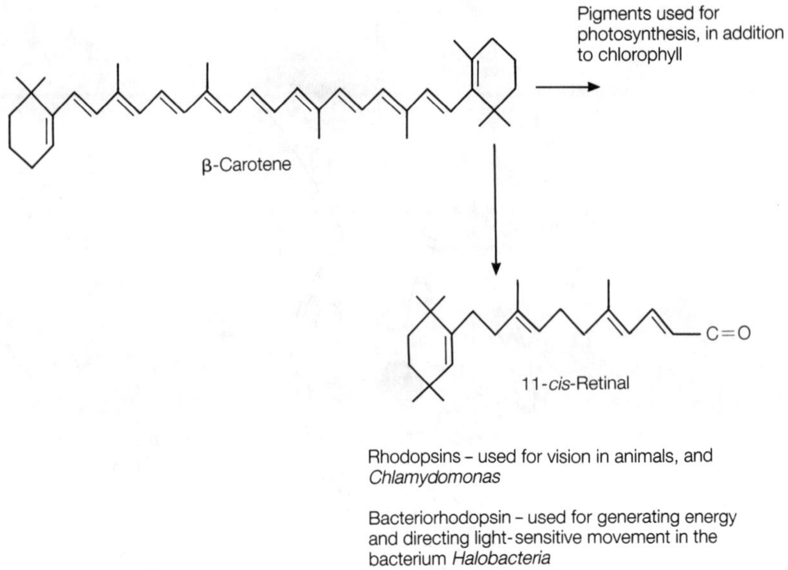

SEEING THE LIGHT

β-Carotene

Pigments used for photosynthesis, in addition to chlorophyll

11-*cis*-Retinal

Rhodopsins – used for vision in animals, and *Chlamydomonas*

Bacteriorhodopsin – used for generating energy and directing light-sensitive movement in the bacterium *Halobacteria*

FIGURE 7.3 Biological light-sensitive molecules have probably evolved from similar molecules.

electrical signals might open up channels in the flagella for calcium to pass through, as happens in the animal-like protozoan *Paramecium* (box 7.1). Little is known about how the alga's rhodopsin behaves, but it is a question that interests animal physiologists since the beauty of studying *Chlamydomonas* is its simplicity compared with animal eyes.

Further clues to the mysteries of algal vision have been discovered in the light-sensitive movements of *Euglena* – a controversial relative of *Chlamydomonas*. Biologists agonize over how to classify it, because *Euglena* has the attributes of both a plant and an animal: it photosynthesizes like a plant, but turns off the photosynthesis when conditions become difficult and reverts instead to an animal-like diet of rotting debris.

The 'plant' is a single pear-shaped cell (figure 7.4(a)). It lives in brackish or fresh water rich in organic matter, often giving ponds a greenish hue. Like all flagellates, *Euglena* swims in a helical, or spiral, path (figure 7.4(b)). Like the blue-greens, it senses the intensity and the direction of light. It does both jobs with a photoreceptor organ bulging out from the base of its tail and stretching along its gullet. Opposite the gullet lies an eyespot. When light shines from the eyespot's side, it casts a shadow onto the photoreceptor, which then triggers a sideways wobble in the beating of the flagellum, pushing the organism over. Eventually, after many wobbles, its front end points straight into the light, eliminating shading of the photoreceptor and stopping the corrective behaviour.

Like *Paramecium* (box 7.1), injecting an electric current into *Euglena* disturbs the beating of its flagellum. Passing currents through a pool of euglenas mimics their photophobic behaviour. So this man-made electricity may be disturbing a natural electrical phenomenon in the little beastie, although that is not conclusive proof.

Thus you can see that 'vision' in these algae is well developed – but is it like our own vision? Jerome Wolken of Carnegie-Mellon University believes that it is. In his book *Photoprocesses, Photoreceptors and Evolution*, he

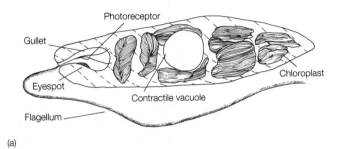

(a)

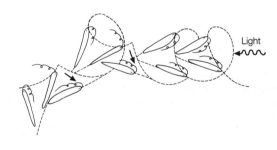

(b)

FIGURE 7.4 (a) *Euglena*'s sense of vision relies on the shadow cast by its eyespot on the photoreceptor, which guides the beating of the flagella propeller. (b) As *Euglena* swims in a spiral, it slowly corrects its direction of swimming until it heads towards a good source of light.
(After Checcucci, 1976)

> Box 7.1: Summary of how *Paramecium* swims
> How does the photoreceptor molecule tell the flagellum what to do? We cannot be sure, but some useful clues come from another single-celled organism, *Paramecium*. This behaves like a pure animal with no signs of plant-like behaviour. This 'animal' protozoan swims along using thousands of tiny hairs – cilia – embedded in the outer plasma membrane covering the creature. The cilia beat in a carefully synchronized wave, but when the *Paramecium* is electrocuted this beating is totally disrupted. Bearing in mind how small this creature is (about 0.1 millimetres long), Roger Eckert and Yutaka Naitoh of the University of California at Los Angeles made some truly remarkable measurements. They inserted electrodes into a single *Paramecium* and recorded nerve-like electrical signals zipping along the plasma membrane. The effects of these impulses on the cilia were astounding. Both the direction and frequency of the cilia beats were controlled by the intensity and frequency of the signals (Eckert et al., 1976).
>
> Eckert and Naitoh even measured the chemistry driving the electrical signals. Calcium and potassium ions carried the electricity across and along the plasma membrane. The calcium also turned on the motors driving the cilia movements – interestingly, the same ion drives muscle contraction (cf. chapter 9). However, the cilia motors are completely different from muscle fibres.
>
> Even though *Paramecium* does not swim in response to light, it is a valuable guide to our understanding of photosensitive algae.

suggests that animal eyes may have evolved from the close union of photoreceptor with flagellum:

> Electron microscope studies of various photoreceptors show flagellar processes associated with all sensory structures, from the flagellum of the protozoan *Euglena* to the flagellum found between the outer and inner segments of vertebrate retinal rods.
>
> (Wolken, 1975)

But does *Euglena* have 'vision'?

> We can then look upon *Euglena* as a primitive 'retinal cell'.
>
> (Wolken, 1975)

Much of his thesis was flawed: he assumed that the *Euglena* eyespot was the actual photoreceptor, instead of acting as a screen for shading the real photoreceptor. And without a brain, how could the alga understand any patterns falling on its photoreceptor? Nevertheless, the electrical excitation of all these algae deserves careful scrutiny, as it may have been the progenitor of much more complex nervous systems – some of which, as we shall see later, have developed in higher plants as well as in animals.

Green Motors

Sliding, twisting, swimming light-sensitive algae are one thing, but what about other plants? *All* green plant cells use their vision for steering movements, borne out of their drive for solar energy. Plants have to perform a tricky balancing act between catching enough good sunlight for powering photosynthesis but not so much light that their chlorophyll is bleached – a plant version, if you like, of sunburn. The dilemma has been partially solved by means of a wide range of light-sensitive movements.

The chloroplasts which house the chlorophyll are accomplished performers. In good light they look like round green pills, but in dimmer conditions they turn into flat tablets. They can also twist and turn, or even move from one place to another somewhat like blue-green algae. All these movements are governed by the brightness of the light that they 'see' (figure 7.5).

The chloroplasts of many lower plants can even point out fingers, like those of amoebae, towards sunlight. For instance, *Funaria hygrometrica* is a moss that lives on damp walls and rocks. In strong light its tablet-shaped chloroplasts roll up into balls, pointing towards the centre of the cell in which they live. Because they are contracted, less of their surface is exposed to the bleaching power of bright light.

Chloroplasts develop from proplastids, and these wriggle even more. They crawl like amoebae, stretching out fingers into their surroundings, feeling their way around and squeezing into all shapes and sizes like a never-ending contortion act. As proplastids develop into chloroplasts they become more rigid and develop another dance.

Like a sunbather caught in the shade, chloroplasts slide round inside their cells to the best lit place. But switch on too bright a light and they line up against the cell's walls parallel to the light, out of harm's way. It is a game of hide-and-seek that can be seen with a very simple experiment. When green leaves that have been kept in the shade are suddenly exposed to bright sunlight, they rapidly turn a brighter shade of green. Their chloroplasts have lined up to face the sun.

Some of the most intricate chloroplast movements in the plant kingdom take place in algae. The single large chloroplast of *Hormidium* slithers along in its long cylindrical cell. It occupies half the cell's circumference, and in dim light is tucked away at the rear of each cell, at the side furthest away

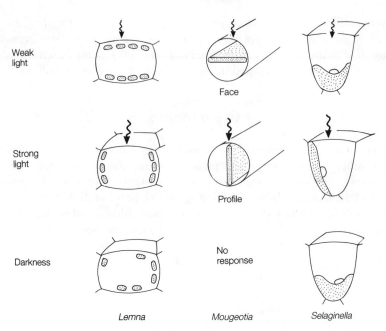

FIGURE 7.5 Chloroplasts move inside their cells, guided by both the direction and the brightness of light. These diagrammatic cross-sections through cells show the chloroplast movements of *Lemna* (duckweed), *Mougeotia* (an alga) and *Selaginella* (a lycopod, which is a group of plants related to ferns and horsetails).
(After Britz, 1979; courtesy of Springer-Verlag)

from the light. There is something wrong with this, you might think — after all, chloroplasts usually gather *towards* the light. A very elegant experiment revealed the answer by changing the optics of the cell (Scholz, 1976). In air the chloroplast slithers off to the back of the cell, but when the cell is immersed in paraffin oil it slopes back to the front. The oil changes the optical properties of the transparent cell walls, focusing the light onto a small spot at the back of the cell where the chloroplast collects. Thus under normal conditions, in water, the cell wall behaves like a lens and focuses light onto the chloroplast.

Plants that live almost entirely in the shade, such as the wood sorrel (*Oxalis acetosella*), have refined this idea further. The transparent skin covering their leaves is shaped into millions of tiny lenses. Like the cell walls of *Hormidium*, these focus the light down into intense spots in the cells below, where the chloroplasts gather.

The chloroplasts of another alga, *Vaucheria*, are entirely different. They are much smaller, and hundreds of them swim around inside long filament-shaped cells. When the light is too strong they rush to the sides of the cell, reacting within 10–80 seconds of illumination. But if only a tiny spot in the cell is exposed to light – using a microbeam – the chloroplasts only gather at that spot. Another interesting thing is that the microbeam does not have to shine directly onto the chloroplasts for the movement to occur. Instead, the surrounding cytoplasm has a sense of vision, and somehow – we do not know how – directs the chloroplast movements. Probably the most dramatic demonstration of this cytoplasmic power is in the diatom *Biddulphia*. In dim light the chloroplasts are spread throughout the cell. But in strong light, or even darkness, they rush to the centre of the cell around the nucleus. Again, full displacement can be triggered by illuminating just a small part of the cell, but why the chloroplasts gather around the nucleus has never been explained – maybe they are shaded by the nucleus.

Two other algae, *Mesocarpus* and *Mougeotia* (pronounced 'moo-gee-oh-shia') also have filament-shaped cells. Their chloroplast movements are, to my mind, the most graceful of the whole plant kingdom. Each rectangular cell holds one rectangular chloroplast, shaped like a sheet of paper and so large that it almost fills the whole side of a cell. Under dim conditions the chloroplast faces directly towards the light. But switch to bright light and it gently swivels around, presenting its thin edge to the light. It is as if strings have pulled the flap of a venetian blind through 90°, and, as we shall see in chapter 9 (pp. 162–3), 'strings' are indeed involved in the movement as well as 'muscles' (figure 7.6).

It only takes about 15 seconds for the movement to start, which is an exciting speed for scientists to study. Their first question was: how does the cell or chloroplast see where the light is coming from? The chloroplast is simply so large that a lens effect can be ruled out, and there is also no eyespot, as in *Chlamydomonas*, for picking up light. Wolfgang Haupt of the University of Erlangen-Nürnberg started by looking for a photoreceptor. He found the common blue-light sensor molecule flavin. But there was another pigment involved as well, and this was sensitive to red light and another wavelength (called far-red) towards the infrared end of the visible light spectrum. Our eyes cannot distinguish between red and far-red, but the plant pigment phytochrome can. This pigment works by changing its chemical structure when excited by red or far-red light. Each time it flips its chemistry, it triggers off all sorts of cell processes. How it does this is very much a burning question in which Haupt, amongst many others, is interested. We'll return to this in chapter 8 (p. 139).

Haupt also worked out the way that the phytochrome 'understood' the direction of light (Haupt, 1973). Rather like polarizing sunglasses, it only sees light polarized (vibrating) in one plane. And as the whole cell is coated with phytochrome molecules carefully arranged in many different planes, only those phytochrome molecules at the same angle as the light plane will be excited, like an envelope only able to pass through the slit of a letter box

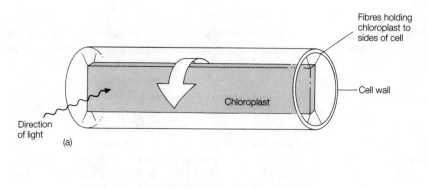

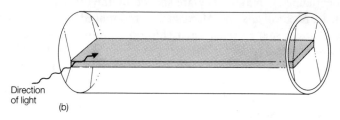

FIGURE 7.6 *Mougeotia* chloroplast (a) facing light and (b) facing away from light. The light is sensed by a photosensor pigment called phytochrome, which sends signals (probably including calcium ions) to the 'muscle' motors made of actomyosin, which actually pull the chloroplast like the ropes on a yacht's sail.

when it's at the right angle. When the phytochrome is excited, it flips into a new shape and sends calcium ions shuttling off to tell the chloroplast at that particular side of the cell to move. In fact this sense of vision is so accurate that when Haupt shone a fine microbeam of red light onto the edge of a chloroplast he could turn its corner over, just like the ear of a book page.

8

Sunbathing, Sleeping and Rhythm

Kissing Mouths

Photosynthesis is the power that supplies plants with energy and sugar; without it green plants could not exist. So it is hardly surprising that plants go to great trouble to trap light, using chloroplast movements, growth movements, swimming movements and, as we shall see shortly, leaf movements. But apart from light, photosynthesis also needs two other vital ingredients – water and carbon dioxide. Flowering plants draw these into their leaves or stems and expel waste oxygen using a sensitive mouth.

On the undersides of leaves are thousands of tiny pores called stomata (figure 8.1), aptly named after the Greek for mouths. Under a low powered microscope each stoma looks like two lips, and their movement opens and closes the mouth. These pairs of so-called guard cells open by inflating with water and bulging out, and close by jettisoning their water and deflating. Each movement takes a few minutes to complete. The guard cells are surrounded by special collapsible cells that act as a reservoir for water and ions needed for the movement. They also absorb the crush of the opening and closing movements, and prevent the tight skin across the leaf surface from crumpling every time the stomata move in unison.

With their stomata open, plants exchange gases with the outside air and lose water by evaporation. The power of millions of stomata all evaporating water together is so strong that a stream of water is sucked up the shoot from the roots, dragging up valuable minerals dissolved in it. A field of grass transpiring between May and July is estimated to lose over 500 tons of water this way. The evaporation even cools the leaf in the same way that evaporating sweat cools our skin. (The bitter gourd or colocynth has leaves which transpire so strongly that they keep about 7°C cooler than the

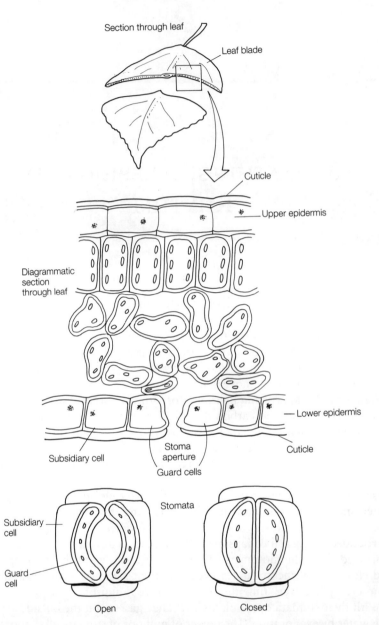

FIGURE 8.1 Stomata are the sensitive 'mouths' of plants. When their apertures open they let water vapour out and vital gases (oxygen and carbon dioxide) in and out. But when they close, the leaf is sealed tight and protected from drought and other dangers. The stomatal movements are governed by various light, chemical and perhaps even touch sensors in the guard cells. These sensors drive potassium ions and water in or out of the guard cells – the same osmotic motor that drives other leaf and flower movements.

surrounding air.) This transpiration stream also keeps the plant watered, fed with nutrients and turgid enough to stand upright if necessary. Unfortunately, this arrangement is gravely flawed.

Generally speaking, most plants open their stomata at dawn, when they start photosynthesizing, and shut them at dusk. But if the weather is too hot, too dry or too windy, the whole plant risks wilting because of loss of water through its open stomata. Most plants respond by releasing an 'alarm' hormone, such as abscisic acid, which warns the stomata to close. In fact, almost any stress – wounding, sudden winds, blows to the leaves, waterlogging – triggers the stomata to close. However, the penalty for closing the stomata is to starve the plant's photosynthesis of carbon dioxide. And that in a nutshell is the dilemma of stomatal movements: wilt or suffocate. This is why the movements have attracted enormous interest from plant scientists, trying to work out how the movements work, what influences them and how to develop crops tolerant of drought and other severe growing conditions.

Because of their control over gas, water and mineral supply, the guard cells have evolved into very sensitive monitors of the conditions inside and outside the leaf. In fact they are some of the most sensitive cells in the plant kingdom, responding to light, moisture, carbon dioxide, temperature, hormones and certain foreign chemicals, and even to touch, shaking, wounding (Martin and Clements, 1935) and electricity (Pallaghy, 1968). They also have their own daily rhythm of movement, which runs quite independently.

Keeping Track of the Sun

Striving for better photosynthesis also often involves much larger movements – the bending of mustard and cress seedlings towards light, or the flickering dimness on a woodland floor produced by the leaves above bending and twisting themselves into a complex jigsaw, each trying to avoid the shade of its neighbours.

But too much heat from the sun can be the death of plants. This is exactly the dilemma facing the shrub *Silphium lacinatum* of the North American prairies. Each day its leaves have to face their high noon, when the full rays of the sun caught face on would force them to close their stomata and stop photosynthesizing. Therefore the plant has evolved a very elegant adaptation – it grows its leaf blades away from the midday sun, and only faces the rising and setting sun. In this way it catches enough sunlight to photosynthesize in the early morning and late afternoon, but avoids wilting in the blistering heat of noon. Its leaves astonished the prairie hunters, because even under a cloudy sky they always stand upright, pointing their faces east and west. So they became known as compass plants, and the sight of wide expanses of their leaves lined parallel to one another looks extraordinary – rather like thousands of sheets of paper all lined up on parade. To get some idea of how useful this adaptation is to survival in the desert, *Silphium* plants with their leaf blades held horizontal instead of vertical lose more water and grow fewer flowers than plants allowed to orient their leaves naturally (Werk and Ehleringer, 1986).

Elegant as it is, the orientation of *Silphium* leaves is only a growth movement, and once fixed they cannot turn again. Other plants, however, have gone a stage further – they have flexible leaf movements that track the sun *throughout* the day, *every* day. Their craving for sunlight is so strong that even at night, before first light breaks, their leaves swing round in anticipation of the dawn.

After one of the desert's occasional rainstorms, the shrublands bloom with a carpet of these short-lived plants. Lupins, sunflowers and many others all soak up enough water to grow for the few weeks or months needed to set the seed that will lie dormant in the soil until the next downpour. Many of these plants track the sun with their leaves, rather like a satellite tracked by a moving satellite dish as a way of fuelling their frantically short growing seasons (figure 8.2). The phenomenon was first catalogued in various plants by Charles Darwin and his son Francis (Darwin and Darwin, 1880), and was then virtually forgotten until some recent work by ecologists interested in the physiology of plants.

James Ehleringer and Irwin Forseth of the University of Utah have been keen followers of the sun-tracking movements of two Californian desert plants, *Malvastrum rotundifolium* and the lupin *Lupinus arizonicus* (Ehleringer and Forseth, 1980; Forseth and Ehleringer, 1983). Their leaves fan out like a peacock's feathers, with their blades facing the sun. As the sun passes across the sky, the leaves twist to face it, and at night they go back to their original starting positions. The mobile leaves are controlled by motor cells in the pulvinus where the blade joins the petiole. Movement of water in and out of these cells is almost certainly controlled by osmotic substances, particularly potassium ions – more details later.

But what advantage is sun-tracking to these plants? Ehleringer and Forseth found that both *Malvastrum* and *Lupinus* collected a good deal more solar energy by tracking the sun than was collected by their neighbours without sun-tracking movements. As a result they grew more vigorously.

Yet the plants still have to face up to the danger of the noonday sun. They have solved this problem using an elegant strategy: as midday approaches, the lupin turns its leaves away from the sun and squeezes its leaf pores shut, but when the sun sinks in the afternoon the leaves twist back to sunbathing and reopen their pores. Exactly how the plant changes from sun-avoiding to sun-tracking action is not known, but the angle of the sunlight is probably the important signal.

On the other hand, *Malvastrum* simply bears the full onslaught of the midday heat, and continues sun-tracking. In fact, *Malvastrum* is so at home in deserts that it can even grow on the floor of Death Valley, California. To explain these different types of behaviour, you have to look at the ancestry of the two plants. *Malvastrum* is descended from a long line of desert plants with experience of coping with drought and a physiology adapted to it.

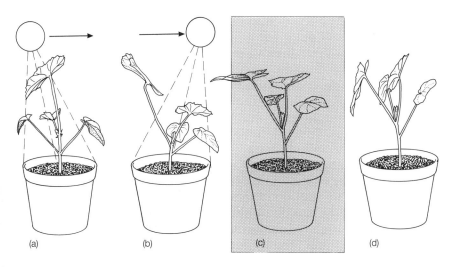

FIGURE 8.2 Leaves like those of *Malvastrum* work at high rates of photosynthesis by swivelling their leaves round to follow the sun through the day, like miniature satellite dishes ((a), (b)). The movements are executed by a pulvinus at the base of each leaf stalk, pumping water in or out of the motor cells much like guard cells open and close stomata. An hour or two after sunset, the leaf blades turn to roughly horizontal (c), but about an hour before sunrise they 'anticipate' where the sun will rise at dawn (d).
(After Salisbury and Ross, 1985; courtesy of Wadsworth Publishing Company)

However, the lupin's ancestors hail from cooler temperate climates where drought is less of a problem. So for this species sun-*avoiding* movements were a relatively recent adaptation to the desert environment. Incidentally, this sort of behaviour, when animals or plants change their behaviour to control their body temperature, is what zoologists call 'behavioural thermoregulation'. Surprisingly, little attention has been paid to this important aspect of plant behaviour.

There are important lessons here for agronomists and crop breeders. A promising range of crops, such as cotton, sunflower, cowpea, soybean and to some extent beans, track the sun with their leaves. Like the desert plants, the leaf movements boost their rates of photosynthesis. And provided that they do not suffer drought, this high metabolism translates into substantially larger harvests at the end of the growing season. Unfortunately, in modern agriculture plants are packed so close together that sun-tracking is inhibited. So better prospects may come from using the vast and increasing expanses of deserts, where space is no problem. Judging by the rising numbers of research papers published recently, sun-tracking movements are now being taken seriously by crop scientists and will probably become an important point for plant breeders to bear in mind when developing drought-resistant crops. Alternatively, we could be using more nutritious plants which already sun-track, such as the cassava.

Cassava is the staple diet of 700 million people throughout the tropics. Its tuberous roots produce higher yields of starch than even maize or rice and make a good flour for bread and tapioca (best known in Britain for the puddings in school dinners!). It grows particularly well in dry areas where the rainfall is too sporadic for most other crops. Scientists at the Centro Internacional de Agricultura Tropical (CIAT) in Colombia have found that cassava's secret of survival lies in two types of drought-avoiding movement.

First, its leaf stomata are so sensitive to changes in humidity that they close as soon as the humidity in the air drops. This conserves water in the plant, but of course stops photosynthesis at the same time. To compensate for this loss, the leaves photosynthesize rapidly during the cooler and more humid times of the day, usually early morning and late afternoon, helped by the leaves tracking the sun.

Second, the whole leaves as well as their stomata sense the moisture in the air, and droop when it gets too hot to avoid overheating. This is a serious risk because when their stomata are closed at noon the leaves are no longer being cooled. The leaves only recover their upright posture again when the humidity of the air rises, usually in late afternoon.

Cassava has a third trick enabling it to survive drought. When water is running short it diverts its growth into roots rather than shoots. So, knowing how cassava tolerates drought, the researchers at CIAT can now help farmers by developing hardier varieties with higher yields. They are also looking for the varieties with the most sensitive stomata and the most efficient use of water for planting in the driest farming regions of the world.

Cassava yields can be improved further by growing it amongst other crops. Because cassava is more sensitive to air humidity than to water in the soil, it is an advantage to grow it between rows of crops such as maize, or even among trees, which increase the humidity of the air around them. All of which highlights the importance of studying leaf movements seriously (box 8.1).

SUNBATHING, SLEEPING AND RHYTHM

At the other extreme of climate to the desert, trees growing in tropical rainforests have recently been found to sense both light and touch. Botanists John Dean and Alan Smith, working at the Smithsonian Institution's tropical research station in Panama, reported tree leaves held out horizontally to catch maximum sunlight, but drawing upwards when rain fell on them (Dean and Smith, 1978). This simply helps drain the rain off the leaf surfaces; otherwise the water can clog up stomata, reflect useful sunlight and encourage the growth of epiphytic plants. This is an interesting example of light-sensitivity and touch-sensitivity in the same leaf, and gives us some clue as to how the movements might have evolved, a theme we will be coming back to later in this chapter.

Box 8.1: Other moisture-sensitive leaf movements

A few other species perform moisture-sensitive movements. The stomata of the epiphyte *Tillandsia*, which grows on rocks, trees and even overhead telephone cables, are capped with a moisture-sensitive plug. When conditions are moist, the plug inflates and pushes out so that the stomata can 'breathe'. But if conditions become too dry, the plug shrinks tight and seals the stomata.

Marram grass grows on sand dunes, where drought is also a threat. During dry periods each long thin vertical leaf folds into a tube, trapping enough moist air inside to let the stomata remain open.

Flowers are also sensitive to moisture. In many species, the stalks bend over so that the flowers droop towards the ground, preventing dew from spoiling their pollen, which is particularly sensitive to moisture. In fact, some species are so sensitive to moisture that they have been used for weather forecasting. It is said that if the flowers of the Siberian sow-thistle (*Sonchus siberius*) shut at night the following day will be fine; if they open, it will be cloudy and rainy. If the African marigold (*Tagetes erectus*) folds its flowers after 7 a.m., rain is supposed to be on the way. If the bindweed *Convolvulus arvensis*, the marigold *Calendula pluvialis* or the scarlet pimpernel *Anagallis arvensis*, also known as the poor man's weather glass, are already open, they will shut at the approach of rain.

Even pollen sacs of flowers can sense moisture. In the plantain (*Plantago*) the anthers squeeze shut on dewy nights or during wet weather to protect the pollen inside. The movement can be fast: the anthers of the unfortunately named bastard toadflax (*Thesium alpinum*) shut within 30 seconds of being moistened.

Solar-heated Flowers

Sun-tracking also comes in useful for plants growing in cold climates. Springtime flowers in the Arctic tundra need to attract insect pollinators when the air is still too chilly for most of them to fly. Some plants have turned into solar energy collectors: they have flowers shaped like satellite dishes which track the sun through the day, collecting its heat in the dish. The warmth collected is enough to encourage any insects brave enough to be flying in the chilly air to seek shelter in the flower, and they pay for this favour by performing cross-pollination.

But seduction by heat is not the whole story. A group of Swedish botanists at the University of Lund studied the heat collected in the flower of *Dryas octopetala*, a favourite plant of rock gardens. Britta Kjellberg, Steffan Karlsson and Ingar Kerstensson wanted to see what effect sun-tracking had on the eventual harvest of seeds following pollination (Kjellberg et al., 1982). They restrained flowers from sun-tracking with a harness. The temperatures of all the flowers under test were monitored using a thermocouple wired to the inside. As the sun rose during the day, all the flowers warmed up. But the flowers which had not been tampered with (sun-tracking intact) warmed up most – by about 1 °C more than flowers restrained from moving. Perhaps the most surprising result of all is that this slight warming had quite a noticeable effect: the warmer the flowers, the heavier are the seeds. It seems that the heat collected by sun-tracking flowers nurtures the developing seeds like a warm incubator. And because heavier seeds tend to germinate better than lighter ones, the sun-tracking plant has a competitive edge over its sedentary neighbours.

Perception

The sun-tracking and sun-avoiding movements all rely on the ability of the plant to sense the direction of the sunlight, and then to turn that information into movement. In addition, sun-tracking plants reorient themselves during the night and *anticipate* where the sun is going to rise an hour or so before it actually does so. Plants can even be trained in only two days to 'learn' the direction of sunrise, even under cloudy skies. How do they do it?

A number of theories have been proposed. The most conventional one only explains how the sun-tracking sensor works, and takes no account of the anticipating move before sunrise. Thomas Vogelmann of the University of Wyoming searched for an 'eye' in the leaves of the lupin *Lupinus succulentus*. He shaded portions of the leaves with pieces of aluminium foil or black graphite, and found that they could only sun-track when the pulvinus was exposed to the sun. Therefore the light sensing must be performed by the pulvinus, and the photodetector was found to be located on the 'underside' near the motor cells (Vogelmann, 1984). The same conclusion was reached by Shinobu Watanabe and Takao Sibaoka of Tohoku University, who shone thin beams of light onto the leaflets of *Mimosa*. Only when the light hit the underside of the pulvinus did the leaf move (Watanabe and Sibaoka, 1973). Other scientists have discovered signals passing from one pulvinus to another in the same leaf (Sheriff and Ludlow, 1985).

However, we are very unsure how the *direction* of light is detected. Perhaps we should go back seventy years to the work of Gottlieb Haberlandt. He found a clear lens of cuticle covering the surface cells of the leaves of some plants. Each lens concentrated the light into a beam inside the cell below. These lenses were so effective that Haberlandt even managed to cut one off a tropical aroid lily, *Anthurium warocqueanum*, and focus the image of a microscope stand with it.

He regarded the lenses as the optical sense organs of the leaves, and tested the idea by defocusing the leaf optics with a thin film of water over the leaf surface. When he shone light onto the wetted part of the leaf its surface lenses were so defocused it no longer recognized the direction of light and didn't bend (Haberlandt, 1914). This is rather like the way the cell wall of the alga *Hormidium* focuses light into its cell, as we saw earlier in chapter 7. Other than that, all we can say for sure is that some plant species use their pulvini and others their leaf blades for detecting the direction of light.

Sleepy Plants

So much for movements designed for collecting solar energy. Many more plants indulge in strange behaviour as night approaches – they go to 'sleep'. Whole gardens can literally change shape at dusk as a wide variety of plants fold up or droop their leaves, flowers or stalks (figure 8.3). And at dawn the next day they reopen to face the sun. As an enthusiastic American naturalist, Royal Dixon, noted in his intriguingly entitled book *Human Side of Plants*:

> What a fantastic fairyland is a garden at night! Here we find many sleepy heads all so quiet and drooping that one wonders whether strange dreams may be forming in their plant minds.
>
> (Dixon, 1914)

The sleep movements of plants are nothing new to us. Androsthenes, one of the generals marching with Alexander the Great 2,300 years ago, noted that the leaves of the tamarind tree (*Tamarindus indica*) folded up at night.

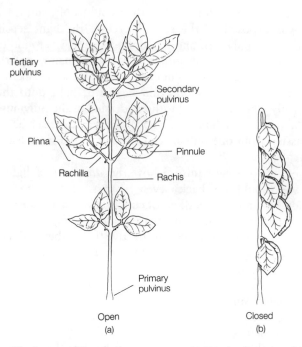

FIGURE 8.3 The leaves of *Samanea saman* open during the day (a) and 'sleep' at night by folding down (b). The movement is driven by fluxes of potassium ions and water in leaf pulvini, controlled by natural rhythms in the plant and by light-sensitive pigments.
(After Satter, 1979; courtesy of Springer-Verlag)

During the same period, the Greek philosopher Theophrastus compared this with the movement of *Mimosa*. In his book *Peri Phyton Historia* (*Enquiry into Plants*) he described a tamarind growing in Bahrain:

> There is another tree with many leaves like the rose, and . . . this closes at night, but opens at sunrise, and by noon is completely unfolded; and at evening again it closes by degrees and remains shut at night, and the natives say that it goes to sleep.
>
> (Theophrastus, 1946)

The great eighteenth-century biologist Carl von Linnaeus (who developed the Latin naming system of all living things) noticed sleep movements in flowers when he saw a lotus plant in his greenhouse come into flower while his gardener was out. When the gardener returned that evening Linnaeus took him to see the amazing flower, but it had disappeared! He assumed that it had been eaten by insects, but next morning the flower had 'reappeared'. In fact the lotus flower had been performing sleep movements, and many other flower species fold up or bend over at night, apparently to prevent nocturnal insects such as moths from pollinating them or, as we mentioned earlier, to guard against dew spoiling the pollen in the flower's stamens.

Throughout history various explanations have been given for sleepy leaf movements, and they are still controversial. Some have claimed that the movements stop the formation of dew on the leaves overnight, which might block the stomatal pores (although there is no evidence that this can happen). Others claimed the complete opposite – that the stomata were deliberately blocked by the leaves folding which stopped them losing water. Again, no evidence was given. But the one explanation that has become increasingly controversial was another brilliant suggestion by Charles Darwin. He prevented the leaves of plants from moving, and noticed that they often died off after a cold night. He proposed an intriguing explanation:

> The blade [of the leaf] is always placed in such a position at night, that its upper surface is exposed as little as possible to full radiation. We cannot doubt that this is the object gained by these movements.
> (Darwin and Darwin, 1880, ch. VI)

In other words, the leaves kept the buds inside warm – another example of 'behavioural thermoregulation' where plant behaviour helps control its own temperature. Darwin candidly admitted that he himself found his own explanation implausible, and doubted whether holding the leaves upright could make 'any sensible difference' to the leaf temperature. Yet the hallmark of a good scientist is to recognize the facts, regardless of any preconceived notions.

Darwin was roundly criticized by botanists then and now for two reasons. First, tropical plants with sleep movements are unlikely to experience cold nights. However, the critics have overlooked the cold nights experienced by plants growing on tropical high plateaux and mountains. Second, and much more damning, nobody seemed capable of repeating Darwin's experiments and obtaining the same results (Schwintzer, 1971). So the idea of sleep movements keeping buds warm fell into disrepute, and is barely mentioned in today's textbooks. But one modern botanist has discovered an extraordinary piece of evidence in the Venezuelan Andes which supports Darwin.

The Andes, as well as the mountains of East Africa and Hawaii, experience wide ranges of temperature throughout each day, from hot days to nights so cold that plants risk freezing all year round. These unusual conditions have resulted in a bizarre alpine landscape resembling the land of Lilliput. Plants that grow as small herbs elsewhere have evolved into fantastic giants. The weedy groundsel has evolved into a tree groundsel,

St John's wort grows up to 12 metres high, and huge lobelias, lupins and heathers have developed. Why these plants have become so incredibly large is not clear, but one way that they overcome the severe climate is by growing their leaves as a rosette, with the older leaves clustered around the younger ones and the single bud that they all develop from in the centre. And some species go even further.

Espeletia is another relative of the groundsel, although it looks rather like a huge artichoke. At dusk its huge rosette of leaves folds inwards to form a dense ball, and in the morning the rosette unfolds again (figure 8.4). Alan Smith, the botanist who discovered raindrop-sensitive movements in tropical trees, wanted to find out whether the older leaves on the outside of the rosette kept the younger ones inside warm at night. He placed thermocouples at different places in the rosette, and found that the temperature of the younger leaves was high enough to stop them freezing.

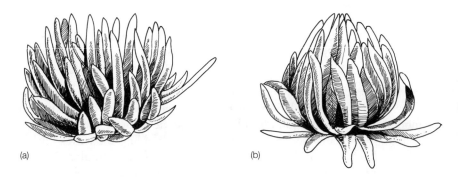

(a) (b)

FIGURE 8.4 The rosette of leaves of *Espeletia shultzeii* (a 'giant' plant of the Andes) open during the day (a) and 'sleep' closed at night (b). Darwin proposed that sleep movements protect younger leaves and buds against cold nights, as borne out by the behaviour of *Espeletia*. Thanks to their sleep movement, older leaves on the outside of the rosette do indeed keep the younger ones inside warmer at night. However, a more recent theory suggests that the sleep movements prevent leaves from absorbing moonlight on bright nights, thus protecting their internal clocks. The arguments still continue.
(After Smith, 1974)

But when sleep movements were prevented by holding the rosettes open, the inner leaves suffered frost damage on cold nights and often died (Smith, 1974). So, just as Darwin had shown a hundred years before, sleep movements do indeed protect young leaves and buds from freezing to death. Smith also discovered that the movements of the *Espeletia* leaf rosettes during the day kept the young leaves warm by acting as a solar dish to concentrate heat into the centre.

Of course, you could argue that giant rosette plants are too strange to have much relevance to most other plants. However, Darwin's theory for temperate plants was finally corroborated by James Enright of the Scripps Institute, California. He restrained the sleep movements of bean leaves and, using thermocouples, showed that the closed leaves kept the young buds about 1 °C warmer than the surroundings. Despite the apparent modesty of this figure it was enough to make a substantial difference to the plant growth (Enright, 1982).

There is, though, a completely different theory to explain sleep movements, which might even be complementary to Darwin's idea. Leaves are more than just solar energy collectors – they also act as clocks, counting off the hours of daylight each day so that they can tune into a 24-hour day, regulating a wide range of chemical and physical processes that run on a daily routine, and also tune into a yearly calendar, regulating a variety of developments which depend on seasons, such as flowering and leaf fall. The latter is particularly important for plants growing in temperate and arctic latitudes, where seasons are very pronounced.

Erwin Bünning, the noted expert on flower movements whom we mentioned earlier in chapter 3, also studied leaf movements for over fifty years. He noticed that sleep movements are particularly common in plants that flower during short days, and this led him and his colleague Ilse Moser to wonder what happened during the long nights. After careful measurements of light intensity, they found that the sleep movements of leaves cut down the absorption of light from the moon, the stars or even street lights. The leaves quite literally hid themselves from light to prevent them from being misled into counting night lights as daylight and upsetting their clocks (Bünning and Moser, 1969). Perhaps Bünning and Darwin are both correct – sleep movements may keep young leaves warmer *and* enable plants to keep time more accurately.

Autonomy in Performances

So far we have mostly looked at movements of plants triggered by cues from the outside world – light, heat, touch, electricity, moisture and chemicals. But many of these movements are also run by clocks *inside* the plants themselves. In other words, they have rhythm, and its discovery dates back to 1729 when the French philosopher and astronomer John De Mairan was perceptive enough to wonder whether the sleep movements of leaves were driven by changes in their surroundings or by something inside the plants. Keeping a *Mimosa* plant in constantly dim light for several days, he found that the leaves carried on moving in a roughly daily cycle, although as time went by the movement grew weaker. So he decided that the movement was controlled by a clock inside the plant. As he succinctly concluded: 'The Sensitive Plant senses the sun without seeing it in any way' (De Mairan, 1729).

In fact, the timing of some movements is so regular that they can be used for timekeeping. Carl von Linnaeus grew a floral clock that showed each hour of the day by selecting different species which opened their flowers at particular times. But be warned about using the flower clocks – they all close up during wet weather!

The nature of this internal clock has been one of the most baffling mysteries of biology since its discovery. It drives not only leaf and flower movements but also a vast range of plant and animal processes. Indeed, all living cells keep their own internal rhythm of metabolism and physiology.

One thing we are sure about is that cell membranes have a strong influence on the rhythms. Upsetting the membranes with chemicals such as lithium also upsets the clock. Current thinking is that molecules in the membranes of plants and animals have a rhythm all of their own, such that the structure, chemistry and 'leakiness' of the membrane are constantly oscillating in a daily cycle. When the membrane leaks certain ions, sugars, hormones or whatever, it triggers a cascade of reactions, some of which will drive the leaf or flower movements.

It was hoped that a convenient system like leaf movements – easy to record, relatively simple compared with animal behaviour and convenient to use in the laboratory – would provide the Rosetta Stone that would unlock the code of all biological rhythms. Although we know that leaf rhythms run a loose 25-hour cycle internal rhythm, which can be reprogrammed with short bursts of light, decades of research have drawn a blank as to what ticks away the minutes and hours inside the plants.

As if that was not baffling enough, some leaf movements keep cycles of only a few minutes, with no apparent function. The telegraph plant (*Desmodium motorium*, also known as *D. gyrans*) has large paddle-shaped leaves with typical sleep movements (figure 8.5). It also has smaller leaves called stipules at the bases of the leaf stalks, and these move in elliptical circles roughly every few minutes like semaphore signals, with the speed increasing as the temperature rises. Wilhelm Pfeffer recorded one revolution of the stipule every 90 seconds at 35 °C (Pfeffer, 1905). It is probably the only rhythmic movement that you can watch with the naked eye, as the little stipules twitch their ceaseless circuits. In pursuit of what? We simply have no idea.

Stephen Britz and Winslow Briggs of the Carnegie Institution, Washington, discovered another unusual movement rhythm. They found rhythmic chloroplast movements in the green leaf-like seaweed *Ulva* (Britz and Briggs, 1976). Towards dusk the chloroplasts in each cell move to the sides of the cell. Just before dawn they return to face the light again. The rhythm

SUNBATHING, SLEEPING AND RHYTHM

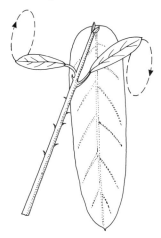

FIGURE 8.5 All mobile leaves have movements of their own accord, in slow rhythms every several hours, and have baffled scientists for centuries. However, the telegraph plant *Desmodium motorium* (*D. gyrans*) performs one of the most mysterious movements known in the plant world. Its small stipules swivel around in ellipses lasting a few minutes, for no apparent reason whatsoever.

persists with a cycle close to 24 hours under constant temperature and in light or darkness. However, there is no obvious reason why the chloroplasts should face the most intense sunlight at midday. Instead, Britz and Briggs suggest that the rhythmic movement may have something to do with establishing the top and bottom polarity of the cells.

How Leaves Move

Leaf pulvini and stomata movements work in basically the same way. They are all driven by water, sucked in and out of motor cells by tides of potassium ions. Wherever the potassium goes, the water follows, so that the motor cells either collapse or inflate, which is what we see as movement. Exactly what drives the potassium fluxes is less clear, but recent research points to electrical changes and the workings of special ion pumps, all of which is uncannily similar to the touch-sensitive movements of *Mimosa*. But there is still much more to be learnt.

The breakthrough in understanding all these movements was the theory of osmosis, described by biologist Ausguste de Candolle in 1837 – the passage of water through a membrane from a dilute to a more concentrated osmotic solution (such as a sugar or salt solution). In 1856 the German botanist Hugo von Mohl of the University of Tübingen applied this theory to explain stomatal movement (Mohl, 1856). When osmotic substances build up in the guard cells, they suck in water from the surrounding cells and the guard cells bulge out, stretch apart like yawning lips and open the slit-shaped pore. When the osmotic substances pass into neighbouring cells the process is reversed – water is expelled from the guard cells which collapse and close the pore.

von Mohl's theory became the cornerstone of all research into stomata, and eventually pulvini, but there has been much disagreement about the nature of the osmotic substances. The traditional theory, dating from the turn of the century, proposed sugar. The idea was that during the day guard cells make sugar by photosynthesis and draw in water by osmosis. At night the sugar is converted to starch, which lacks the water-pulling power of sugar, and the cells lose water and deflate. One of the most compelling pieces of evidence for this theory was the way that starch granules disappeared (presumably as they broke down into glucose or other sugars) in the guard cells as they opened. But, as we will see below, this was a red herring.

The sugar–starch theory was completely debunked in the mid-1960s. There was no evidence that the levels of sugar rose as expected. The spotlight has now turned from sugar to potassium ions (Raschke, 1976). When potassium enters the guard cells it drags water in with it by osmosis, and when it leaves it takes the water out again. Using stains, dyes and microscopes equipped with apparatus for chemical analysis, you can actually see potassium migrate into or out of guard cells after they have moved. The potassium ions carry a positive electrical charge which must always be balanced, otherwise the cell risks being blown up by electrical short-circuits. This is done by negative ions shuttling in the same direction. Usually these are chloride ions, but malate (a product of photosynthesis) can also be involved in some species. The process is shown schematically in figure 8.6.

Now that we know what drives the water movements, the hunt is on for what triggers the movement of potassium ions in the first place, and recent breakthroughs have just recently revealed yet another startling similarity between plant and animal chemistry.

Cells are usually reluctant to let messengers from outside penetrate inside. Instead, when a signal like a hormone arrives at a cell it locks onto a special receptor on the cell's surface and its message is conveyed into the cell by another so-called 'second messenger' (figure 10.4). One of the best known second messengers in animal cells is inositol 1,4,5-triphosphate ($InsP_3$), and when released it liberates calcium ions previously locked away inside the cell, which then triggers a cascade of cell reactions.

There is now some evidence that $InsP_3$ is also at work in guard cells, where it keeps calcium under its thumb. When the hormone abscisic acid locks onto a guard cell it releases $InsP_3$, probably triggering a rise in calcium

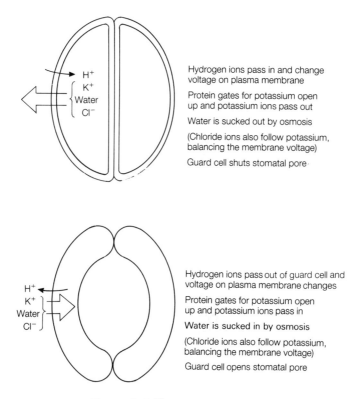

FIGURE 8.6 How stomata move.

inside the cell. The calcium could then tell the plasma membrane to open its potassium ion channels, leading to the swelling of the guard cell and closure of the stomatum (Hedrich and Neher, 1987; Schroeder and Hagiwara, 1989; Blatt et al., 1990; Gilroy et al., 1990; McAinsh et al., 1990). Whether guard cells use $InsP_3$ in sensing light or other stimuli (as happens in animal cells) is not known, but $InsP_3$ is probably important in conveying many other signals in plants (Morse et al., 1989), and could well be involved in telling plant cells about touch stimulation, as we'll be mentioning in chapter 12.

But hydrogen ions are also thought to fire the potassium shuttle into action. The first sign of this came from a completely unexpected story. Parasitic fungi are constantly trying to find their way into plants. Those that infect leaves often enter through the stomata, but of course they have to wait for the pores to open. Not so with the fungus *Fusicoccum amygdali* which has the power of 'open sesame' and actually makes the stomata open for it. This power greatly intrigued scientists, and ten years ago they identified a potent chemical – aptly called fusicoccin – in the fungus which made the guard cells throw out hydrogen ions, making the cells more negatively charged and leading to potassium ions flooding in and opening the stomata. And that is why the fungus eventually makes the leaves wilt.

There is more evidence pointing to an electrical factor at work. When stomata are illuminated after they have been kept in the dark, there is a spontaneous electrical 'shock' wave in the guard cells. It occurs in less than a thousandth of a second, and is the first measurable reaction of stomata to light (Zeiger et al., 1977; Moody and Zeiger, 1978). It is possible that the chloroplasts in the guard cells 'wake up' and start photosynthesizing – itself a highly charged process. Or perhaps a light-sensitive molecule in the guard cell becomes excited and pumps out hydrogen, creating an electrical force across the plasma membrane. This would drag in potassium ions. As with pulvini, stomata use the blue light sensor flavin and the red–far-red light sensor phytochrome. The electrical work of Racusen and Satter (1975) on *Albizzia* pulvini and Takuma Tanada's research on mung bean root tips (Tanada, 1968) suggest that phytochrome does change electrical voltages on cells, and this can have other bizarre effects (as we shall see in chapter 10, p. 183). It has also recently been shown that electrical forces can open special gates in the membrane, allowing certain ions such as potassium through (Schroeder et al., 1984) – for more details see pp. 184–8.

(Incidentally, the hydrogen ion story helps explain why earlier botanists were fooled into thinking that glucose and starch were the driving forces in stomatal movement. The starch in opening guard cells disappeared not into sugars (there is no evidence for this) but into acids such as malic acid. These acids split up into positive hydrogen ions and negative malate ions, which are both used in the ionic fluxes of guard cell movement.)

The electricity could have other effects. Waller (1925) detected a light-triggered wave of electricity zipping across the surface of leaves. The signal was so strong that it could travel through parts of the leaf kept in darkness or lacking chlorophyll altogether – clearly photosynthesis was not needed. The great stomata scientists Oscar Heath and J. Russell of the Rothamsted Experimental Station found that one group of stomata kept under poor light conditions could tell other groups of stomata kept under bright light on the same leaf to start closing (Heath and Russell, 1954). It is unfortunate that this work has not been followed up. By far the most promising recent research on stomatal signals has been the discovery of a hormone(s) released from wounds that triggers an electrical signal through a shoot, and apparently signals stomata to shut. We shall look at this phenomenon again in chapter 10 (pp. 181–2) when we examine the nature of these hormones in more detail.

Whole leaf movements follow a similar scheme of events to those of the stomata. They generally move from the pulvini at the base of the leaf stalk, although some leaves simply twist their leaf stalks. When water is pumped from one side of the pulvinus to the other, or is expelled altogether into the surrounding leaf stalk and stem, the leaves droop. When water passes back again, the pulvinus inflates and the leaves rise up. This explains the mechanics of leaf movement, but what actually drives the water pump?

First, the light-sensitive movements have to be stimulated by a light signal. All the sun-tracking, sun-avoiding and sleep movements are controlled by two familiar types of light sensor – the blue light detector flavin and the red light detector phytochrome. Most work has been done on sleep movements, and these are strongly controlled by phytochrome as it detects the subtle changes in red and far-red light. But what, you might ask, is the use of sensing red and far-red light?

Although we cannot see far-red radiation, daylight none the less contains an almost equal balance of far-red and red light. But changes in the quality of sunlight, like a passing cloud or shade from a nearby tree, will swing the balance towards far-red. On the other hand, as the sun nears the horizon at dusk, the balance tips more towards red light and then reverses as the sun rises again at dawn. So the plant can glean a lot of information about its environment simply from the shifts between these two types of red radiation.

But elegant as this theory is, other ideas have been suggested. The simple cooling of the evening air or a passing cloud also affects phytochrome, and could set in train many of the processes that it controls. Another theory is that phytochrome itself virtually 'goes to sleep' in the evening, and remains dead to the outside world (Lunsden and Vince-Prue, 1981).

We have already seen that the light sensor controlling many leaf movements is located in the pulvinus. It is believed to be embedded in the plasma membrane envelope sheathing each motor cell and behaves like a frontier post, controlling the movements of ions such as calcium between the inside and outside of the cell (figure 8.7). (It's interesting that light-sensitive leaf movements also probably involve the $InsP_3$ messenger we saw earlier in stomata movements (Morse et al., 1987), in which case $InsP_3$ could also unlock calcium ions from their stores inside the pulvinus motor cell.) The calcium ions in turn trigger fluxes of hydrogen, potassium and chloride ions across the plasma membrane. The potassium then draws water into or out of the motor cells by osmosis. Each cell inflates or deflates, making the pulvinus turgid or floppy. Because the pulvinus behaves like a hinge, it either pulls the leaf up or lets it flop down. More recently, positively charged hydrogen ions have also been implicated in the ionic traffic. They may be one of the first ions to start moving, but we don't know how they tie in with the movements of calcium ions. The sequence of events leading to pulvinus movement might go something like this (Satter and Galston, 1981):

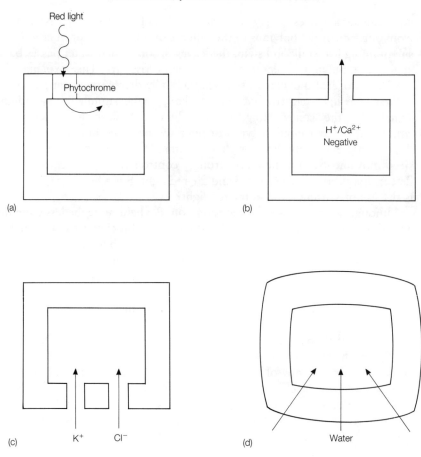

FIGURE 8.7 How leaf pulvini might move: (a) red light excites phytochrome in the motor cell membrane; (b) phytochrome somehow opens a gate for pumping out hydrogen or calcium ions, leaving the cell more negative; (c) protein gate opens to potassium (and choride ions, to balance the electric charge); (d) water is sucked in by osmosis, and the motor cell swells up.

Hydrogen is pumped out of cells using metabolic energy, and/or $InsP_3$ triggers a surge of calcium.

As hydrogen ions pass through the plasma membrane they create a pH and electrical forces.

The electrical forces open special gates in the membrane, which allow certain ions such as potassium to pass through.

Other ions simply leak whichever way they are dragged by the electrical and pH forces.

Negatively charged ions such as chloride can also be swapped with hydrogen ions as they are pumped in or out of the cell.

At the tissue level electricity could help one cell signal to its neighbour to start moving. At the pulvinus level it could both signal the start of and help to direct the massive fluxes of ions from one side of the organ to the other.

What are so remarkable are the similarities between all these leaf movements and the touch-sensitive flower and leaf movements. The English botanist John Ray was close to the truth when he suggested in 1693 in his book *Historia Plantarum* that sleep leaf movements were caused by a withering of their parts. He rather astutely explained the touch-sensitive *Mimosa* movements in the same way (Ray, 1693).

One aspect that we have not touched on so far is the role of hormones. Traditionally, plant physiologists have explained many forms of plant behaviour in terms of hormones. In the light-sensitive movements, phytochrome is known to affect plant hormones such as IAA (indoleacetic acid) and gibberellic acid. These chemicals promote the opening and inhibit the closing of darkened *Mimosa* leaves. They work by latching onto each cell's membrane, where they probably trigger some of the ion fluxes. Certainly, IAA triggers a rapid release of hydrogen ions. The inhibition of the evening movement is also sensitive to a little-known hormone called jasmonic acid, which antagonizes the activity of IAA in a number of plants (Tsurumi and Asahi, 1985). As we'll see later in chapter 10 (pp. 179–80) jasmonic acid is also closely related to a group of animal hormones called the prostaglandins.

Perhaps the greatest surprise in the chemistry of leaf movements in recent years has been the discovery of a completely new group of hormones, never found before in any plant. They were found in *Acacia* and several other plants by the organic chemist Hermann Schildknecht of Heidelberg University. He believes that these hormones activate both sleep and rhythmic leaf movements in a wide variety of plants. They stimulate both their own leaf movements and those of *Mimosa*, and he called them turgorins because of their effect on the turgor of plant pulvini (Schildknecht, 1984). We will look at these hormones in more detail when we examine the effects of related chemicals on plants in chapter 10 (pp. 180–1). One puzzling feature of this work is that plant physiologists have paid the turgorins very little attention, and their activity still remains something of a mystery. Furthermore, no one is even sure how important hormones may be in pulvini movements – just because they affect pulvini in the laboratory does not necessarily mean that they are significant in the natural life of a plant. What we need to show is that a hormone *signals* a movement.

One consistent feature of all leaf movements, whatever their cause, is a change in membrane voltage. Sir Jagadis Bose, the maverick Indian scientist who worked on the electrical signals of *Mimosa*, made another interesting discovery. He found that the rhythmic movements of leaves were always accompanied by changes in the electrical potential on the surface of the pulvinus, as in the telegraph plant. Some sixty years later, Richard Racusen and Ruth Satter of Yale University found that at the onset of the natural rhythm of *Samanea* its electrical potential grows stronger, and the potential oscillates daily as long as the plant is not kept in darkness for too long (Racusen and Satter, 1975).

This is about as far as we can go in the physiology of leaf movements without getting bogged down in myriads of theories. However, the interested reader is recommended to study the review by Satter and Galston (1981), in which they set out their ideas of how leaf movements might be controlled.

How then do the motor cells themselves behave during all this activity? The pulvinus of the *Samanea* leaf looks similar to the pulvinus of *Mimosa* – a slender core of vascular tissue (the 'plumbing' which carries water, salts, sugars and other chemicals) surrounded by a thick wodge of parenchyma tissue wrapped in a layer of epidermis (figure 8.8). The parenchyma cells are well connected with each other through microscopic pores called plasmodesmata, particularly in the motor cells closest to the core of vascular tissue. These pores provide a rapid network for sending signals and carrying water and ions between vascular tissue and the motor tissue in the outer region.

The motor cells of *Albizzia* and *Samanea* both contain dozens of vacuoles (bubble-like packages) of various sizes. These seem to split up when the cells shrink, and re-form when the cells swell. Indeed, Darwin (1875) noticed a similar phenomenon in the motor cells of sundew tentacles when they bent towards meat. No one knows what importance this might have. Tannins – the substances that give tea its bitter taste – have also been found in the vacuoles, although they are not as abundant as in *Mimosa*. The size of the tannin bodies changes shape during the transition from day to night, so they may have a role to play although we cannot be sure what it is. Our main clue is that in *Mimosa* the tannin seems to be an ion reservoir.

Another intriguing puzzle is reminiscent of the *Mimosa* 'muscles'. *Albizzia* looks remarkably like *Mimosa* but it only performs the light, sleep and rhythm movements – it is not touch sensitive. Pierette Fleurat-Lessard and Ruth Satter, working at the University of Poitiers, found numerous fibrils rather like those of *Mimosa* in *Albizzia* motor cells (Fleurat-Lessard and Satter, 1985). Could they contract like the *Mimosa* fibrils? So far there is nothing else to suggest that they might contract. However, if they did, how would that help the motor cell? Fleurat-Lessard and Satter noticed that the cell walls of *Albizzia* were thicker than those of *Mimosa* or the stamen filaments of *Berberis*, so could the fibrils help contract the cell? We simply do not know yet.

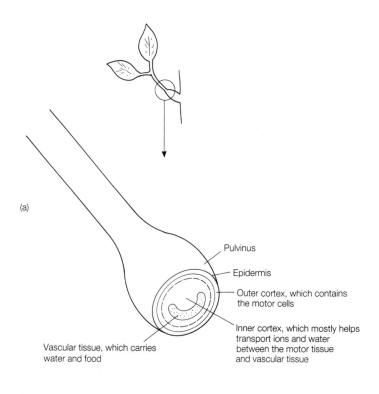
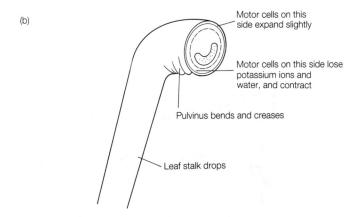

FIGURE 8.8 Leaf movement of *Samanea saman*: (a) leaf open; (b) leaf closed.

To summarize this mass of information, imagine yourself for a moment actually inside the motor tissue of a pulvinus when it performs a sleep movement (figure 8.9). It would probably hit you like rush hour in a city. First, as the sun sinks in the outside world, photosensors could suddenly open up holes in the motor membranes. Hydrogen or calcium would come flooding in or out. The environment becomes electric as the voltages on the membranes are suddenly jolted, and this is followed by a torrent of potassium ions and water streaming through every available space in the pulvinus, quickly succeeded by chloride ions travelling in the opposite direction. Motor cells would rapidly collapse or inflate in unison, and with them the tissues would wrinkle up or bulge out. And then the whole pulvinus would lift up or down.

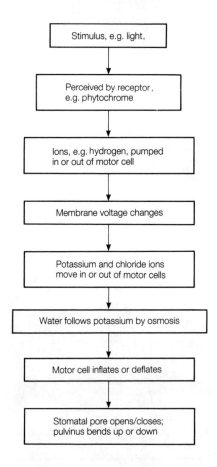

FIGURE 8.9 General scheme of events leading to stomata and pulvinus movements.

It is an enormous exercise, fuelled by considerable amounts of energy from respiration and photosynthesis. Clearly, it must pay off for the plant. Of all the plant performances described in this book, leaf movement has been one of the most intensively studied in recent years. And yet we still know so little about it.

Evolution of Leaf Movements

Now take one step back from the minutiae of all these pulvini and stomata movements. It seems to be asking a great deal of coincidence that so many different movements should all work with the same sort of ion and water pumps and their accompanying electricity. It's stretching credibility a bit far to imagine that they have all evolved completely independently of each other, particularly as many of the species with touch-sensitive movements are closely related to those with light-sensitive movements. It seems far more likely that, with so much in common, the touch-sensitive leaf movements evolved from the other pulvini movements. Consider these similarities.

They both rely on water acting as a hydraulic engine to move the pulvinus.
The water is moved by enormous potassium fluxes.
The current of potassium ions carries positive electric charges, balanced by
  negative chloride ions travelling in the same direction.
The electrical potential of the membrane changes, and probably helps the
  ion fluxes.
Tannin bodies in the vacuoles are believed to be ion reservoirs.
Caffeine and a number of other chemicals that upset calcium ion transport
  in the plant cells also block movement.
The structure of the pulvinus is basically similar – a thin core of vascular
  tissue surrounded by parenchyma tissue connected by plasmodesmata.
The mechanics of the pulvinus are basically the same – a lever pumped up or
  down by changes in turgor in its two halves. Motor cells change shape as a
  result of the changes in turgor and their inherent flexibility.

However, Ruth Satter, a noted authority on leaf movements, disagreed (Satter, 1979). She argued that there are too many differences between, say, *Mimosa* and its close relative *Albizzia*. The motor cells of *Mimosa* have thinner walls and the bulk of movement occurs on the underside of the pulvinus. In *Albizzia* both halves of the pulvinus are involved, and have to co-ordinate their movements more closely. Also, a number of chemicals that affect the movement of *Albizzia* have no effect on *Mimosa* touch-sensitive movements.

To my mind the similarities outweigh the differences. The *execution* of the movements is basically the same, but the *perception* of a stimulus is different. When an insect touches a *Mimosa* leaf the plant has to react rapidly, and so electrical signals are vital for speed. Time is not so crucial to the other types of movement, and so the same sort of electrical waves are not found.

Probably the greatest unknown factor is the possibility of muscle-like fibrils in *Mimosa*. Much more needs to be done to find out whether they do indeed help the motor cells contract, and whether non-touchy plants use them as well. This remains a thoroughly intriguing area of research.

Leaf movements also have much in common with the movements of stomata – water is channelled in and out of motor cells by osmosis, driven by potassium and balanced by other ions. So, imagine how all this sensitivity and motion may have evolved.

All leaves grow towards sunlight by bending their stalks and blades, but the movement is restricted by their limited capacity to grow. More flexible leaf stalks may have evolved to make bending easier. They relied less on growth and more on special motor cells with thin walls, which would let them stretch and contract according to the amount of water that they held. Eventually a pulvinus developed as a flexible hinge, allowing leaves to move backwards and forwards without relying on any growth at all. As for the movement itself, the motor cells of the pulvinus developed a light-sensitive hydraulic/ion motor similar to that already well established in stomatal movements. With a well-developed network of plasmodesmata between motor cells, the movements of the pulvinus then became co-ordinated.

It is probably no great jump in evolution to go from light-sensitive to touch-sensitive leaf movements. First, all the touch-sensitive species also exhibit light-sensitive movements. Second, we can see a gradation in touchy movements from those that need repeated and heavy blows before they move (like *Oxalis acetosella*) to those that respond with the slightest touch, such as *Mimosa pudica*. If only we could find something in the plants that correlates with their level of touchiness, we could point to the vital ingredient that has evolved in touch-sensitive movements. There is only one anatomical feature that even remotely correlates – the tannin bodies. Species with the strongest touch-sensitivity have the largest and most numerous tannin vacuoles. And the touch-sensitive species have more tannin than those plants that only perform sleep, light-sensitive and rhythmic movements. Could tannin have been a crucial key in helping plants respond to touch? We need to find out more about what tannin does before anyone can say for sure how important it might have been in the evolution of touch-sensitive movements.

One of the crucial phases in the evolution of touchy movements was the development of action potentials to send rapid messages. Rapid communication allows rapid movement, and it is interesting that Philip Applewhite and Frank Gardner of Yale University reported how *Mimosa* can respond rapidly to light as well as touch stimulation (Applewhite and Gardner, 1971). They switched a light on and off every few minutes. When the periods of darkness lasted less than a few minutes the leaves did not respond, but after 5 minutes of darkness they suddenly closed after 90 seconds. The longer the darkness, the quicker their reactions, with the fastest recorded reaction at 49 seconds. Therefore rapid light-sensitive movement might have been the precursor of rapid touch-sensitive movement.

Touch-sensitive pulvini may have evolved first as a way of defending leaves against rain. Tropical leaves that could orient to sunlight folded down during heavy rainfall, preventing a harmful film of water building up on the leaf surface. The relatively weak touchiness of these leaves probably helped them stay open for as long as possible to catch sunlight. But touchy plants such as *Mimosa pudica* grow in open sunny habitats where they could afford to close during much lighter rainfalls. As the leaves became faster at responding and moving to rainfall, so they could also defend themselves against animals. For successful tropical plants like *Mimosa* the movements have evidently paid off. But one vital question is: how did the perception of touch evolve? The answer to that lies in 'ordinary' plants in general, and that's an intriguing area that we'll look at more closely in chapter 12.

9

PLANT MUSCLES

Power in the Cell

Cells have no hands or feet, yet nearly every living cell is seething with movement. Look inside a cell and you'll see a chemical factory running a 24-hour production line to manufacture the life support systems vital for a living being. Each of the cell's bodies, its organelles, runs its own business (see box 9.1) and even lives a semi-independent life. You can actually see much of this activity. For example, the cytoplasm in a plant cell looks like a race track, incessantly whirling around and often dragging the organelles with it. This was a mesmerizing sight when the Italian biologist Bonaventura Corti, professor of botany at Bologna University, first noted it over 200 years ago. He commented on the animal-like movements of the cytoplasm in the fungus *Tremella* as follows:

> that *Tremella* are endowed with movements said to be spontaneous in animals are considered characteristics of animals. And here we have plants that by now are confused with true animals.
>
> (Corti, 1774)

Box 9.1: The components of cells and their functions
The organelles of the cell consist of mitochondria (for respiration), chloroplasts (for photosynthesis), the nucleus (repository of most of the cell's genes) and so on. The cytoplasm itself is roughly divided into two zones – the layer of cytoplasm lying next to the plasma membrane is rather jelly-like, but the rest is fluid.

He thought that the streaming was the cell sap moving around – in those days they did not know about cytoplasm. But one of his favourite specimens, the stonewort alga *Chara*, has one of the best studied of all cell movements because of its relatively large size – a tube almost a millimetre in diameter by several millimetres long (figure 9.1). The cytoplasm in *Chara* is a granular ribbon usually sandwiched between the plasma membrane and the tonoplast, but sometimes it sends slender fingers across the vacuole. At speeds of up to 19 micrometres per second, the cytoplasm's streaming behaves like a rudimentary blood system, supplying the cell with its food, dissolved gases, enzymes, hormones, storage products and so on. But whether it also serves as a conduit for passing chemicals from one cell to another is not certain.

The rate of cytoplasmic streaming varies enormously (Kamiya, 1959) from 1.35 millimetres per second in *Physarum polycephalum* up to 19 millimetres per second in pollen tubes. However, the rate depends on various conditions – temperature, the age of the cell, light, the time of year, ions present in the surrounding medium and many other factors. Even magnetism has been reported to affect it, but the significance of this in the natural situation is unclear (Ewart, 1903).

The discovery of the driving force behind cytoplasmic streaming took the world of biology by storm – and botanists are still reeling from it. The first hints of something unusual started with work on the humble slime mould fungi. As we have already seen in chapter 4, the slime moulds are a peculiar half-way house between animal and vegetable kingdoms. They slither around searching for food, or for each other's company, or away from substances that they find distasteful.

Many theories have been proposed to explain the movement. Some said that the stiff jelly-like cytoplasm dissolved at one end of the cell and the fluid cytoplasm gelled at the other end. There were ideas that the surface tension on the cell continually changed with the metabolism inside, and this somehow drove the cell along. There were ideas involving the fluid cytoplasm pushing itself. Others believed that the particles suspended in the fluid cytoplasm somehow solidified. But none could fully explain the driving force behind the movement.

However, in 1952 Ariel Loewy, a graduate botany student at the University of Pennsylvania, made an astonishing discovery. 'I was convinced that the time had come to stop talking about the colloidal properties of protoplasm', he now remarks. So he took a lead from the new protein chemistry that was being developed by Albert Szent-Gyorgyi and others, who were unravelling the engine that drives muscle contraction. They had found two proteins – actin and myosin – which together formed a dynamic duo called actomyosin. This protein partnership has the clever knack of being able to contract whenever the energy-rich compound ATP (adenosine

PLANT MUSCLES

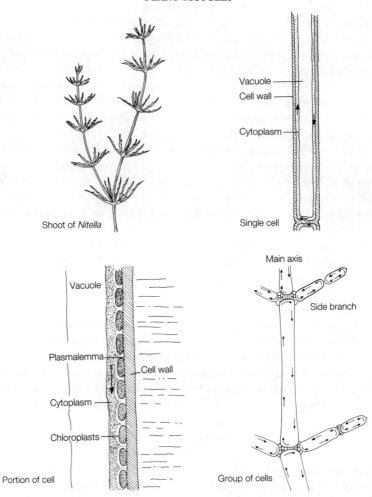

FIGURE 9.1 Plant cells have primitive 'muscles'. The cytoplasm inside the cells of *Chara*, *Nitella* and other so-called giant algal cells constantly churns around in never-ending circuits (the arrows indicate the direction of streaming). The engine driving the movement is actomyosin, the protein complex responsible for contracting animal muscles. In fact, the movements of *Chara* and *Nitella* can even be controlled by electrical signals and calcium ions, rather like the controls of muscle fibre contraction.

triphosphate) is added to it. We now know that actomyosin behaves as a catalyst – an enzyme – and splits the ATP molecules, unlocking much of the energy inside them (box 9.2). This energy fuels the contraction of actomyosin, and when the supply of ATP dries up the protein relaxes and returns to its original length. Loewy marvelled at the possibilities:

PLANT MUSCLES

My own scientific awakening began when, as a graduate student, I read the marvellous little volume by Szent-Gyorgyi, *Chemistry of Muscular Contraction* and when I first saw the wonderful little organism *Physarum polycephalum* move briskly under the dissecting microscope of my Professor William Seifriz. Naturally actomyosin and *Physarum* had to be put together!

(Loewy, personal communication)

Loewy added ATP to *Physarum*, and it made the cytoplasm less viscous in much the same way that Szent-Gyorgyi had found for muscle actomyosin. It was the first, albeit rather crude, evidence that actomyosin could exist in cells other than muscles. Loewy even proposed a theory to explain how the filaments of actin and myosin could slide over one another and create movement in *Physarum*, and also in chloroplasts, nuclei and chromosomes (Loewy, 1952). Several years later Hodgkin and Huxley proposed a similar scheme to explain muscle contraction – and they received a Nobel Prize!

Several years after Loewy's discovery, and after intense work, the Japanese biochemists Sadashi Hatano and Fumio Oosawa of Nagoya University eventually isolated actin and myosin from *Physarum* (Hatano and Oosawa, 1966). Under the electron microscope they saw that the slime mould's actin and myosin behaved like that in animal cells, forming filaments 10 nanometres (a hundred-thousandth of a millimetre) long which contracted by a quarter when ATP was added.

As techniques for studying actomyosin became more sophisticated, the story of fungal muscle power became clearer. It turned out that the actin and myosin in *Physarum* is so closely related to that in animal muscle that if purified actin from a fungus is added to myosin from muscle then plant and animal proteins join together and actually *work* together! The same happens in reverse if muscle actin is added to fungal myosin. To appreciate this astonishing similarity between fungi and animal muscles fully you need to look at how actin and myosin work (see box 9.2); it is powerful testimony to the preservation of these motor proteins over hundreds of millions of years of evolution, from amoeboid cells to mammals.

The prospect of a weight-lifting slime mould is still in the realms of science fiction though. For one thing, the slime mould organizes its 'muscles' differently. Instead of having actin and myosin arranged in long permanent bundles, the fungus makes temporary 'muscle filaments' wherever the cell needs to move. We now know that slime moulds slither along by squeezing columns of protoplasm back and forth inside their cells at speeds of up to 1.5 millimetres per second. It behaves rather like toothpaste pressed in and out of a tube. The engine driving the squeezing is a core of actomyosin fibres anchored to the plasma membrane. When the core contracts, the protoplasm is squeezed.

So far we have only looked at the motor and the fuel which powers the slime mould movement. And yet there must be something controlling the motor, otherwise the slime mould would be in perpetual convulsions, not knowing which way or when to turn. The answer, once again, is remarkably similar to the control of animal muscle.

When pieces of *Physarum* protoplasm are treated with caffeine (to make them more leaky to calcium ions), a concentration of only 0.000,000,1 molar (0.000,002 grams per litre) of calcium in the solution bathing it is enough to trigger movement. This sensitivity is almost the same as in muscle. And like the way that the muscle's sarcoplasmic reticulum stores reservoirs of calcium, the slime mould stores its calcium in special little bubbles (vesicles). Bursts of calcium can even be detected just before contraction begins (Ridgway and Durham, 1976), but what controls the movement of calcium?

Physarum also contains proteins similar to tropomyosin–troponin (described in box 9.2), which are so close to their animal counterparts that they can sensitize rabbit muscle (Kato and Tonomura, 1975). But tropomyosin and troponin are only two of many other actin-binding proteins, and hardly a month goes by without one of the science journals reporting another actin-binding protein. Some are simply the same protein with two different names because they were discovered in two different laboratories.

Box 9.2: How actomyosin works

Actomyosin forms long thin filaments which have the canny knack of shortening when fed with energy from ATP (adenosine triphosphate). When the ATP runs out, they relax and return to their original length. A muscle only a few millimetres in diameter can generate a force of a few newtons, which would be enough for you to pick up this book.

The muscles of animals basically consist of the fibrous actin and the paddle-shaped myosin. They are both arranged in a highly ordered striped pattern: the actin forms thin stripes, whereas myosin builds by side-to-side bonding, thick stripes lying parallel to the actin. The complex of actin and myosin is called actomyosin. Myosin itself acts as an enzyme, an ATPase, liberating energy from ATP, and is helped along by actin. Driven by the energy supplied by splitting ATP, the thin and thick filaments interact and slide over one another, contracting the whole muscle.

In many types of muscle, control of contraction is helped along by calcium ions and the proteins troponin and tropomyosin. These proteins are located in the actin filaments and confer calcium sensitivity on the contractile process. In the absence of calcium, they prevent the actin and myosin filaments from touching and moving against one another by inhibiting the stimulation of ATPase activity by actin.

PLANT MUSCLES

When the muscle receives an electrical signal (action potential) through its own nerve supply, the message is relayed through a network of membranes – the sarcoplasmic reticulum – running throughout the interior of the muscle. As the electricity from this signal passes, it unleashes calcium previously stored in the sarcoplasmic reticulum.

The calcium ion does not take part in a chemical reaction straightaway. It first binds to a collaborator molecule – in muscle this is the protein troponin – and then fires off one reaction which triggers a cascade of other reactions among the muscle proteins, eventually releasing energy needed for muscle contraction from the ATP.

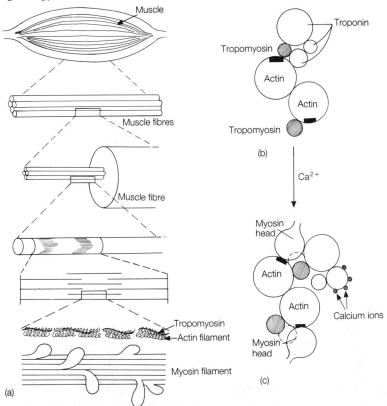

How muscles work. (a) Organization of muscle, down to the level of actin and myosin molecules. (b) At rest (low levels of calcium), the troponin bonds in such a way with the actin and tropomyosin as to interfere with the actin. (c) A rise in calcium ion levels allows troponin to bind Ca^{2+}, causing a shift of the tropomyosin. This allows myosin to bond to actin, and so the muscle contracts.

> Box 9.3: Diatom 'muscles'
> One type of alga, the diatom, also relies on actomyosin for its movement.
> Its rigid cell wall is only exposed to the outside world through a slit running along the bottom of the creature. Since the last century botanists have realized that the movements probably involved the slit, but there was one baffling problem – nothing could ever be seen moving there. The lack of any structure which could drive the movement only deepened the mystery. Only recently has the secret of this cleft been revealed.
> Botanists Lesley Edgar and Jeremy Pickett-Heaps working at the University of Colorado looked at the diatom *Navisula cuspidata* under an electron microscope and discovered some strange features (Edgar and Pickett-Heaps, 1983). They found that tiny filaments lining the slit contained actin. These filaments drive strands of slime down through the diatom's body and onto the ground below, pushing the diatom along like a punt pushed along a river with a pole. They found actin in the filaments in the slit, and these may drive the diatom along.
> Even more intriguing evidence came from another laboratory at the University of Miami. Barbara and Keith Cooksey found that calcium was needed for another diatom, *Amphora coffeaeformis*, to move. This again suggests that calcium may be involved with the actin protein (Cooksey and Cooksey, 1980).

This complex machinery keeps control of the contractions, probably in both muscle and non-muscle alike (figure 9.2). Incidentally, by studying the simple cell of *Physarum* and other primitives we get a good insight into our own muscles. But all this begs the question: what controls the controllers? What, for instance, controls the tropomyosin–troponin?

Given how similar the fungal cells are to a muscle, the obvious thing to look for is a nerve-like signal. But the results here have been conflicting. The most exciting discoveries have been made not in fungi but in two large-celled algae, *Chara* and *Nitella*, which, as we'll see later in this chapter, use nerve-like impulses to stop their cytoplasm streaming (figure 9.3). They trigger their electrical signals with calcium, potassium and chloride ions, and control their actomyosin contractions with calcium ions – remarkably similar to our animal muscle model.

Returning again to the slime moulds, over thirty years ago the veteran protoplasm expert Noburo Kamiya and his colleague Abe discovered pulsating voltages on the plasma membrane of *Physarum* (Kamiya and Abe,

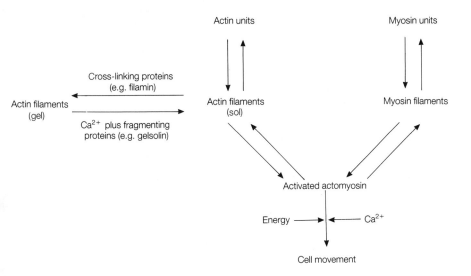

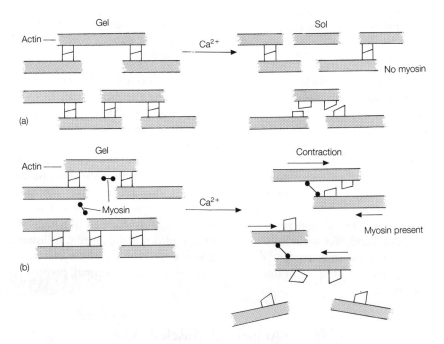

FIGURE 9.2 How actin and myosin make slime mould plasmodia and amoebae move. (a) At low levels of Ca^{2+}, actin is bridged by gelatin factors. When Ca^{2+} levels rise, the gel breaks down. (b) When myosin is present, contraction occurs when Ca^{2+} levels rise.

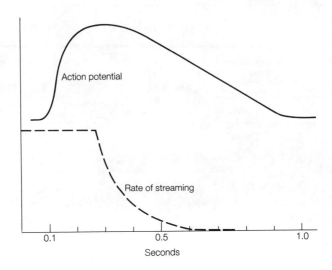

FIGURE 9.3 Touching a cell of the alga *Chara* triggers an action potential (electrical signal) which tells the cytoplasm to stop moving. The response is rapid, but the movement recovers within seconds or minutes.
(After Seitz, 1979; courtesy of Springer-Verlag)

1950). But although the electrical rhythm followed the rhythm of protoplasmic pulsations, it lagged slightly behind the movement. In other words it appeared to be the *effect*, and not the *cause*, of streaming movement. Whether this discrepancy was real, or simply due to the way that the electrical measurements were made with the electrodes outside the beast, is now a contentious issue. Ridgway and Durham (1976) correlated the fluctuations in voltage with the levels of free calcium inside the protoplasm. Achenbach and Weisenseel (1981) used a highly sensitive vibrating electrode to measure electric currents flowing around *Physarum*. Current always left the advancing front of the cell and returned through the retracting end. Yet what is cause and what is effect: is the current following or leading the movement? We get a much clearer idea of how electricity could be controlling plant muscles from the single-celled 'animal' protozoan amoebae.

Amoeboid Muscles

Amoebae come in a variety of sizes, but they are all microscopic single cells and therefore a great deal smaller than *Physarum*. They appear to move by means of the same finger-like extensions, called pseudopodia ('false feet'), pushing out in front of the creature in advance of the bulk of the protoplasm. In the wonderfully titled amoeba *Chaos chaos*, vibrating electrodes held just outside the surface of the cell have revealed electric current entering the portions of cell about to form a pseudopodium. In fact, studying the movement turned out to be surprisingly easy. The amoeba thought that the vibrating probe looked a lot like its usual protozoan prey, and sent out a feeler to investigate it immediately.

Following the discovery of actomyosin in slime moulds, it was not long before a similar motor was discovered in *Amoeba proteus* by Thompson and Wolpert (1963). And as in the fungus, ATP also fuels the amoeboid motor. However, there is some argument as to how the contractions actually propel an amoeba. One idea is that the rear of the cell contracts and pushes the protoplasm out in front – again, like squeezing a tube of toothpaste. Alternatively, the protoplasm may be pinched at the 'nose' of the cell and pull the rest of it behind. Or the two mechanisms might even work simultaneously.

One thing is certain. There is now overwhelming evidence from many different laboratories that calcium ions are again the key signal for regulating amoeboid movement.

Cellular slime moulds, such as *Dictyostelium discoideum*, form amoeba-like structures, and the work on animal amoebae has been immensely useful for studying their movement. (Recall that the cellular slime moulds have one nucleus per cell, and tend to be much smaller than the acellular slime moulds which contain many nuclei.) Although the two differ slightly, their mechanisms of movement are fairly similar: actomyosin fuelled by ATP, under the thumb of calcium control. The outstanding difference is the way that actomyosin is put to use. In the fungal amoeba it pushes the protoplasm out of the rather rigid rear of the cell (figure 9.4). Because the sides of the cell are also rigid, the protoplasm has nowhere to move except forwards – probably the closest in nature to the toothpaste squeezing analogy – whereas in animal amoebae the whole cell is much more flexible.

Dictyostelium also contains a troponin-like protein, conferring calcium ion sensitivity upon the ATPase of myosin in the actomyosin complex (Mockrin and Spudich, 1976).

Nature does not waste a good opportunity. So successful was the movement of the protozoan amoebae that they became incorporated in higher animals, including ourselves. One type of human white blood cell called a neutrophil behaves exactly like an amoeba. It roams through our blood and lymph vessels searching out foreign invaders such as bacteria and fungi. When these are found the neutrophil deals with them in true amoeba fashion by totally engulfing them, wrapping them up in plasma membrane, digesting them with enzymes and absorbing them. The amoeboid movements of the white blood cells are now attracting the attention of many

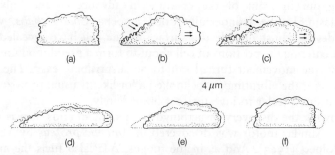

FIGURE 9.4 Sketch of how an amoeba of Dictyostelium discoideum moves in response to a chemical stimulus. (a) Outline of a resting amoeba. (b) A stimulus of cAMP diffusing from right to left triggers movement. (The broken line shows the position of the cell in the preceding figure. Arrows show the direction of flow of cytoplasm. Stippling shows the contraction of cytoplasm.) (c) Continual contraction at the rear of the amoeba pushes the amoeba forward, sending out a 'pseudopodium' (false foot) at the front. (d) The rear of the amoeba begins to relax. (e) The amoeba comes to rest. (f) The amoeba is ready to begin another surge forward.

(After Poff and Whitaker, 1979; courtesy of Springer-Verlag)

biologists, including Frederick Southwick and John Hartwig working at Harvard University (Southwick and Hartwig, 1982).

They found that like the slime moulds and amoebae, actomyosin also drives the white blood cell movement. However, they went further. Actin filaments are made up of many identical globules. As long as the globules remain separate, the actin flows like a fluid, but in a suitable salt solution they join together into a long chain – the actin filament. Conversely, when the actin threads are no longer needed, they are broken down into their component globules. In this way the amoebae, slime moulds and many other non-muscle cells use their actomyosin like portable motors, by building up and breaking down actin filaments wherever they are needed. The Harvard laboratory's breakthrough was finding how the actin assembles and disassembles itself in slime mould amoebae.

The key discovery was a protein called gelsolin, which has two important properties. First it likes to grab hold and cap the ends of actin filaments, preventing them from growing longer. Its second important property is that it can only do this at rather high levels of calcium. This is extremely important because many cellular signals, including of course the signal for muscle contraction, are transmitted by changes in the level of calcium. Therefore a local change in calcium levels would activate gelsolin to loosen the actin network, and at the same time the calcium would probably be triggering the contractions of the actin with the myosin filaments. The chances are that a similar gelsolin also exists in the primitive animal and fungal amoebae.

What I find astonishing about these discoveries made in these seemingly 'primitive' cells is that they tell us a great deal about the movements of our own human cells: not only white blood cells, but also the traffic of young cells sliding around in an embryo as they find their right locations in the foetus; the amoeboid movements of young nerve-endings forming new connections with their neighbours; new cells sealing off a wound; cancer cells migrating through the rest of the body. Above all our white blood cells reflect our primeval past; they are, if you like, simply roving amoebae with their own 'muscles', and their independence has recently been confirmed when it was discovered that they also possess a keen sense of taste and nerve-like electrical signals (McCann et al., 1983). Yet they are still part of us. Clearly, the life of an amoeba is a highly successful adaptation, yet ironically the amoeboid cells of a cancer can be the very downfall of an animal.

Plant Cytoplasm

Turning now to plant cells, amoeboid movement is fairly unusual. The corset of cellulose wrapped around most plant cells restrains the cytoplasm inside from expanding outwards like an amoeba. Therefore, when the cytoplasm moves, it has nowhere to go but round and round the central vacuole. This ceaseless merry-go-round of movement is called cytoplasmic streaming, and is common to all plants. Even single-celled animals such as *Paramecium* have streaming cytoplasm. But the movement does not come cheap; plants have to spend a considerable amount of energy on their streaming movements. Therefore it is quite clearly very useful.

So how does streaming work in the higher plants? The best evidence comes from algae. In 1976, Yolande Kersey and her colleagues at Stanford University used a heavy meromycin from animal muscle to reveal actin filaments in the cells of *Chara* and *Nitella* (Kersey et al., 1976). The actin filaments themselves don't move, but stay anchored in a stiff layer of cytoplasm near the edge of the cell, reaching out to myosin threads passing by in the adjacent layer of loose cytoplasm. The actin and myosin behave like countless tiny muscles, building and breaking arms between each other and so pushing the cytoplasm along.

What controls the ordinary day-to-day streaming of the algae is not clear, but one useful insight comes from their dramatic response to being touched. A slight tap or prod is all that is needed to make the cytoplasm stop almost instantly (figure 9.3), although after a short rest it restarts again. It has long been known that touching a cell of *Chara* or *Nitella* triggers an action potential inside it, and if the stimulation is strong enough the electrical signal fans out into neighbouring cells (Hope and Walker, 1975). There is no doubting the importance of the electrical signal though. The start of the action potential always precedes the sudden brake on the streaming (figure 9.3), and if the electricity of the cell is artificially lowered that will also trigger the stoppage.

Recently, a quite bizarre discovery was made: electrical signals were found to jump across into neighbouring *Chara* plants if their cells were lying parallel, and to make their cytoplasm stop streaming as well. This is probably the only known case of one plant communicating with another by electricity (Ping et al., 1990). Quite what significance this would have for algae living in the wild is a mystery.

The actomyosin 'paddles' that drive the streaming cytoplasm of these two relatively simple algae also power the streaming in the cells of the much more highly evolved flowering plants (Hepler and Palevitz, 1974). We know far less about the control of streaming in flowering plant cells, largely because *Chara* and *Nitella* are so much easier to study, but we do know that calcium is involved (Kohno and Shimen, 1988).

On the face of it, the 'muscles' of *Nitella* and *Chara* seem to work uncannily like animal muscles. They both need an action potential to make them contract by releasing stores of calcium which interact with the actomyosin. But there are some subtle differences. First, there is no evidence that an electrical signal is required to control the regular streaming of cytoplasm; only the stoppage needs an action potential. Second, in contrast with animal muscles, the surge of calcium ions in plants inhibits movement. Having said that, though, the differences are insignificant compared with the similarities. There seems little doubt that both animal and plant muscles evolved from the same primeval ancestors.

Before actomyosin was discovered in plant cells, David Fensom, professor of botany at MacAllister University, Canada, made a truly astonishing proposal. He suggested that plants might be using contracting fibres to pump solutions through their sugar-conducting tubes (phloem) (Fensom, 1972). Phloem tubes are relatively long plant cells, capped at each end by sieves. The cells are empty except for one feature – strands of material containing proteins (figure 9.5). The function of these strands is one of the great enigmas of plant science. The conventional view of phloem transport is that it works by simple water pressure – a rich supply of sugar in a phloem tube sucks in water, and the pressure forces the sugar to move to a more dilute region. There should be no need for sieves or strands, and most textbooks are content to write them off as safety devices for plugging the phloem shut when they are under stress. So Fensom's idea has never really caught on, yet the circumstantial evidence improves with time (always a good sign of a correct theory!). Over the past decade, several scientists have discovered actin in the phloem, or even attached to the phloem strands (Ilker, 1975; Kursanov et al., 1983). The idea that contractions in these fibres might push or pull material along is nothing to scoff at. As we will see shortly, nerve fibres contain a network of microtubules coated with contractile proteins that transport small bodies along.

Actin may also help move the organelles. A good example of this has been found recently in pollen tubes, where actin filaments carry the nucleus along its journey that eventually culminates in fertilization with an egg cell (Heslop-Harrison and Heslop-Harrison, 1989). Rings of actin may also help some organelles divide by splitting into two halves (Hasezawa et al., 1988). Most dramatic of all, the insides of some algae shrink into a ball or dozens of balls when they are wounded. These contractions are also driven by actomyosin (La Claire, 1989).

Chloroplast Choreography

The most accomplished organelle movements are perhaps those of the chloroplasts. As we saw in chapter 7, they have their own rhythmic and

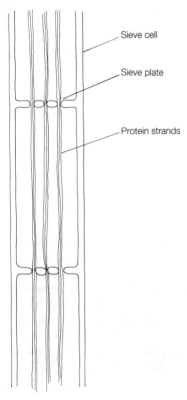

FIGURE 9.5 Phloem tube cells.

light-sensitive performances – rolling, sliding or turning to find the best available light.

Heavy meromycin from animal muscle has been used to identify actin near the chloroplasts, but the precise location in the cell was difficult to find. The problem was the tough cell wall envelope, which gets in the way of the treatment. So Karin Klein, Gottfried Wagner and Michael Blatt of the Institute for Botany and Pharmaceutical Biology, Erlangen, Germany, stripped the cell wall of *Mougeotia* away with enzymes and then applied the heavy meromycin. What they found were perfect tiny plant 'muscles': 5–7 nanometre long filaments of actin identical with the strands found in the untreated cell, anchoring the chloroplast to the surrounding cytoplasm. Assuming these were the same filaments, then their contraction will pull the chloroplast like the rigging of a yacht's sail (Klein et al., 1980).

Again we return to that familiar question: what controls the movement? Recent breakthroughs now show even more similarities with those in muscle contraction. The chloroplast 'sees' the direction and intensity of light using the light-sensing pigment phytochrome and an unidentified blue light sensor. Calcium ions help to regulate the contraction of chloroplast actin at similar concentrations to those in muscle contraction. American botanists Bruce Serlin and Stanley Roux of the University of Texas even showed where the calcium acts (Serlin and Roux, 1984). They used a special chemical called an ionophore, which punches a hole through the plasma membrane allowing calcium to leak in or out. As a result the chloroplast turned. But Serlin and Roux carried out an intriguing test. They applied the ionophore to very small patches of the cell's plasma membrane, and saw that the chloroplast turned only at that spot nearest the area of membrane which had been treated. So calcium ions are important in chloroplast movement.

Sigrid Jacobshagen et al. (1986) at Justus-Liebig University, Giessen, Germany, have narrowed down the behaviour of calcium even further. They believe that calcium is stored in vesicles in the cell. Phytochrome then releases the calcium from these vesicles. Furthermore, the calcium may then bind to a special protein called calmodulin (see box 9.4), which has also recently been isolated from *Mougeotia*. They found that drugs which interfere with calmodulin inhibited the phytochrome-controlled light response. Jacobshagen and colleagues also found that phytochrome activates the calcium-binding behaviour of calmodulin, releasing calcium.

ATP fuels chloroplast movement, and may even help to control it. The blue light photoreceptor may directly affect the myosin or perhaps the supply of ATP energy. Other chloroplast movements are simply passive – they are dragged along in the cytoplasm stream, e.g. chloroplasts of *Elodea* and *Vallisneria* are located in the 'fluid' part of the cytoplasm.

To complicate things further, different chloroplast movements may use different motors. Steven Britz and Winslow Briggs of the Carnegie Institution believe that microtubules may be involved in the rhythmic movements of the chloroplasts of the green alga *Ulva* (Britz and Briggs, 1983). They prevented the chloroplasts from moving from the sides to the ends of the cell by using the drug colchicine which breaks up microtubules. The effect of colchicine is reversed in the presence of ultraviolet light – characteristic of a microtubule-controlled movement. And yet the same chloroplasts can move from the ends of the cell to the sides unhindered. These movements are driven by actomyosin.

Microtubules have also been implicated in many other plant and animal cell movements: chloroplast movements in *Bryopsis* and in the alga *Micrasterias*, migration of vesicles in the animal *Hydra*, the movements of coloured pigments in the camouflage behaviour of some fish and various other organisms, and the movements of mitochondria (the cell organelles concerned with respiration). But we will now look at one cell component whose microtubule behaviour is being intensively studied throughout the plant and animal kingdoms, and which might one day help unlock the mysteries of cancer development.

Dancing the Cell Division Jive

If there's one single feature that makes living things different from the rest of the world, it's their ability to reproduce themselves. This extraordinary feat needs two vital steps: a set of instructions to work from, and the power to tear a cell into two new cells. The instructions are kept in a chemical language DNA (deoxyribonucleic acid), stored largely in the nucleus of the cell. What we're concerned with here, though, is how the cell rips itself apart in two separate operations: dividing the nucleus into two new nuclei, and then dividing the rest of the cell into two new daughter cells.

Yet these two fundamental pillars of life are still one of the greatest mysteries of biology. The whole process of cell division involves an enormous upheaval in the cell's architecture and chemistry, precisely co-ordinated and demanding a great deal of energy. Yet all organisms above the level of bacteria perform this dance (bacteria do not have nuclei as such). So what makes a cell divide? What is the programme that choreographs the exquisite dance of a cell duplicating and then tearing itself into two identical copies? (We shall only deal here with the commonest division, mitosis, in which the DNA is divided into two equal portions. In the other form of cell division, meiosis, the DNA divides into unequal portions. This is only performed by cells that produce gametes (sex cells).)

Pursuing these questions, particularly the quest to find what devastating switch turns ordinary healthy cells into anarchic cancers, has become the

PLANT MUSCLES

> Box 9.4: Calmodulin: the calcium partner
>
> Work from numerous laboratories now shows that calmodulin is ubiquitous in living organisms. Calcium has such a special liking for it that it clings to it and distorts its shape, and so changes its properties. The conversion is almost like Clark Kent changing into Superman – the calmodulin with its new shape can take on a dazzling range of jobs: turning on enzymes which release energy, turning on other enzymes which bind phosphate onto proteins, opening up channels in the cell membrane to specific ions and even affecting the transport of calcium itself. Again, there are parallels with our own muscles: when smooth muscle contracts (the type of movement that we do automatically, such as breathing), the calcium ions released from the sarcoplasmic reticulum reservoirs bind to calmodulin in the muscle. This activates a protein kinase.
>
> (Kinase is an enzyme which adds phosphates onto other proteins – often a key step in the regulation of other physiological or metabolic processes. One kinase that adds phosphate to tubulin, making it aggregate into microtubules, has been found in nerve cells. These tubules may then interact with the cell plasma membrane and release a nerve-transmitting chemical, one of the substances that relays a nerve impulse from one nerve cell to another. Kinases, and hence calmodulin, are involved in many other reactions.)
>
> Because calmodulin binds so tenaciously to calcium, it also regulates the concentration of calcium ions in a cell, and this in turn affects cell growth, cytoplasmic streaming, photosynthesis, and a wide range of other plant processes (see box 5.2).
>
> Calmodulin behaves the same in animals. It has long been known that a variety of external stimuli can cause an enormous release of calcium into animal cells. The calmodulin molecule mops these up, binding with the calcium ions. In so doing, calmodulin itself becomes active and interacts with the many key regulators of the cell's vital processes.

Holy Grail of biology and medicine. There must be something that tells the cell when and how to divide. A breakthrough in this field might reap enormous benefits, and yet one discovery often only leads to another problem. The medical world was taken by storm when cancer-causing genes – oncogenes – were discovered a few years ago. Yet they only give us a tantalizing glimpse of cures or treatments to come. The story I'm about to relate is quite different: scientists have discovered remarkable parallels between dividing nuclei and the movements of cilia and flagella.

When cells divide they usually split into two daughter cells with equal numbers of chromosomes. The division follows a precise timetable of events (figure 9.6). Loose strands of DNA inside the nucleus condense and wind themselves into dense chromosomes, like candyfloss wound on a stick although much more complicated. Meanwhile, the membrane wrapped around the nucleus disintegrates, and a mass of microtubule 'scaffolding' inside the cell breaks out and forms into a beautiful diamond-shaped dance floor – known as the 'spindle' – on which the sausage-shaped chromosomes perform an exquisite dance.

The chromosomes attach themselves to the microtubules, and at a precise moment they all pull up to the equator of the diamond, like pairs dancing a barn dance. At a given signal all the pairs split down their lengths into equal halves, each partner moving off to its own pole at the apices of the diamond. The chromosomes then wrap up into two new nuclei, membranes enclose each nucleus and the spindle disappears. So much for the division of the nucleus. The rest of the cell then divides into two new daughter cells, each with their own nucleus. All this usually takes several hours, but the fastest spindle movement known, in juvenile stomatal guard cells in onions, takes only 10–15 minutes.

The Chromosome Motors

The spindle tubules behave like cables, guiding and moving the chromosomes so that each of the two new daughter cells receives an identical copy of each chromosome. The movements of the chromosomes are linked with the microtubules of the mitotic spindle. There are several conflicting theories of how the microtubules might work, but the two ideas sketched out here have the bonus of being complementary to each other and also having very recent evidence on their side.

Chromosomes can be plucked from their spindles and their movements recreated in the test tube by replanting them on microtubules. The microtubules shorten by falling to pieces at one end, as they do in the spindle, pulling the chromosomes with them. This behaviour is surprisingly like that in natural cell division, where you can actually see the two arms of the chromosome flapping behind as the middle of the chromosome is pulled away to the poles of the spindle (Koshland et al., 1988).

Shortly after the chromosomes are tugged apart by shortening the microtubules, the spindle itself paradoxically grows longer. This is caused by the growth of other groups of microtubules at the equator which push the two halves of the spindle apart. Thanks to new techniques which can float spindles free of their cells – a very tricky operation, given how small and delicate the whole system is – we now know that the two halves of the

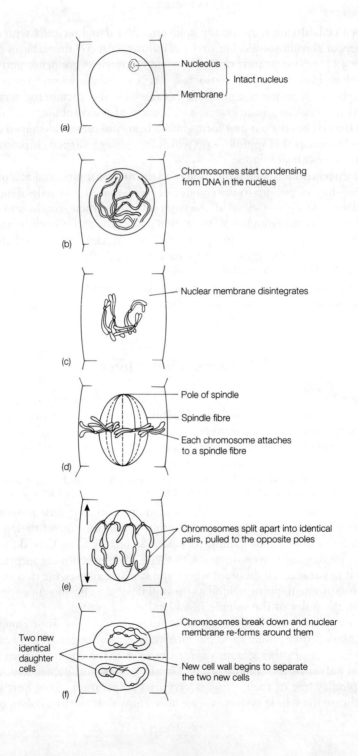

spindle overlap. Using this technique, botanists Zacheus Cande and Kent McDonald of the University of California at Berkeley found that the spindles from a diatom alga stretched apart (Cande and McDonald, 1985). The microtubules overlapping at the equator grew apart like a chest-expander stretching out, seemingly pushing the chromosomes apart. So what drives the microtubule movement?

The microtubules move uncannily like the wriggling of sperm, spores or other cells with cilia and flagella. Dynein has recently been identified in the spindles of sea urchin eggs and other animal cells by means of an elegant new technique – a fluorescent molecule attached to an antibody tailored only to fit dynein. It is a neat partnership: the antibodies only recognize and grip onto dynein, and the light from the fluorescent tag lights up the dynein under the microscope like a small torch (Mohri et al., 1976). As further proof that dynein is involved, an antibody that blocks sperm dynein from working also blocks the chromosome movement (Sakai et al., 1976).

The dynein motor works in the following way. Each chromosome is pinched into a tight 'waist' known as the kinetochore, which attaches the chromosome to the mitotic spindle. Over the past few years we have come to realize that the kinetochore is more than just an attachment zone. More and more evidence points to the kinetochore as the chromosome motor, holding the dynein engine and running along the mitotic spindle like a cable car (Pfarr et al., 1990; Steuer et al., 1990). Since dynein propels the sliding of the microtubules in the beating of cilia and flagella, so the spindle may behave in the same way as sperm and similar cells.

There is also some evidence that actin may be involved in moving chromosomes at some stage of cell division. In recent years tremendous progress has been made in pinpointing actin in plant cells, using antibodies specially designed to fit them exclusively. This new development is quite clearly reflected in a flurry of papers published over the past three years which have reported evidence of actin in the mitotic spindle (Menzel, 1989), but whether the actin is driving the chromosome movement remains to be seen. The issue of which engine drives chromosome movements is further complicated by the recent discovery of several other potential microtubule-based motor proteins which are implicated in movement (Menzel, 1989). Therefore we must be prepared for more discoveries, and even surprises, in the mitosis engine story in the coming years.

FIGURE 9.6 What controls the exquisite choreography of chromosomes as they divide into two new daughter nuclei is one of the greatest mysteries of biology. Protein motors and microtubules similar to those in the cilia or flagella are probably involved, but how they are controlled is not clear. Microtubules are also involved in creating a new cell wall, by apparently laying down 'tram-lines' for building materials to travel along ((d), (f)).

The Conductor of the Chromosome Orchestra

Yet something must tell all these chromosome motors when to move, and there is now compelling evidence that the conductor of the cell division orchestra is calcium. In fact, calcium is probably involved in triggering the very start of cell division, before the chromosomes are even formed.

Positively charged electric currents of calcium ions and possibly hydrogen ions start flowing before cell division starts, when the membrane around the nucleus degenerates (Hesketh et al., 1985). And even before this, when the nucleus prepares for division by positioning itself in the cell (see box 9.5), tiny electric currents pass through the cells of the moss *Funaria hygrometrica* nearest the nucleus. At the critical moment just before the cell divides, the strength of the current doubles (Saunders, 1986).

Box 9.5: The mobile nucleus

Nuclei also move of their own accord, without having to divide. For instance, they are sometimes very sensitive to wounding. If a *Tradescantia* (spiderplant) leaf is wounded, within 15 hours the majority of nuclei nearby will move to the wound. You might assume that the nuclei are simply sucked towards the wound by the drop in pressure, but metabolic energy is spent on the movement because cyanide, a potent inhibitor of respiration, blocks the movement.

Nuclei also move during the final stages of mating, just before the male and female sex cells finally fuse together. Many fungal nuclei, in particular, have to go through elaborate manoeuvres before they can fuse. One nucleus has to pass from one cell to the next, through a tube specially grown for the occasion, at precisely the right time before it fuses with the nucleus opposite. In flowering plants, pollen tubes carry two sperm nuclei into the female ovule. There they break free: one moves across to enter the egg cell and fuse with her nucleus, whilst the other travels across to the female's so-called central cell, where it fuses with two nuclei, eventually forming endosperm, the nutritious food reserve of the seed.

Cell division can even be turned on and off at will; by injecting sea urchin eggs with calcium the cells can be induced into premature division. And the division can be stopped with chemicals which block calcium activity, and then re-induced by injecting extra calcium (Izant, 1983; Hepler, 1985; Steinhardt and Alderton, 1988).

Calcium also affects both the sliding movements and growth of the spindle's microtubules. A group of physiologists headed by Martin Poenie of the University of California at Berkeley has made an astonishing breakthrough that will no doubt be used in many future cell studies (Poenie et al., 1986). For the first time ever, they peeped inside a cell during its division and continuously monitored its calcium levels using a new fluorescent dye called fura-2. The greater the levels of calcium, the more the dye fluoresced, which was recorded with a video camera mounted on a microscope and then analysed using a computer. It revealed a burst of calcium levels over the entire spindle, to up to six times their normal values for 20 seconds, signalling the start of the chromosomes splitting apart.

Calcium might also trigger the other stages of mitosis (Wolniak et al., 1983; Hepler, 1985). It seems to be released from special stores, which might inject bursts of calcium into the spindle. A recent report by two cancer researchers – Christian Petzelt and Mathias Hafner of the German Cancer Research Centre, Heidelberg – shows that the poles of the spindle contain small vesicles of calcium, and given the correct cues they could supply the spindle with its calcium trigger (Petzelt and Hafner, 1986). The calcium-binding protein calmodulin is also concentrated in the spindle itself, and this substance is known to help construct microtubules from their component parts. Or it might well behave like the calmodulin in sperm – by controlling the activity of enzymes involved in releasing energy from ATP.

No doubt in the future we will learn much more about calcium, calmodulin, pH, microtubules and electric currents – and many more regulators of the division of plant and animal cells. And for all the euphoria about oncogenes, the control of microtubules and their calcium master might eventually prove to be a more tangible weapon for fighting cancer.

Muscle Protein Again

But nuclear division is only half the story. Once the nucleus has divided into two new nuclei, the cytoplasm divides into two. Only here do we see a fundamental difference between plants and animals, with most fungi behaving like plants. The cytoplasm of animal cells divides by the action of a ring of actomyosin around the edge of the cell which garrottes the protoplasm into two halves. Plants, on the other hand, divide their cytoplasm less violently by partitioning the cell into two halves with a wall made up of plasma membrane and cell wall.

Now we come full circle once again, back to actomyosin. As cell division gets under way, the streaming of the cell's cytoplasm comes to a halt and so we might expect the actomyosin motor behind the streaming to disappear. But not so – actin fibres remain throughout the cell's division. In a recent review of this puzzle, Clive Lloyd of the John Innes Research Laboratory, Norwich, has mapped out another possible role for actin (Lloyd, 1988). He believes that the actin fibres could hold the nucleus in position before it starts dividing. As the spindle forms, the actin is closely wrapped up with the microtubules – not to drive the chromosomes apart, as once thought, but preparing the groundwork for the next stage, when the cytoplasm splits into two halves.

But before tackling the enigma of how a plant cell cleaves its cytoplasm into two halves, it's well worth taking a look at how animals solve the problem. They garrotte and sever their own cytoplasm into two completely independent cells with a constricting ring of actin fibres around the periphery of the cell. By some strange quirk of evolution, yeasts use the same technique in their division (Watts et al., 1985). If yeast is a forerunner of animal-type division, then the alga *Spirogyra* may be the bridge between plant and animal cytoplasm divisions. It uses both plant-like cell plates and animal-like constrictions, and most importantly it also seems to use actin fibres (Gotto and Ueda, 1988).

Plant cells partition themselves by spreading out two plates: one from the edge of the cell growing towards the centre, and one growing out from the centre to the edges. Even though they seem completely isolated from each other, the two plates still somehow meet and join up accurately to make a completely neat partition dividing the cell into two halves. The actin in *Spirogyra* could be the evolutionary bridge to the rest of the plant kingdom. We now know that a web of actin filaments connects the growing plates of a dividing plant cell, like the spokes of a wheel holding the hub to the rim. Could the actin fibres direct the growth of the new cell wall, even though it clearly doesn't garrotte the cytoplasm, as in an animal cell?

In the past few years scientists in the United States have made fundamental claims for actomyosin, with implications for all growing cells. To start this story off it is worth recalling the nineteenth-century naturalist and microscopist Joseph Leidy. He found a curious animal no larger than a grain of sand growing in a pond near his home. It had a remarkable network of filopodia, or 'false feet', stretching out from its body like tentacles, with which it moved and fed. It is called *Gromia*, and is a fresh-water relative of the amoebae. Leidy described the streaming of the cytoplasm inside the tentacle-like filopodia in a report to the US Geological Survey:

> in incessant motion along the course of the threads, flowing in *opposite* directions in all except those of greatest delicacy.
> (Leidy, 1879, cited in Allen, 1987; emphasis added)

Flow in both directions *within* a single thread of protoplasm seemed to defy the laws of physics, and some hundred years later attracted Robert Allen of Dartmouth College, who confirmed that particles flowed at different speeds as well as in different directions within a single filopodium. In 1971, after ten years' work on *Allogromia*, a close relative of *Gromia*, he and Samuel McGee-Russell of the University of New York at Albany identified the threads described by Leidy as microtubules (Allen and McGee-Russell, 1971, cited in Allen, 1987). Touching a microtubule was enough to spur the acceleration of previously stationary particles. The microtubules themselves also moved. Could the microtubules be moving the particles?

William Burdwood, a graduate of Allen's, discovered a heavy traffic of organelles moving in both directions inside nerve cells, at up to 200 millimetres a day – quite a pace on a microscopic scale. Allen then looked at the giant axon (nerve stalk) of the squid using a video enhancement technique, and saw an 'express railway' of filaments carrying mitochondria, vesicles carrying membranes and vesicles carrying substances ready to be made into transmitter hormones, all travelling to the terminals of the nerve where they are needed to carry nerve impulses from one nerve to another (Allen, 1987). It was a momentous revelation – after all, it had been believed that nerve cells only conducted waves of electrical excitation. Now they had found that nerves also acted as conduits for transporting materials by cable.

When the dust had settled, they analysed their findings. Not only did different particles travel in different directions, but larger ones moved more slowly than smaller ones. And the traffic seemed independent of nerve impulses. When the nerve cells were given electrical shocks to stimulate electrical excitement, the traffic continued unaffected. The next problem was to find out what the filaments were that carried the organelle traffic.

Using antibodies that would only latch onto tubulin (the building block of microtubules), the decisive evidence was obtained – the fibres were microtubules. So here at last after years of exasperating, and very delicate, work was the evidence that a single microtubule was responsible for the enigmatic two-way movement in *Allogromia* and the squid nerve axon, and for Leidy's streaming in *Gromia*.

The next question was how the microtubule railway worked. Any directed movement in cells needs chemical energy and some kind of 'force-generating enzyme' to convert the energy into motion. ATP seemed the most likely fuel, since the movement ground to a halt when the cells were deprived of it. As for the engine driving the microtubules, the most we can guess, and it is a wild guess, is that they might work like spindles or cilia and flagella, sliding past one another (Allen, 1987).

What significance could this have to plants? Any dividing cell needs to be supplied with vesicles to build new cell walls and organelles to manufacture the wall materials. Organelles also need to be moved into one of the two daughter cells before the wall is completed. Botanists at the University of Michigan used fluorescent stains to pinpoint the locations of both actin and microtubules in yeast cells during their division (Adams and Pringle, 1984;

Kilmartin and Adams, 1984). Where the new daughter cell grows out from its mother, its bulging cell wall is a hive of activity (figure 9.7). The new cell wall is built up of components brought up to the growing tip in small

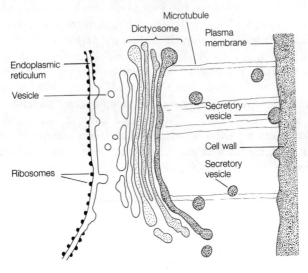

FIGURE 9.7

membrane-bound vesicles from protein factories (endoplasmic reticulum) lying deeper inside the cell. These vesicles appear to travel along tramways of microtubules like the vesicles travelling inside nerve cells. When the packets arrive at the tip, they dump their cargo of cell wall components into the building site. The growth of fungal hyphae tips and seaweed eggs is thought to proceed in the same way. Actin at the building site may be involved in the vesicle movement. Furthermore, the new daughter nucleus may also run along microtubules as it migrates from the tip of the spindle into the daughter cell itself. Similarly, microtubules are thought to direct the movements of the nucleus of another fungus, *Aspergillus nidulans*.

We will probably learn much more about actin, calcium, calmodulin and microtubules, and no doubt many more regulators in the divisions of plant and animal cells, in the future. The chemicals controlling the different stages of division, and the ways that they themselves are controlled, will have important implications for understanding how organisms grow and develop, and what makes them revert to a juvenile state in cancers.

10

EXCITABLE CHEMISTRY

Anaesthetics: Waking Plants Up!

On 16 October 1846 the prominent surgeon and teacher John Collins Warren removed a tumour from the neck of Gilbert Abbott at the Massachusetts General Hospital. The operation made medical history, but not because of the surgery, the surgeon or the patient. The event only proved historic when a dentist called William Morton arrived, 15 minutes late, and Warren remarked: 'Well, sir, your patient is ready.' This was to be the first public demonstration of anaesthesia. After a few minutes of giving the patient ether, Morton replied: '*Your* patient is ready, sir.' Although Morton had successfully used ether a month previously on one of his own patients, this surgery at Massachusetts General Hospital, in front of medical witnesses, proved to be a milestone (Davis, 1982).

Following the successful demonstration of surgical anaesthesia, it was tried out on literally any living thing that moved. So it was no surprise when the French physiologist Paul Bert, the German plant physiologist Wilhelm Pfeffer and many others gave anaesthetics to the Sensitive Plant (*Mimosa pudica*). To their great pleasure, the plant succumbed to both ether and chloroform – the leaves became totally unresponsive to all stimulation. Even more impressive, when the anaesthetic wore off the sensitivity returned, like an animal waking up. The story of plant anaesthesia has recently taken another surprising twist. Anaesthetics might very well prove useful in agriculture, horticulture and other areas of study involving plant growth, but more of that later. First, was the anaesthesia of *Mimosa* truly anaesthesia?

A few years ago French scientists confirmed the nineteenth-century experiments using ether and chloroform on *Mimosa*. They showed that the motor apparatus of *Mimosa* obeys the same laws of narcosis as are observed in animals. Comparable concentrations and durations of anaesthesia, or withdrawal of the anaesthetic, affected plants and animals alike (Gaillochet et al., 1975).

Mimosa is not the only plant that can be put to sleep like this. The touch-sensitive bendy stamens of the garden *Berberis* are paralysed by doses of ether, chloroform or even 1 per cent morphia. Charles Darwin found that giving a whiff of ether or chloroform to the Venus's flytrap stupefied it, even when the sensitive trigger hairs on the trap were flicked vigorously (Darwin, 1875).

Even the simpler single-celled creatures are susceptible to anaesthesia. Chloroform and ether suppress the sensitivity of swimming bacteria, fungi and algae to various stimuli such as chemicals, air or light (Rothert, 1904). So do anaesthetics really behave the same way in both plants and animals?

To answer this question, we need to look at how anaesthetics work on the nervous systems of animals. Unfortunately, all is not clear. As one of the leading medical experts, Sir William Paton FRS, confessed recently, neither anaesthetists nor anaesthesiologists have any idea how general anaesthetics work!

The excitatory nerves – including the vital central nervous system – generally become inhibited. In humans, the amount of general anaesthetic circulating in the blood produces several distinct stages of anaesthesia. Starting with analgesia – an absence of sense of pain – it moves on to mental or physical excitement. In the third or surgical stage complete loss of consciousness is accompanied by muscle relaxation. If too much anaesthetic is given, respiration and circulation may become so depressed that they stop completely. Therefore the science of anaesthesia is very much concerned with controlling the amount of anaesthetic given to the patient.

The most contentious issue is the exact way that anaesthetics affect the physics and chemistry of the nervous system. However, it is certain that they dampen the electrical impulses of nerves (in fact anaesthesia can be induced in a specific part of the brain by passing alternating current over the skull – a technique known as electronarcosis).

How, then, are the electrical signals blocked? The curious thing about general anaesthesia is the astonishing variety of chemicals that produce this effect. They include the inert gas xenon, chloroform and the barbiturate thiopentone. Furthermore, their effectiveness seems to be completely unrelated to their structure, size or shape. All they seem to share in common is a rather vague preference for 'fatty' over watery solutions. This has led to the idea that they burrow their way into the middle of nerve cell membranes. The resulting bulge would then squash the ion channel proteins embedded in the membranes, blocking any flow of electric current. But a new discovery by Nick Franks of Imperial College London and Bill Loeb of

King's College London has suggested an answer. They found that general anaesthetics bind onto protein molecules, but which ones in the nervous system is not known (Franks and Loeb, 1984).

Returning to plant movements, we can see an obvious parallel with animal nerves. *Mimosa*, *Berberis* and the Venus's flytrap all control their movements with electrical impulses. As in their animal counterparts, these signals are driven by shuttling of ions across excitable cell membranes. The resulting waves of electrical excitation are very similar to an animal's nerve impulses, but strangely no one has yet measured what effect anaesthetics might have on them. Instead, we must turn to work from the 1920s on 'ordinary' plants such as the kelp *Laminaria*. The American physiologist Winthrop Osterhout of the University of California at Berkeley found that concentrations of around 1 per cent of ether, chloroform and even caffeine (although this is not an anaesthetic for animals) lowered the electrical resistance of the kelp's fronds (Osterhout, 1922). But when the fronds were washed clean in sea water, their electrical resistance returned to normal.

As far as modern botanists are concerned, the arena has shifted to an even more unlikely topic – the germination of seeds. Consider seeds as dehydrated packets of all the chemical goods needed for life: DNA (deoxyribonucleic acid), enzymes, food and so on. Because of the almost total dessication of the seed, the dry packet is held in near suspended animation (so-called dormancy), in some cases for up to hundreds of years. If you think of this dormancy as sleep, then the seed needs to 'wake up' and germinate. The 'alarm call' comes in various forms: hormones, light and temperature combined with water are all ideal candidates. When the seed finally wakens, the membrane wrapping its dormant packet of chemicals leaks in enough water, oxygen and nutrients to fire off the germination bandwagon, particularly the enzymes and hormones needed to spark the seed's chemistry into life.

But the dormancies of some seeds are notoriously difficult to break. Ray Taylorson and Sterling Hendricks of the US Department of Agriculture have used growth hormones and light in attempts to break the stubborn dormancy of the seeds of the Fall panicum grass *Panicum dichotomoflorum* (Taylorson and Hendricks, 1979). They realized that ethanol used as an anaesthetic upset animal cell membranes, and so it might help upset seed membranes. They gave seeds an ethanol supplement on top of their usual hormone or light treatments, and it worked. Then, against all their expectations, the ethanol broke seed dormancy on its own without the help of either hormones or light.

Knowing that they had hit on something important, they tried out chloroform, and this was also a good germination trigger, not only for Fall panicum but for many other grasses as well. Even more surprisingly, the doses of ethanol or chloroform needed to break dormancy matched those needed to anaesthetize animals. Furthermore, the seeds share another feature of anaesthesia with animals. When seeds treated with chloroform are put under pressure they stop germinating, in the same way that pressure releases anaesthesia in animals (Taylorson and Hendricks, 1980).

There is yet another surprising effect. Seeds that need light to break dormancy are able to germinate in the dark if an anaesthetic is used. How can an anaesthetic substitute for light? Seeds detect light using the light sensor phytochrome which is attached to the outer membrane. Growth-regulating hormones also have a liking for this particular membrane, and bind onto special receptor sites there at the onset of germination. Thus, anaesthetics probably break seed dormancy by affecting the same membranes as phytochrome and hormones.

This leads to a much more difficult question: can anaesthetics affect fully grown plants, perhaps by substituting for light or hormones? The pioneering nineteenth-century animal physiologist Claude Barnard realized that anaesthetics affected more than just nerves:

> Let us remember that chloroform does not act solely on the nerve tissues. . . . It anaesthetises all the cells, benumbing all the tissues, and stopping temporarily their irritability.
>
> (Barnard, 1875)

With this in mind he went on to try anaesthetizing plants, and found that chloroformed water plants stopped bubbling oxygen by photosynthesis, but recovered when the chloroform was washed out (Barnard, 1878). This discovery sparked off a spate of work on the effect of anaesthetics on plant photosynthesis and respiration. Algae, pond weeds, crop plants and even trees were all tested. Even though the results were rather contradictory – probably because of the wide variety of and amounts of anaesthetic given – low levels of anaesthetic appeared to inhibit photosynthesis but boost respiration. On the other hand, high concentrations stopped both types of metabolism.

In the 1920s it was claimed that ether upset the flow of auxin hormone in plant shoots, possibly by disrupting the incessant streaming of cytoplasm in individual cells, but not much supporting evidence was provided. There is much better evidence for narcotics, close relatives of the simpler anaesthetics which also depress the central nervous system.

The barbiturates amobarbital and secobarbital slow down cytoplasmic streaming, stalling the germination and growth of pollen tubes (Kordan and Mumford, 1978), stunt the growth of rice seedlings and make them 'anaemic' (they lack their normal quota of green chlorophyll) (Kordan, 1977, 1978), affect seed germination (Kordan, 1984), impair the production of anthocyanin (colouring pigment) (Kordan and Rengel, 1988) and disorient the gravity-sensing mechanism of roots (Kordan, 1980). The drugs probably make their mark by locking onto the plant cell membranes. More than that is conjecture.

Given that they are properly applied, anaesthetics and narcotics may prove valuable tools for plant growers in breaking awkward seed dormancies or perhaps even controlling the plant's growth. And there is another surprising group of medical drugs with promising applications for plant growers – the painkillers.

Plant Aspirin

When American Indians had a headache, they eased the pain by strapping mashed extract of willow tree bark across their foreheads. Today we use a more refined version of that extract – it is called aspirin, and its active ingredient, salicylic acid, is named after the willow tree, *Salix*. The story of plant aspirin is remarkable (indeed, it is still unfolding), and although the connection with nervous plants won't be immediately obvious, bear with me.

Plants themselves are very sensitive to artificial aspirin. According to folklore, popping an aspirin into a vase of cut flowers helps keep them fresh, and this has recently found remarkable scientific support. Synthetic aspirin triggers a plant's own natural defences against microbes, as well as promoting leaf bud formation, stimulating leaf growth, cutting down the plant's water demands by closing the stomata, making roots leak ions, stimulating various species to flower and stopping tobacco mosaic virus from multiplying (Simons, 1982, 1991).

These impressive feats have inspired plant scientists for the past twenty years to find out what the natural aspirin may be doing in plants, but the work hasn't been easy. When salicylic acid, the active component of aspirin, was detected its activity didn't fit any obvious pattern of plant behaviour.

The first breakthrough finally came in 1987 in a very bizarre plant: the Voodoo lily, *Sauromatum guttatum*. What makes this species particularly interesting from our point of view is that it makes its own heat – a rather unusual behaviour for a plant (figure 10.1). The temperature of its bloom can suddenly rocket up to 15 °C above the temperature of the surrounding air, fuelled by a rate of respiration as great as a hummingbird's in flight.

The heat, incidentally, is part of an elaborate hoax designed to trick carrion flies into believing they've found rotting detritus to feed or lay their eggs on. Instead, they get trapped into cross-pollinating the flowers. The heat comes from a large crimson poker wrapped in a sheath open at the top and pinched into a chamber around the small flowers below. The heat of the poker boils off volatile and distinctly dung-like scents to attract the flies, but

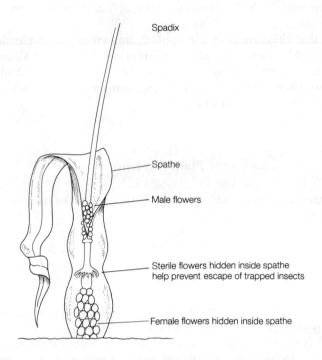

FIGURE 10.1 Salicylic acid, a close relative of aspirin, fires the voodoo lily (*Sauromatum guttatum*) into heat, quite literally. The poker-shaped spadix protruding from the bloom heats up to 15 °C warmer than its surroundings. This helps to vaporize foul-smelling chemicals which attract the pollinators: carrion flies and beetles. The insects become trapped in the base of the cornet-shaped spathe, where they collect and deposit pollen onto the real flowers hidden there. Eventually the spathe withers and the flies are released to cross-pollinate other blooms.

once they've landed the insects fall into an ambush: they crawl down into the warmth of the floral chamber where downward pointing hairs prevent their escape. Only after the flowers have been pollinated do the hairs wither and release the flies, hopefully to pollinate another plant.

The biochemistry behind the heat is even more bizarre. The plant pays such a heavy metabolic cost for its hot love-making that it almost starves to death. Almost a third of its calorific value is burnt up in a type of respiration unusual in plants. Normally, plants respire by breaking down starch and sugars to release energy, which is then temporarily mopped up in ATP (adenosine triphosphate) molecules. In the Voodoo lily, the plant switches to aerobic respiration in which the energy normally trapped in the ATP molecule is released directly as heat (which is why we too get hot during aerobic exercise).

So to avoid starving itself to death, the Voodoo lily carefully rations its heating to coincide with the few days when the flowers are fertile. But what sort of signal it is that tells the plant to switch its metabolism from cold to hot so dramatically has been a mystery for the past fifty years. Only now has the signal been identified as salicylic acid.

Using a sensitive assay technique, Bastiaan Meeuse of the University of Washington in Seattle and Ilya Raskin of Du Pont in Delaware monitored salicylic acid throughout the flowering period. First, they found that salicylic acid is carried into the poker as the inflorescence matures (signalled by the length of days which turn on flowering). Then, the day before flowering, salicylic acid concentrations explode a hundredfold, sending the plant off into its aerobic drama. A second salicylic acid pulse in the flowers themselves signals their own heat-up. All these observations are powerful evidence that salicylic acid is a plant hormone (Raskin et al., 1987; Meeuse and Raskin, 1988).

Building on this substantial discovery, Raskin continued the work with Jocelyn Malamy and her colleagues at Rutgers University in New Jersey. Prompted by the earlier work on synthetic aspirin, they looked at salicylic acid in tobacco plants infected with tobacco mosaic virus, and their discoveries have now established salicylic acid as a powerful hormone in all plants, with important implications for agriculture.

Before any signs of infection or disease resistance were detected in the tobacco plants, salicylic acid levels surged almost fivefold through the plants. This surge acted like an alarm call, setting off the production of special disease-fighting proteins to combat the viral attack (Malamy et al., 1990).

At the same time, another group of biologists, led by Jean Pierre Métraux from Ciba-Geigy's laboratory in Basel, Switzerland, reached similar conclusions. They identified a salicylic acid signal that triggers disease resistance in cucumber plants infected with a fungal disease (Métraux et al., 1990). The possibilities for exploiting these discoveries in agriculture are fascinating. Infected crops might be sprayed with aspirin to trigger their own natural disease-fighting systems, or the plants could be bred with greater levels of salicylic acid. And it is highly likely that salicylic acid controls other important phenomena in plants; Raskin already has preliminary evidence that salicylic acid helps to toughen plants against environmental stresses. All of which shows these to be important discoveries of a remarkably powerful hormone.

Exactly how salicylic acid triggers its effects in plants is another interesting question. In animals, aspirin relieves pain by knocking out pain-triggering hormones called prostaglandins. Similar prostaglandin hormones, and a closely related hormone called jasmonic acid, are also found in a wide variety of plants (Miyares and Menendez, 1976; Vick and Zimmerman,

1984), where they stimulate growth, close stomata and affect night-time leaf movements and a wide range of other phenomena (Larqé-Saavedra, 1979; Saniewski et al., 1979; Roblin and Monmort, 1984; Tsurumi and Asahi, 1985). Most remarkable of all, methyl jasmonate (a close relative of jasmonic acid) also triggers a plant's defences against insects or diseases, and as we'll see later in chapter 12 salicylic acid blocks a similar wound response (Farmer and Ryan, 1990). But it is far too early to say whether a plant's salicylic acid behaves as it does in an animal, by blocking prostaglandin signals or something similar.

There is an alternative way it could work in plants, and there's a very strange twist to this tale that brings us back to electrical signals. Unbeknown to the scientists working on salicylic acid, their hormone is closely related to another painkiller, gentisic acid, which has recently been discovered as an active ingredient of the 'excitable' hormone Ricca's factor. As we saw in chapter 6, Ricca's factor triggers electrical signals in wounded *Mimosa* plants, telling the leaves to move. It is far too early to say whether salicylic acid behaves like gentisic acid, but they are chemically so close it is a tantalizing prospect.

The gentisic acid discovery was made independently in two laboratories. Barbara Pickard of the University of Washington, St Louis, found gentisic acid amongst *four* other ingredients. The gentisic component could on its own trigger the leaf movement, but it also interacted with the other four factors to produce an even stronger response.

Meanwhile, and unknown to Pickard, a similar discovery had been made in another laboratory. Organic chemist Hermann Schildknecht of Heidelberg University published a detailed chemical paper on Ricca's factor in an obscure German science journal (Schildknecht, 1978). He described how the gentisic acid in *Mimosa* is attached to a sugar 'companion' and triggers the leaf movement. But, surprisingly, Schildknecht failed to comment on its significance to electrical signals. Since then he has characterized a family of closely related chemicals which trigger *Mimosa* movement. He has called them the leaf movement factors or turgorins, after the effect that they have on the turgor of the leaf pulvini which move the leaves. Many of these chemicals are also present in other plants, where they induce sleep or rhythm leaf movements. Schildknecht also identified another factor in the *Mimosa* leaf movements – glutamic acid, which is a common amino acid in plants and animals and also functions as a nerve transmitter in animals.

But how does this relate to the rest of the plant kingdom? Karl Umrath found that Ricca's factor extracted from *Mimosa* had some astonishing effects on other plants: it provoked the movement of touch-sensitive stamens in the barberry (*Berberis*) flower and triggered stomata to close in cucumber leaves (Umrath 1943, 1966). Clearly, this substance(s) was not exclusive to *Mimosa*, but if it existed in other plants, what did it do in them?

Several years ago Pickard and her postgraduate student Jerome van Sambeek discovered that *all* plants probably contain Ricca's factor. They found that burning the leaf of, say, a tomato plant released a burst of Ricca's factor into the plant's transpiration stream. This in turn triggered an electrical signal not unlike the wound-induced signal in *Mimosa*. Soon after the electrical signal had passed through the wounded plant, the rate of photosynthesis dropped, the rate of respiration rose and the stomatal leaf pores squeezed shut (van Sambeek and Pickard, 1976a, b, c). Eric Davies and Anne Schuster of Nebraska University also found a mysterious signal triggered by wounding which collected the protein-building bodies (ribosomes) inside cells into an organized chain. They reported that wounding pea, soybean, maize or tobacco plants triggered a sudden and massive recruitment of ribosomes into bead-like chains known as polysomes (Davies and Schuster, 1981b). This assembly cranks the cell up into high speed production of protein, which is essential for repairing the damage inflicted by wounding. Most exciting of all, Davies and Schuster concluded that the message to form polysomes travelled up and down the plant like an electrical signal. Another group of scientists, in Washington State University, could only explain a sudden weakening of cell plasma membranes in wounded plants by a rapid signal released by the wound (Walker-Simmons et al., 1984).

There is yet another possible function for Ricca's factor. Pollination can be thought of, in a rather twisted way, as the wounding of the female carpel by the burrowing pollen tube. A.W. Spanjers at Toernooiveld, The Netherlands, reported measuring electrical disturbances in the styles of various flowers following pollination. Self-pollination triggered electrical signals distinct from those triggered by cross-pollination (Spanjers, 1981). The remarkable thing about the cross-pollinating signal is how similar it is to the signal made by a wounded *Mimosa* leaf. Is Ricca's factor responsible for these electrical signals? Perhaps, but the signal that Spanjers reported only travelled at 3 centimetres per hour – far slower than the electrical signals recorded in the leaves of the crops. Maybe other hormones are involved as well.

We're on the threshold of a fascinating new field of plant science. Ricca's factor probably behaves as an alarm call in all plants, triggering electrical signals which warn the plant to prepare their defences against attack. Salicylic acid, too, can behave like an alarm signal, telling the plant about sudden changes in its environment or attack from microbes. But although it's far too early to say whether salicylic acid behaves like gentisic acid by triggering electrical signals, we may yet discover its involvement in plant excitation.

Nerve Transmitters in Plants

There are even closer similarities between the chemistry of plant and animal excitability and the first hint of something extraordinary started with a painful experience.

If you have ever brushed against *Urtica dioica* you can feel the closeness of plant chemistry to ours – it's the sting in the common stinging nettle! In 1947 two physiologists, Nils Emmelin and Wilhelm Feldberg, then working at Cambridge University, were interested in the various allergies caused by histamines, and turned to nettles because, as Feldberg explains, 'After finding histamines in so many things, it seemed natural to look at the nettle.' But application of histamine to the skin didn't produce the same burning sensation as a proper stinging nettle. So Feldberg and Emmelin tested two other chemicals purely on the basis of wild speculation: 'I have only worked on three substances in my whole life – histamines, acetylcholine and 5-hydroxytryptamine, so we tried out acetylcholine.' The combination of acetylcholine and histamine mimicked the stinging sensation (Emmelin and Feldberg, 1947). They then tested for the two substances in the stinging hairs. The histamines irritate the skin and acetylcholine brings on a burning sensation. In fact the burn is so strong that the Germans call the plant *Brennessel*, the 'burning bush'. What made the discovery all the more remarkable is that acetylcholine is a nerve transmitter hormone, used in nervous systems to relay messages from one nerve ending to another.

A third substance in the stinging hairs was later identified as 5-hydroxytryptamine, another replica of an animal nerve transmitting hormone. There are even plant chemicals in the broad-leaf dock which inhibit 5-hydroxytryptamine, which explains why rubbing dock leaves on skin stung by nettles is so soothing (Brittain and Collier, 1956).

Perhaps there is nothing surprising about finding acetylcholine in a plant. Plants have evolved all sorts of chemicals, such as curare, that mimic the chemistry of animal nerves to repel predators. Yet Emmelin and Feldberg (1949) found acetylcholine not just in leaves, but throughout the rest of the plant shoot. It was a discovery with far-reaching implications.

Some twenty years later, acetylcholine and its accompanying enzyme acetylcholine esterase (which mops up any unwanted acetylcholine) were found in a wide range of 'ordinary' harmless plants from beans to duckweeds, and cucumber leaves to potato tubers. Furthermore, many of the drugs that interfere with acetylcholine in animals, such as atropine and scopolamine, are derived from plants in the potato and bean families in which acetylcholine has been discovered. So what use is a nerve chemical to a plant with no sting?

The first clue came as the result of another accident, when Japanese plant physiologist Takuma Tanada was studying the red–far-red light-sensing pigment phytochrome. Tanada noticed that the tips of bean roots he had been studying clung to glass beakers washed in detergent. Under red light the tips came unstuck, but on switching to far-red light they became stuck again. This red–far-red counterplay is a classic hallmark of phytochrome at work (Tanada, 1968). He suspected some sort of electrical effect on the roots, because the detergent had left a negative charge on the glass. Later, Racusen and Etherton (1975) proved that electricity was involved by mimicking the red–far-red movements with electrodes.

In a wild flight of fancy Tanada also reasoned that a nerve transmitter like acetylcholine, which alters the electrical charges on nerve cells, might also affect the electrical charge on the root surface. So he dipped the root tips into acetylcholine, and to his delight the roots stuck to the glass as if they had been treated with red light and detergent. The work was continued by Mordechai Jaffe at Ohio University, who discovered that the roots had their own natural supply of acetylcholine which bumped up their respiration, depleted the reserves of the energy-carrying molecule ATP and stimulated the release of hydrogen ions (Jaffe, 1968).

From these somewhat obscure pieces of research came a new wave of experiments which revealed a much stronger link between the nerve transmitter acetylcholine and phytochrome. Acetylcholine has now been shown to mimic the effects of red light on a wide range of phenomena: helping seeds to germinate, helping duckweeds to flower and many others (Jones and Stutte, 1986). The evidence for the link between acetycholine and phytochrome was further strengthened when it was found that red or far-red light affects the levels of natural acetycholine in bean roots – in some way as yet unknown red light and phytochrome raise the levels of acetycholine in plants. There's even evidence that plant acetylcholine binds to two different kinds of receptors on the cell plasma membrane: a nicotine-sensitive receptor and a muscarine-sensitive receptor, as is the case in animal cells (Tretyn et al., 1990).

If the behaviour of acetylcholine does depend on upsetting the plant electricity, should we be looking for nerves and nerve-like signals? We already know that plants do not have nervous tissue, and there is as yet little to suggest that electrical signals are involved with plant acetylcholine. Any electrical disturbance might simply be localized in individual cells. Yet recent research has highlighted another property of acetylcholine that animals and plants share – telling the time.

From migrating birds to hibernating bears, almost all animals in temperate latitudes tune into the seasons. They tell the time of year by counting the hours of daylight – for example, the testes of the male golden hamster shrink during short days (i.e. wintertime) and expand during long days (i.e. summertime). Recently, David Earnest and Fred Turek of

Northwestern University, Evanston, Illinois, have prevented the shrinkage of hamster testes during short days by flashing light or injecting the animals with a chemical that antagonizes acetylcholine (Earnest and Turek, 1983). Other evidence shows that rat brains make acetylcholine in cycles during the day, peaking in the light and dipping in darkness. The light is clearly important, as flashes during the night raise the acetylcholine levels. Plants behave in much the same way: light promotes synthesis of acetylcholine in leaves, but this process stops in darkness (Miura and Shih, 1984).

The acetylcholine clock in animals and plants helps to send messages as well as counting the hours of daylight. Acetylcholine sends instructions from the brain in animals and from leaves in plants to the tissues that actually respond to the changing seasons. The way in which these instructions are sent is not yet known, although it may be as electrical signals.

One other avenue worth exploring is the fertilization of eggs. Plants and animals share the same electrical phenomena during fertilization. When a sperm touches an egg cell, it releases a wave of electrical disturbance around her plasma membrane before it starts burrowing into her (Hagiwara and Jaffe, 1979). Research published recently shows that chemicals on the sperm lock onto acetylcholine receptors on the egg surface, and that this triggers the wave of voltage changes. In some eggs this wave blocks other sperm from trying to penetrate the egg. It may also prepare the egg for its development (Gould and Stephano, 1987; Kline et al., 1988). Because many aspects of egg fertilization in animals and plants are similar, a phenomenon involving acetylcholine and electrical disturbances may also be at work in plant egg cells.

Whether plants contain other nerve transmitters is not certain. Philip Applewhite (1972) of Yale University found the nerve transmitters serotonin (also known as 5-hydroxytryptamine) and norepinephrine (noradrenaline) in a number of plant species, such as *Mimosa*, with the ability to move. Furthermore, he found higher levels of norepinephrine in the pulvinus than in the rest of the plant, although its role, if it has one, is a mystery. But what is becoming increasingly obvious is that plant cell membranes exhibit electrical behaviour very similar to that observed in animals.

Opening and Closing Gates

So far in this book we have covered almost everything you can conceivably think of about plant movements. In each chapter we have outlined how plants control their movements using electrical signals, ions, hormones and so on. In this chapter we have seen how drugs and the plant's own natural excitable chemicals influence their movements. But now let me show you the

EXCITABLE CHEMISTRY

detailed biophysical nuts and bolts of plant and animal excitability, an area which has boomed over the past few years with some very exciting discoveries.

A nerve carries electrical signals by means of sodium and potassium ions scuttling in and out of the cell (see figure 10.2). This is a clever feat, because any living cell's plasma membrane behaves like an electrically insulated wire, coated with a layer of fat (which cannot conduct electricity) sandwiched between layers of protein. The only way that sodium and potassium ions can pass across the fat is through special gates which burst open.

The whole phenomenon depends on a bizarre piece of contortionism. The gates, which are made of protein, straddle the width of a plasma membrane and are sensitive to electricity. When the voltage across the membrane reaches a critical level, the proteins suddenly change shape and allow selected ions to slip through. When the voltage returns to normal, the protein gates squeeze shut and the ions are once again blocked.

Such voltage-sensitive ion channels are now known in organisms as simple as bacteria. In the single-celled animal *Paramecium* they control the nerve-like electrical signals which, as we discussed in chapter 4, drive the cilia motors. Ion gates in other animals are sensitive to other stimuli: light, hormones, stretching or increases in the level of calcium ions. The shuttling of ions creates an electrical disturbance and discharges an action potential in nerves.

However, plant scientists are far behind their zoologist colleagues in this fascinating work, partly because of the special problems of the small size of plant cells and their tough cellulose walls. But we can now track down the tell-tale electrical signatures of ion gates opening and closing. Because the proteins are sensitive to electricity, we can turn them on by passing electric current through a plasma membrane, using an exceptionally narrow tipped electrode (usually less than 1 micron (0.001 millimetre) across) adjusted until a crucial voltage is reached which makes the protein gates suddenly switch open, allowing the ions to surge through the membrane. The ionic flood is picked up as a sudden change in the membrane's electric charge.

Previously, only the extraordinarily large cells of *Chara* and *Nitella* could be studied. These are large enough to inject with current and monitor for protein gates by measuring the traffic of ions in and out of the cell. A number of biophysicists have used this technique to find voltage-dependent potassium gates in these algae (Smith, 1984). Similar gates have been found in a wheat leaf cell by a group consisting of biophysicists from the National Institute of Neurological Disorders, Bethesda, Maryland, in collaboration with plant scientists from the Weed Science Laboratory, Beltsville, Maryland. Moran et al. (1984) overcame the practical problems by stripping the plant cells of their walls and then using a 'patch electrode'. With this technique, the electrode first sucks up and breaks off a patch of membrane in a tight clinch. The recordings rely on a simple law of electricity

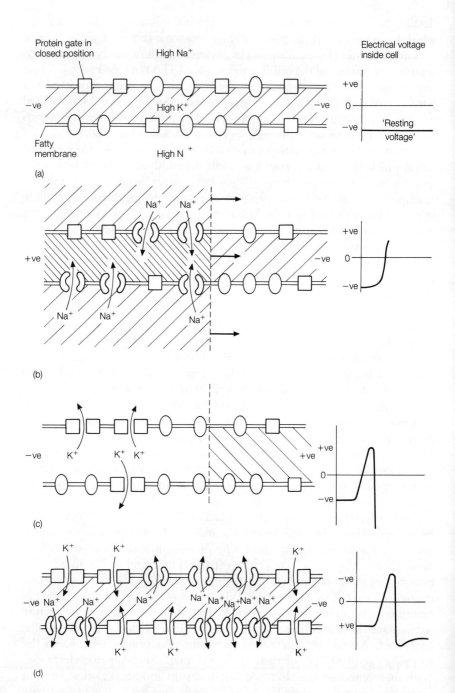

called Ohm's law, whereby a tiny current I is shown up by a large resistance R and a small voltage V applied to the membrane: $V = IR$. Therefore exceptionally small fluctuations in current which represent the opening and closing of individual protein ion gates can be detected. Moran et al. recorded various types of ion channels opening and closing, although it was not clear which ions were carried in each ion channel. In fact, not all the ion channels were turned on by the voltage, and since then other channels only sensitive to certain ions have been found. And another surprising thing was found: the recordings revealed 'hot spots' containing hundreds of voltage-sensitive channels in a small patch (1–10 microns square). Interestingly, clusters of the same sort are also found at nerve–muscle junctions. The functions of these 'hot spots' and of each type of gate are not known. However, another research team used the patch clamp technique to probe one particularly important ion channel.

The week after Moran et al. reported their success with wheat protoplasts, a German group at the Max-Planck Institute, Göttingen, published an account of potassium ion channels in the plasma membranes of bean stomata guard cells (Schroeder et al., 1984). As we saw earlier, potassium ions drive the water pumps in the opening and closing of stomata. When the voltage on the plasma membrane changes, as happens at dusk or dawn, potassium is drawn into the stomata cells through their own private ion channels and water is dragged with it by osmosis (Schauf and Wilson, 1987; Schroeder et al., 1987).

Over the past couple of years there has been an explosion of work on ion channels in plants, particularly in stomata, but the significance of electricity in this process is hotly contested. Voltage-sensitive chloride and potassium channels have been found, but this does not square with a report that the hormone abscisic acid opens potassium channels in guard cells without an appreciable change in the cell voltage (Blatt, 1990). The picture is now even more complicated as we begin to realize that calcium ions inside the guard cells may also be turning the ion channels on or off (Schroeder and Hagiwara, 1989).

FIGURE 10.2 How animal nerves pass action potentials. (a) Nerve cell ready to fire. The inside is rich in potassium and poor in sodium. The voltage inside is slightly negative. (b) The wave of the action potential sweeps along inside and outside the membrane, opening the protein gates to sodium ions. The sodium rushes in, making the cell turn from electrically negative to positive. (c) The change in electric charge opens up the potassium gates, and when potassium ions flood out the cell voltage swings negative. (d) The original voltage is restored as the potassium ions flood back in again through their gates, and sodium ions pass out again through their own gates. The cell is then ready to fire another action potential.

EXCITABLE CHEMISTRY

Voltage-sensitive channels have also been found in *Samanea* pulvini (Morse and Satter, 1987). In chapter 11 we shall see how ion gates can respond to touch, and probably many other signals. Much of this behaviour is very similar to that of ion gates in animal neuromotors. This then is an exciting field of plant biology which could unlock many of the mysteries of plant movements: how signals from the outside world are decoded into messages that the plant can act on, how electrical signals pass through plants and how enormous surges of ions are triggered in motor cells.

Plant and Animal Hormones

The similarities between plant and animal physiology do not end there. Like animals, plants control much of their growth and development using hormones, which are made in one sort of cell and carried to another where they are used. Most botany textbooks describe only a few groups of plant hormones (strictly speaking, 'plant growth regulators'): auxins, gibberellins, ABA (abscisic acid), ethylene and cytokinins, although as we saw earlier in this chapter many more are being discovered, such as salicylic acid and jasmonic acid. But our concern here is that many plant hormones have much in common with their animal counterparts. Thus, we have already mentioned the similarity between jasmonic acid in plants and prostaglandin hormones in animals.

Take the auxins. One outstanding feature of the auxin IAA (indoleacetic acid) is its uncanny similarity to a nerve transmitter hormone called 5-hydroxytryptamine (the Americans call it serotonin) (figure 10.3). This chemical is particularly common in the parts of the brain concerned with consciousness, and may help to control dreaming. In fact, commercial auxins, such as those used for horticultural rooting compounds, are made from animal urine, since their auxins are the broken down waste products

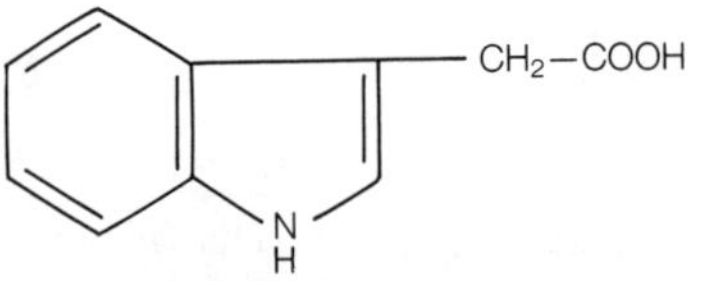

Indoleacetic acid (IAA) 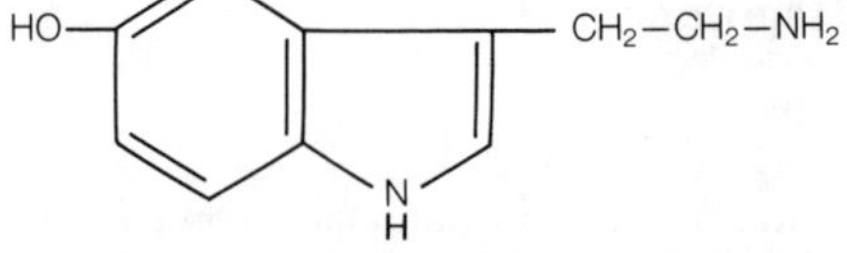   5'-Hydroxytryptamine (serotonin)

FIGURE 10.3 Some plant and animal hormones share similar chemical structures, but do they behave the same? Evidence shows that hormones from one kingdom can affect the behaviour of organisms in the other kingdom, which leads to the interesting question: how much is our behaviour affected by the plants we eat?

of 5-hydroxytryptamine. A wide variety of plants also contain 5-hydroxytryptamine – as we saw earlier, it makes dock leaves a useful remedy for nettle stings. But 5-hydroxytryptamine is of little use to a plant as an antidote to human skin rashes. A much better clue to its function is in walnut seeds, where its levels are high and rise further during seed germination (Lambeck and Skofitsch, 1984). It may behave as a hormone in plants, as it causes release of 'excitatory substances' in Mimosa pudica (Umrath and Thaler, 1980) and is found in unusually high concentrations in the pulvini (Applewhite, 1973). It also increases the root length, root weight, coleoptile weight and cell division activity of barley seedlings (Csaba and Pal, 1982).

The similarity between IAA and an animal nerve transmitter hormone is uncanny. In fact, one plant hormone has even given a lead to a new animal hormone. The French biochemist Michel Ladzunski and his colleagues recently discovered a hormone in pig and rat brains which has the same structure as the plant hormone ABA – a key regulator of the movements of stomata, shedding of leaves in autumn (hence its name) and many other plant responses to stress (Le Page-Degivry et al., 1986). But did the pigs really make their own ABA, or could they have simply eaten and absorbed it with their food? Not according to control experiments on rats, of which three were fed on diets high in ABA and three were fed on diets containing no ABA. Amazingly, the amount of ABA in the brains of rats fed the zero diet was almost double that in the brains of rats fed the high ABA diet. Therefore they must have synthesized their own supplies. But just to show how similar ABA is in animals and plants, Ladzunski and his colleagues found that plants could not tell the difference between their own ABA and that from pigs.

What use ABA is to the animals isn't certain, but there is one intriguing hint. When the hormone was supplied to muscles it strengthened their contractions. The effect was blocked by nifedipine, a drug used for treating angina and high blood pressure. As nifedipine blocks pores in the muscle cell which pass calcium, it is possible that ABA affects the same pores, in which case it could have the same power in plants.

With 200 nerve transmitters already known, there are probably many more similarities between plant and animal hormones. So is it coincidence that plant and animal hormones resemble one another, or do they work in similar ways? Because zoologists are far ahead of botanists in their understanding of hormones, the comparisons here are rather difficult. Nevertheless, there are some striking parallels. Like animal hormones, plant growth regulators have to lock onto special partners ('receptors'), rather like a plug and socket, before they can work. When a hormone locks onto its receptor, it triggers a second messenger to carry a signal inside the cell – usually an ion such as calcium or a molecule such as cAMP (cyclic adenosine

EXCITABLE CHEMISTRY

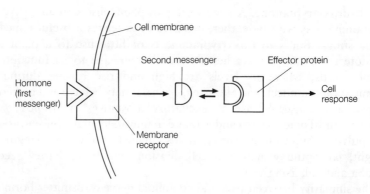

FIGURE 10.4 Simplified outline of how a hormone works. Hormone binds to a receptor on the surface of the cell. The receptor releases another, second, messenger inside the cell. The second messenger binds to, and activates, an enzyme or other effector protein. The effector protein triggers a cell response, possibly by opening an ion gate in the membrane.

3′,5′-monophosphate). The second messenger triggers another reaction, and eventually a whole cascade of reactions follow (figure 10.4). As in nerve impulses, any shuttling of ions across the plasma membrane upsets the delicate electrical balance across the cell, so that we can measure the activity as a change in voltage.

Acetylcholine is a good example of how a hormone works. As we saw previously, it is released from the end of one nerve cell, passes to another across a tiny gap, binds to a special receptor on the nerve/muscle/gland cell opposite, opens up ion gates and changes the voltage across the membrane of the cell there. Whether the same thing happens with acetylcholine in plants has not been investigated, but there is some evidence in support of this with the common plant hormone IAA.

George Bates and Mary Helen Goldsmith of Yale University found a characteristic electrical signal whenever they applied the auxin to oat coleoptiles (Bates and Goldsmith, 1983). The way that the signal developed coincided with a rise in acidity caused by a flood of hydrogen pouring out of the cell. Hydrogen stimulates growth of the corset of cellulose wrapped around each plant cell – one of the characteristics of auxin at work. But which came first? Did the auxin create the change in electric charge, or did the electricity change the auxin? This is a hot potato in botany at present, particularly in the tropic growth movements of plant shoots towards light or of roots towards gravity. Both these tropisms are too complex to describe in detail here, but the important point is that once a shoot or root is stimulated by gravity or light one of the first reactions we can measure is a change in voltage across the tissues. At the same time there is a change in auxin concentration across the same tissues. Arguments are now raging as to whether electricity is involved in the bending movement. Does the auxin trigger electrical signals that switch on ion pumps in neighbouring cells? Or does the electricity drive a current of auxin through the cells? Or is it simply a by-product of auxin, with no real use to the plant? Or does it amplify the movements of auxin? Because the changes in electricity and auxin begin at almost the same time, separating one from the other has proved very difficult. One intriguing line of evidence is that a very weak electric current can prevent auxin from moving.

Plants manufacture IAA in the tips of their stems and leaves and pass the hormone through the stems into most of the root system. David Morris of Southampton University passed electric currents of a mere 15-20 microamps along pea stems – the size of current you might expect in a growth movement (Morris, 1980). By labelling IAA with a radioactive marker, he could apply the hormone to the tip of the stem and follow its passage downwards. The electric current behaved like an invisible plug, completely blocking the migration of IAA and locking it up in the top of the stem. But when the current was switched off the movement of IAA was rapidly and fully restored. How the electricity works is not known, but the hormone does not seem to be repelled or attracted by the current, since reversing it made no difference to the blockage. This suggests that electricity is important for carrying IAA through plant tissues.

Incidentally, plants themselves are also affected by animal hormones in nature. There's recent evidence to show that earthworm casts promote root growth and development, crop yield and flowering, and that at least part of the reason lies in the hormones in the casts. Indeed, roots can sometimes reverse their growth and turn upwards towards casts on the surface of the ground. The hormone activity may come from bacteria living in the earthworm gut, rather than from the animals themselves (Tomati et al., 1988). However, it's a powerful demonstration of how hormones affect creatures from different kingdoms.

Animals Turned on by Plants on Steroids

The hormones of plants and animals can be so similar that it's hardly surprising that they influence each other. But plants are superbly inventive chemists, and over the past decade scientists have uncovered a sophisticated war between them and leaf-eating insects that capitalizes on chemical mimicry by subverting the hormone system of the insects.

Insects reach sexual maturity only after passing through a series of stages punctuated by moults. The moulting is controlled by an ecdysteroid hormone known as ecdysone, but the outcome of a moult depends critically on a second hormone called juvenile hormone. Because the level of juvenile hormone falls at successive moults, the insect gradually matures into the adult. Anything that the plant can do to upset this balance between the two hormones hastens or delays development and causes lethal abnormalities.

Over the course of evolution, this hormonal network has been penetrated many times. One of the earliest cases came to light in 1964 when entomologist Karel Sláma moved from Czechoslovakia to the United States, bringing with him his favourite insects – colourful creatures called firebugs. The insects immediately became incapable of proper development, as though life in the West had proved too much of a distraction. Bugs that had moulted normally in Prague remained stubbornly immature in Cambridge, Massachusetts (Sláma and Williams, 1965). The secret of their protracted youth turned out to reside in the paper towels carpeting their living quarters. Various other brands of paper were substituted in an effort to pinpoint the chemical culprit. The London *Times* allowed them to moult properly; the *New York Times* held them back. The trail eventually led to a substance that occurred naturally in the balsam fir tree from which the US newspapers had been made. The substance was christened juvabione.

This substance delayed Sláma's insects because it mimics juvenile hormone. Yet, like many similar substances discovered since then, it has this effect on only a narrow range of closely related bugs. Insects from other families are simply not taken in, possibly because its chemical structure bears only a faint resemblance to the real thing. This state of affairs obviously limits the utility of juvabione as an insecticide and prompted a search for plants that make more accurate mimics.

Chemists eventually unearthed a plant that does make an exact copy of the juvenile hormone. After a detailed and lengthy piece of detective work, Yock Toong of the University Sains Malaysia, Penang, and two scientists from Sandoz Crop Protection in California discovered the substance in a Malaysian relative of the sedge family called the grasshopper's cyperus (Toong et al., 1988).

For good measure the creative cyperus also makes a related material called methyl farnesoate, which is thought to play the role of juvenile hormone in certain crustaceans.

Grasshoppers raised on a diet of cyperus experience a range of defects attributable to hormonal mayhem. Twisted wings and altered wing patterns are common. As if to add insult to injury, adult females must carry the additional handicap of underdeveloped ovaries. The cyperus imposes a harsh penalty on herbivores.

Analysis reveals that the cyperus is a rich source of juvenile hormone. Weight for weight, it contains over a thousand times more hormone than the average insect. One day the genes of this industrious plant may be of interest to genetic engineers attempting to develop pest-resistant crops.

However, the question is: who is the victim and who the vanquisher? The cyperus chemical defence may have evolved from something that originally favoured the insects.

Saralee Visscher, professor of entomology at Montana State University, has taken a close look at how conventional plant hormones might affect insect fertility. When she fed the plant hormone ABA to grasshoppers, the fertility of the female insects was badly affected – they laid fewer eggs over a longer time (Visscher, 1980).

Feeding juvenile hormone to the grasshoppers had much the same affect as feeding them ABA, and when Visscher looked at the chemical structures of the hormones she found that they belonged to the same family of terpenoid chemicals. This may be more than just fortuitous. In the real world insects may have tapped into plant hormones to regulate their own development. For instance, when plants are suffering drought, disease or some other stress, they make more ABA. Juvenile insects feeding on the leaves will eat enough of the hormone to stall their own development until conditions improve. Then the ABA level drops and the insects can turn into adults and start breeding. ABA may be only one of the cues that insects use. Visscher (1982) also found that low levels of other plant growth hormones – auxins, gibberellic acid and kinetin – affected locust reproduction.

Therefore a balance of plant hormones could help insects control their own development. The twists and turns of this plant versus insect battle are so complex that there is plenty more left for us to unravel. How, for instance, do you explain why the plant hormone cytokinin has recently been discovered in newly laid locust eggs (Tsoupras et al., 1983)? It was discovered chemically bound to the ecdysone hormone, but no one knows what its function might be.

Insects also plundered plants for another part of their chemistry – their ecdysteroid hormones. Insects depend entirely on plants for ecdysteroids, because they cannot make the hormones themselves. Over fifty ecdysteroids from plants have been found, and just a few kilograms of dried yew leaves or *Podocarpus* roots contain as much of the steroid as 500 bumblebee larvae. However, there is controversy at present as to whether plants use ecdysteroids for their own growth. The results from several laboratories have been conflicting, although one aquatic fungus which goes by the ignominious title of *Achlya ambisexualis* uses chemicals closely related to ecdysteroids to control its own sexual reproduction on the basis of 'you scratch my back and I'll scratch yours'. The female of the fungus secretes a steroid called antheridiol, which turns the male strain on to produce the male sex organs. At the same time, the male makes oogonial steroid which turns on the female sex organs.

In the past decade, a completely new class of plant hormones called brassinosteroids has been discovered; these resemble but also inhibit ecdysteroids (Lehmann et al., 1988). Brassinolide was first isolated from pollen grains collected by bees from oil seed rape (*Brassica napus*) (Grove et al., 1979). It has since been found in dozens of other plants where it promotes growth by encouraging cells to divide. In young plants it helps to elongate stems and has a wide number of other effects.

The traditional five groups of plant hormones – ethylene, ABA, gibberellic acid, auxins and cytokinins – are probably only the tip of a hormone iceberg. Apart from brassinolide, no less than four other plant steroids have recently been found to promote the growth of plants: dolicholide in seeds of the lablab bean (*Dolichos lablab*), the rather disconcertingly named castasterone from a chestnut insect gall (a gall is an aberrant growth in plants caused by insects), typhasterol from pollen of the bullrush common cat-tail (*Typha latifolia*), and stigmasterol which is widespread in plants and, together with vitamin D3, encourages root cells to divide. It will be interesting to see whether any of these hormones, and possibly many more yet to be discovered, could be exploited for agriculture.

Apart from insects, plant chemistry also subverts other herbivores. Certain types of clover, for example, reduce the reproduction of sheep by making a substance that mimics the fertility effects of oestrogen, one of the mammalian sex hormones. Our own reproduction is not immune from vegetable hormones either. The earliest oestrogen oral contraceptives were made from yam tubers which had long been used as a form of birth control by South American Indians. Plant oestrogens may also explain why hungry women who ate tulip bulbs in wartime Holland suffered a range of menstrual abnormalities, including a failure to ovulate. In effect, they were eating oral contraceptives.

But even the simplest organisms could be dispensing us with exact replicas of our own hormones. David Feldman, chief endocrinologist at the Stanford University School of Medicine, has spent most of his career studying steroid hormones in humans. But a few years ago he and his colleagues wondered how far back in evolution mammalian-like steroid hormones and receptors could be traced. 'We felt it was worthwhile to make a big leap back and look in really simple organisms,' he explains. There was already evidence from Jesse Roth and others for peptide hormones in yeast, and so Feldman and his colleagues David Loose, a graduate student, and David Schurman, an orthopaedist interested in joint infections, decided to look for steroid hormones and receptors in *Candida*, a well-studied yeast that frequently infects humans, especially when their immune systems are suppressed. 'To our amazement, we found a binding site for corticosteroids that looked somewhat like the receptors we see all the time in mammalian cells,' Feldman recalls. 'We got very excited and we asked, "What is it there for? They must have something like a hormone as well"' (cited by Kolata, 1984).

The group went on to discover that the yeasts indeed made hormones for their receptors to lock onto. Mammalian-like receptors and steroid hormones were found in at least three different yeast species. What is extraordinary is that simple organisms and humans make chemical messengers that can bind to each other's cell receptors. When Feldman and his associates looked at the common bakers' and brewers' yeast *Saccharomyces cerevisae*, they found that it also had receptors that bind oestrogens (Feldman et al., 1982). They also found that it actually makes the human female sex hormone 17β-oestradiol (Feldman et al., 1984). 'The startling thing is that it is identical to the human hormone', remarks Feldman, and this raises the alarming possibility that people may be eating and drinking small quantities of oestrogen. It certainly has important implications for some infectious diseases.

The yeast *Paracoccidiodes brasiliensis* causes a devastating disease in South America, infecting seventy-five men for every woman. At first it was assumed that the unusual sexual ratio occurred because more men than women work in the fields where they inhale the organism. But research showed that men and women have equal contact with the fungus. The Stanford team suspected that the yeast may prefer men because it has sex hormone receptors that bind sex steroids. And indeed the researchers discovered oestrogen receptors in *Paracoccidiodes*. However, this only deepened the mystery, because men clearly have far less oestrogen and yet many more men were infected with the fungus. They then discovered that oestrogens inhibit the conversion of the harmless filamentary form of the yeast which is breathed in by the victims into the invasive form of the fungus that creates the infection (Loose et al., 1983). Therefore a woman's own oestrogens were her protection against disease, whilst the lack of oestrogens in men permitted the fungus to run its full infectious course.

The implications of this work are extraordinary. It shows that the sophisticated steroid hormone system existing in such a simple organism must have developed very early in evolution. The 'hormones' probably behave more as pheromones: it is well known that the peptide hormones are critical for the mating of yeast.

But what effect could oestrogen have in the yeast? The simple answer is that we don't know. In mammals oestrogens act on the central nervous system, turning on mating behaviour in females. Oestrogens also stimulate the growth of the reproductive tract and breasts. Yeasts are only single-celled organisms and have none of these exotic features, but there is at least one case where a yeast uses the oestrogen from humans.

Steroid glycosides are steroids to which a sugar molecule is attached, and they too are used by plants to combat animals. For example, the cardiac glycosides stimulate the activity of the mammalian heart. The best known, such as digitoxigenin, come from the foxglove (*Digitalis purpurea*) and are

> Box 10.1: The truffle story
> Another interesting case of sexual attraction by fungi is the search for truffles. It has long been known that sows easily detect underground truffles by scent and dig vigorously for them. No one knew why the sow was so keen on the scent until the story was unravelled by Maugh (1982). Truffles make a sterol that is also made by boars as a strong sex pheromone that attracts sows. The same sterol is formed by human testes, and Maugh noted results indicating that both men and women who were exposed to the sterol rated pictures of normally dressed women more beautiful than others not treated with truffle odour!

the source of useful drugs for stimulating weak hearts. Various *Acokanthera* trees such as bushman's poison (*Acokanthera oppositifolia*) contain a toxin in their wood which African bushmen boil into a thick tar and use as an arrow poison. The active ingredient is a cardiac glycoside called ouabain, and it has attracted the attention of physiologists studying nerve impulses because it inhibits the sodium–potassium ion-transporting protein in the plasma membrane. Botanist Herbert Jonas of the University of Minnesota applied ouabain, digoxin and digitoxin to the pulvinus of *Mimosa pudica* (Jonas, 1976). All three glycosides paralysed the leaf movement, suggesting that they had interfered with the movements of ions in the *Mimosa* cell membranes. Enid MacRobbie of Cambridge University reported that these drugs inhibited ion transport in many other plants as well (MacRobbie, 1971).

Cardiac glycosides have also been used by animals and plants to subvert each other's chemistry for their own ends. The milkweeds (*Ascelpias*) produce several bitter-tasting cardiac glycosides that protect them against being eaten by most insects and even cattle. However, monarch butterflies have built up a resistance to the glycosides, and the caterpillars contain such a high concentration of them that they induce vomiting in bluejays. The experience is so traumatic for the birds that they avoid other monarchs on sight alone. Thus the butterflies acquire considerable protection against predation from the plant's glycosides.

Self-defence by chemical subversion is probably why many plants make morphine, a chemical with similar structure and properties to our own endorphin and encephalin brain hormones. Apart from the well-known effects of opium poppies, plant morphines could have other interesting effects on us. Eli Hazum of the Wellcome Research Laboratories, North Carolina, discovered traces of morphine in the sap of lettuce and grass. This becomes concentrated in the mammary glands of cows and humans and is passed on in milk (Hazum et al., 1981). Similarly, cocaine from coca leaves looks and behaves like atropine, another type of nerve transmitter hormone.

Animals possess only two groups of hormones: the peptide hormones (such as insulin) and the steroid hormones, which include the sex hormones and a wide variety of others.

In 1978 Jesse Roth and his colleagues at the US National Institutes of Health were very surprised to find that the brain makes insulin. This was unexpected because insulin is normally only made in the pancreas. Roth's group made another remarkable discovery when they found insulin in many other parts of the body. But then they went much further – they found it in worms, fungi, single-celled protozoa and even the *Escherichia* bacterium which lives in our own digestive tract. Wherever they looked the answer was the same – all the organisms produced insulin that was 'closer to insulin than any other known substance' (Le Roith et al., 1980). Roth's results were later confirmed when receptors for insulin were found in bacteria and yeast (Dietz and Uhlenbruck, 1989).

So it was perhaps no surprise when Roth's research group also found insulin-like substances in plants (Collier et al., 1987), backing up previous reports that insulin affects plant growth (Csaba and Pal, 1982) and binds to plants (Legros et al., 1975). Insulin may well behave like a hormone in plants.

Roth and other scientists then turned to several other peptide hormones which had also been assumed to be specialized products of advanced animals. The story was much the same. Several single-celled organisms produce hormones similar to human somatostatin, adrenocorticotropin, beta-endorphin, arginine vasotocin, cholecystokinin, glucagon and chorionic gonadotrophin (Csaba, 1980).

From an evolutionary point of view, these discoveries are astonishing. Take insulin for example. Previously, most experts assumed that insulin first appeared in animals at the same time in evolutionary history as the pancreas, the gland in which it is made. But its presence in primitive single-celled organisms, with fossil records dating back some 3 billion years, suggests that insulin's origin is much more ancient than was first thought. In vertebrate animals, insulin acts as a chemical messenger released by the pancreas into the blood to regulate the uptake of small molecules such as glucose and amino acids for protein-building by cells elsewhere in the body. But for insulin to have retained its chemical structure so exactly over hundreds of millions of years of evolution, its basic effects on cells must also have remained the same.

Jesse Roth is one of the few modern biologists who takes an overview of all these bizarre results and gives an answer based on evolution. He believes that the earliest single cells secreted messenger molecules to communicate with each other – as we saw earlier with the chemotactic movements of plant sperm to the female's pheromones, or the *Achlya* steroid hormones. These molecules bind to specific receptor proteins on the surface of their target cells, inducing them to mate, multiply, clump together or otherwise behave in biochemical harmony.

As single cells evolved into organisms containing many cells (all living in physical as well as biochemical harmony), the same messengers were still used for communicating between the cells, although they were passed around inside the organism, in the blood, between nerves, through the phloem or whatever. They have clearly been very good at their job, because their chemical structure has hardly altered over millions of years. As a result, hormones from one organism can easily affect another organism, even if they are totally unrelated. For example, an opioid peptide from the brain of a vertebrate animal can bind to receptors on the surface of amoebae and alter their feeding pattern. Or a peptide pheromone responsible for mating in a yeast is the same as a hormone serving a key function in mammalian reproduction, controlling the testis and ovary (Loumaye et al., 1982).

The discoveries about these hormones have shaken biology to its fundamentals and are so potentially important that the relationships between plants and animals is being re-examined. For ecologists, the way that plants and animals control themselves and each other with their hormones will throw a new light on their relationships. For zoologists, the interest is in how the vegetable diet of herbivores affects their growth, development and behaviour. Botanists are trying to find out how plants try to fight off herbivores with hormone mimics. Pharmacists may find new or more easily extractable drugs in relatively simple organisms. We may even find that our own growth, development and behaviour is affected by the plants and their hormones that we eat.

11

Good Behaviour

Exploding flowers, tactile leaves, snapping carnivorous traps, cells that swim around – the list of these and all the other plant performances is impressive. Yet we've only been talking about reflex actions – movements already programmed into the plants in which learning or decision-making aren't needed, in some ways like a human knee jerked by a sharp tap.

But there *is* some intriguing evidence that plants remember events from their recent past. The first sign of a vegetable memory was recorded by Sir John Burdon-Sanderson during his pioneering work on Venus's flytrap, although he didn't call it a memory. The Venus's flytrap refused to shut unless the trigger hairs on the trap lobes were flicked at least *twice* within about 35 seconds (Burdon-Sanderson, 1873). The trap somehow remembered the first flick before responding to the second one. In fact, the trap remembers much more.

If the second flick is delayed longer than 35 seconds, the hairs need a third prod. And the longer the intervals between flicks, the more flicks are needed (figure 11.1). The trap 'remembers' some fifteen hits delivered every 18 minutes, although by the time it responds it is very sluggish (Brown, 1916). There is probably a good reason for the trap's memory. A chance blow from a falling twig or perhaps a piffling little midge can be ignored, so that the flytrap does not waste its energy on needless movement.

The reasons for other, more humble, plant memories are very obscure. Mordechai Jaffe, one of the few experts on touchiness in ordinary plants, found that a pea plant's tendril remembers being stroked (Jaffe, 1977). Normally the tendril starts coiling around a support only minutes after being rubbed, but when he chilled a tendril immediately after stroking it, the tendril became paralysed. And the astonishing thing is that, when it was rewarmed over an hour later, it coiled up as if nothing unusual had

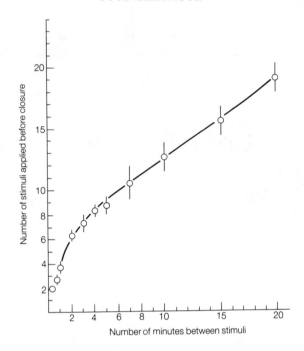

FIGURE 11.1 Evidence of a memory in the Venus's flytrap. It usually takes two flicks of the trigger hair(s) within 35 seconds to make the trap close. But if the flicks are delayed, the trap needs more stimulation before it decides to close. This graph shows that the longer the delays between flicks, the more flicks are needed to close the trap. In other words, the trap somehow 'remembers' how frequently it was stimulated.
(After Williams, plotted from the data of Brown, 1916)

happened. Clearly the rubbing sensation was stored and later retrieved when conditions improved.

More recently, a group of botanists led by Marie-Odile Desbiez of Clermont University published a provocative paper about plant memory in the well-respected scientific journal *Planta* (Desbiez et al., 1984). They removed both the leaves and the top bud of a bur-marigold seedling, allowing its two side buds to sprout open and grow a symmetrical pair of shoots. It's a common phenomenon in most plants known as apical dominance. Then they tested for a memory by pricking one of the pair of leaves with a needle just before stripping the leaves off. The side shoot nearest the trauma grew stunted – it had apparently 'remembered' the wound, even though it had been injured several days before.

So how does a primitive recall system like this work? Some of the best clues come from the Venus's flytrap. We saw in chapter 4 that the trigger hairs need a sharp blow before the trap closes. But if the trap is given a series of weak taps it eventually closes. The reason is that one weak tap on its own is not enough to trigger an action potential, but it does raise the electrical excitation in the trigger hair – the so-called receptor potential. Thus, if the trigger hairs are given enough weak blows, the receptor potentials build up until eventually they trigger an action potential. What is more, the greater the number of action potentials triggered, the more easily they pass through the trap. All this tempts one to plump for an electrical answer to the mystery of the plant memory. The trap may be sprung only when the electrical excitation reaches a critical level. Or perhaps some accompanying chemical activity, such as a build-up of calcium ions, is also involved. Interestingly, bur-marigold seedlings also fire electrical impulses when they are wounded (Frachisse et al., 1985), but what role these signals play in bud growth is unknown, and the possibility that they control the memory is pure speculation.

The trouble with all this is that we can only guess. There is so little research in this field that any item of information is like gold. One particularly interesting nugget that has recently been unearthed is that lithium – a tranquilizer in humans, and an agent that upsets primitive animal memories – also disturbs the bur-marigold's memory. Perhaps it's far too early to draw parallels of this type between plants and primitive animals, yet there is another peculiar aspect of plant behaviour like that in animals which I shall discuss in the next section.

Plants in Training

On 13 August 1960 a typhoon hit Tokyo, and apart from the damage to people and buildings it led to a strange discovery. A leading expert on sensitive plants, Hideo Toriyama of the Womens' University, noticed that *Mimosa pudica* plants growing outdoors first folded up in the typhoon, and then several hours later reopened, even though they were still buffeted by the wind. Somehow they had grown used to the wind and lost their sensitivity, whereas plants kept indoors were completely unaffected. So Toriyama tried recreating the phenomenon by buffeting *Mimosa* plants in front of an electric fan and found that they too lost their sensitivity (Toriyama, 1966). (Perhaps it is only a remarkable coincidence, but it took 10 hours for the desensitization to take effect, the same period as burmarigold takes to memorize the asymmetry of its wounding treatment.)

Somehow the *Mimosa* plants had *learned* to adapt to the constant wind and 'ignored' it. In fact, the French botanist Augustin de Candolle found that the sensitivity of *Mimosa* plants became depressed if they were repeatedly hit or gently electrocuted (de Candolle, 1832), and this was later confirmed by Wilhelm Pfeffer (1875). The leaflets first closed, but after prolonged stimulation eventually reopened, although they no longer responded to any stimuli. The sensitivity returns only after a suitable rest period.

Perhaps the most animal-like behaviour was discovered in 1965, when the neurophysiologists Eric Holmes and Gail Gruenberg of the University of Southern California found that once *Mimosa* had 'ignored' one type of stimulus, say being constantly touched, it could still respond to another type of stimulus, such as an electric shock. In other words, the plant distinguished between different sorts of stimuli (Holmes and Gruenberg, 1965). This opened up astonishing possibilities – could you in fact train the plant? The results have been mixed. Training is given by first applying two different stimuli together – the leaves react more vigorously than usual – and then testing the plant's reactions with only one of the stimuli. If training is successful, the response to one stimulus alone should be unusually strong, as if it had received two stimulations together (suggestions for training experiments are given in chapter 14).

But *Mimosa* is far from alone in this behaviour. Fungi also perform learning-type responses, as Joseph Ortega and Rustem Gamow of the University of Colorado showed for the spore-bearing stalk of *Phycomyces*. The stalk bends towards light and away from solid objects. They saturated the stalk with so much light that it eventually lost all its light sensitivity. Yet when they tested the ability of the fungus to avoid objects, it still remained sensitive (Ortega and Gamow, 1970). Therefore the fungus had modified its behaviour, ignoring light but remaining sensitive to other types of stimuli. It was a classic test of what psychologists call habituation – another way of saying that plants can be trained. The sort of habituation found in *Mimosa* is also known in simple animals like flatworms, and the physiologist Philip Applewhite of Yale University wondered how far back in evolution this behaviour might go. To his astonishment he discovered that he could habituate single-celled animals and even bacteria to different stimuli (Applewhite, 1975). Memory and learning may well be fundamental properties of all living cells.

Plant and Animal Behaviour

The question is: how closely do plants behave like animals? Simple memory and learning is one thing, but over the past fifteen years there have been staggering claims that plants have emotions and can even read human minds. The public's imagination was caught, and made books like *The Secret Life of Plants* into best-sellers (Tompkins and Bird, 1974). Their appealing blend of journalistic stories together with fringe and conventional science gave the feeling of credibility and a pleasant read. It was also a clear-cut snub to the orthodox scientific community, who were outraged when the reputable journal *Science* appeared to lend some credibility by publishing a review, albeit critical. Yet, despite the uproar, there has been little public debate on the issue, so let me put another side of the case here.

The authors made much of the wilder claims of Sir Jagadis Bose, whose work we have discussed previously. They also drew heavily on the contemporary work of Cleve Backster (1968), a lie detector expert working for the US Federal Bureau of Investigation. His credentials for research on plants are minimal (in fact non-existent), yet by clamping electrodes usually used for human hands onto plants, he claims to have measured electrical signals which give accounts of their 'feelings'. They can even feel the thoughts of nearby people.

The test of any experiment is whether it can be repeated by others, and this is where Backster came unstuck. Qualified scientists have been unable to reproduce his results. Kenneth Horowitz and his colleagues at Cornell University took great care to follow Backster's exact research accounts and apparatus, and even got in touch with him to ensure that everything was correct. They randomly dropped brine shrimps or distilled water into tanks of boiling water and recorded electrical activity on the leaves of rubber plants. According to Backster, the plants became very agitated whenever the shrimps were killed, as distinct from merely dropping water into water. But as Horowitz and his colleagues reported:

> We believe that we matched, and in several instances improved on, Backster's experimental techniques, such as controls, [external electrical field] shielding, number of observations, methods of analysis, and number of shrimp killed per injection. We obtained no evidence of primary perception in plants. While the hypothesis will remain as an intriguing speculation one should note that only the limited published data of Backster support it.
>
> (Horowitz et al., 1975)

Two leading plant physiologists, Arthur Galston and Clifford Slayman, performed a careful analysis of Backster's extrasensory plant research and reported:

> It is the lack of any plausible anatomical substratum, rather than any single experimental fact or flaw, which – in our view – drives the final nail into the coffin for the Backster, Tompkins and Bird [authors of *The Secret Life of Plants*] view of plant 'sensory perception' ... nowhere in the plant kingdom is there a gross anatomical structure that approximates the complexity of insect, or even worm, nervous systems, much less the mind-boggling intricacies of the cerebral cortex in higher primates.
>
> (Galston and Slayman, 1979)

So much for the more fantastic claims. What we still haven't decided is how far plant behaviour has evolved. In fact, even the pinnacles of plant behaviour, like the Venus's flytrap, can barely match primitive animals such as jellyfish, sea anemones, hydra and other coelenterates – aquatic creatures with no brains but only a simple network of nerves for relaying messages from sensors to cells that make stinging, feeding, swimming/crawling and defensive movements. Even then, 'these have an array of neural responses which makes any plant in the Droseraceae [Venus's flytrap and sundew family] look simple by comparison', comments Stephen Williams in his review of the evolution of the sundew family's sensitivity (Williams, 1976). In fact, Williams may have been too flattering to the plants. The even less specialized sponges are slightly nearer the mark – colonies of cells with no nerves.

12

THE ORDINARY PLANT

Solomon Islanders can reputedly kill trees by creeping up on them at dawn and suddenly uttering piercing yells close to the trunk. The tree is supposed to die a month later. Perhaps that seems incredible, but 'ordinary' plants are surprisingly sensitive to touch, although what use this is to plants living in the wild is not always obvious. There are extreme examples, like plants with obvious tactile movements such as the Venus's flytrap and *Mimosa*. But there are far more mundane cases: plants swaying in the wind often knock into themselves and one another; insects and other small animals bend, vibrate, knock, rub and scratch plants as they move on them; roots and young shoots rub against particles of soil as they grow. Indeed, plants are so sensitive to mechanical sensation that Canadian biologists were able to improve plant growth by means of strong sound vibrations (Weinberger and Measures, 1968), perhaps giving some credibility to the Solomon Islanders' feats.

With another one of his unnerving insights, Charles Darwin made a pioneering discovery in this field. He was fascinated by the way that plant growth responds to the direction of light, gravity, temperature and water. To him, touch was simply another influence to take into account in the everyday life of plants. He stimulated the touch sensitivity of pea roots by attaching a tiny piece of mica to one side of the root tip and observed that the root curved away from the mica (Darwin, 1865). In nature this is a useful way for roots to grow round solid obstacles such as stones. But Darwin's discovery did little to inspire further interest in touchiness in ordinary plants, and that's more than a pity – it has left a gaping hole in our understanding of plants. Only recently has the orthodox botanical world begun to realize the profound influence of touch sensitivity on the ways that plants grow and develop.

THE ORDINARY PLANT

The renaissance in interest probably began in 1971, when forest ecologists at the University of California at Davis discovered that trees responded to shaking. If young *Liquidambar* trees were shaken for just 30 seconds every day, they grew up to a third shorter than normal and developed thicker trunks (Neel and Harris, 1971). This makes them stronger and better able to stand up to strong winds. Trees prevented from swaying by wire supports showed none of these responses. Thus swaying is an important stimulus to the adaptation of plants to wind, and this carries an important lesson for tree growers. If saplings are tied to a post to stop them swaying, their trunks will be thinner and more likely to snap.

But it came as more of a surprise when most of the shaken trees also dropped their leaves, as if preparing for winter. This was confirmed in other studies. Manhandling other plants, such as cocklebur, maize or cucumber leaves, for only a few *seconds* each day also slows their upward growth by a third and triggers premature ageing (Salisbury, 1963). The same phenomena have been recorded in dozens of other ordinary plant species. In fact they are probably ubiquitous – as common as phototropism and gravitropism (Jaffe, 1985). It amazes me that it has taken so long to acknowledge this sensitivity to touch.

Plant physiologist Mordechai Jaffe at Ohio University, Athens, Ohio, has become one of the world's experts in this field, and he's been trying to make more sense of how and why plants respond to touch. Thirty minutes after rubbing bean plants their girth increases, as the surface and woody tissues expand sideways instead of upwards. Normal growth eventually recovers in about four days, and so it is only a temporary phenomenon (Jaffe, 1976a). He also showed that touching plants or letting them sway in the wind stimulates stomata to close, and this cuts down on water lost by the plant through its stomatal pores. Flower production can also be delayed and leaf fall triggered (Giridhar and Jaffe, 1988). Touching can even toughen up some plant species, such as beans, to drought, frost or chilling, but, for reasons that we don't understand, has no effect on other species like lettuce and cauliflower (Jaffe and Biro, 1979).

Natural history film-makers are aware of the problems that plant touchiness can present. World-renowned Oxford Scientific Films tried repeatedly to film the opening of dandelion heads, but as soon as they got the dandelions into a studio the plants grew in circles which prevented them getting a good shot. Yet when they tried clamping the dandelion flower stalks to stop their circular growth the plant growth suddenly shot upwards and sent the flower heads out of shot! The cause was probably touch-sensitivity in the flower stalk. In their natural habitat this phenomenon probably helps the flower head stand up above its surroundings to advertise itself to passing insects.

So what lies behind all these reactions? One of the answers came via two strange stories: from street-lights and from florists. At the turn of the century, Russian physiologist Dimitry Neljubow found that plants growing near street gas lamps grew thicker and shorter stems than other plants, and most characteristic of all, they drooped their leaves. The cause lay in ethylene pollution from the lamp gas. Decades later, florists were finding that poinsettias wrapped in paper or polythene sleeves to protect them during transit drooped their leaves, just like the lamp gas phenomenon. The culprit was again tracked down to ethylene, although in this case the plants themselves made the ethylene when they were rubbed. We now know that ethylene is a plant hormone released by plants under stress; when any flowering plant is touched it releases tiny invisible puffs of ethylene gas. As for florists, they can prevent leaves from drooping by treating plants with an ethylene inhibitor before wrapping the plants.

Ethylene is one of the plant's first lines of defence against drought, wounding, waterlogging and almost any other environmental stress. It affects the elongation of plants and the onset of flowering, helps shed leaves and flowers, ripens certain types of fruit and even stimulates the flow of latex in rubber trees. It's no surprise, then, that touch stimulation shows many of the same effects: interfering with flowering, promoting leaf shedding, and causing thicker and stunted growth.

Only a few parts per million are enough for ethylene to weave all this magic, working in a number of ways. It collaborates closely with auxins, which lead to the sideways growth of cells, and affects a number of enzymes, which is probably why it produces a sharp rise in the rate of respiration. It also blocks up the sugar-carrying channels (phloem) in the plant, again probably by influencing enzymes. Once the phloem is plugged, the tissues that it nourishes with sugars and hormones are effectively starved. This is a good way to shed leaves and flowers or fight viruses or seal wounds – a quick response to isolate potential trouble spots. In fact, the blocked phloem is the first sign of a plant's response to touch or even to high frequency vibration. The sideways growth then appears and lasts as long as the blockage remains in place; afterwards both normal phloem activity and growth resume. The correlation between the two events tempts one to assume that the blocked phloem helps sideways growth, but we've no concrete proof – it could simply be that they both rely on a continual supply of ethylene.

We can get another useful insight into how touch and ethylene work together during the infant life of seedlings as their young shoots push up through the soil to reach the daylight above. Plants such as peas grow a thick hook in their shoots, like a crooked finger. This acts like armour, driving the soil aside as the shoot elongates upwards, protecting the more delicate leaves tucked underneath from damage. Ethylene is responsible for developing this armour. As the shoot grows it scrapes past the particles of soil, and this rubbing sensation stimulates the release of ethylene. The gas in turn makes the hook grow thicker and stronger (Goeschl et al., 1966). The question is: what stimulates the release of ethylene? Barbara Pickard may have an answer. She detected small action potentials in the pea hook as it pushed against glass beads representing particles of soil. She suggested that the scraping of the hook against the soil stimulates action potentials, and these may release the ethylene (Pickard, 1971). However, it is still very difficult to tell which comes first, the chicken or the egg – do the electrical signals trigger the ethylene or vice versa, or perhaps the two are totally unrelated? We just don't know because the events happen so quickly.

Nevertheless, we've already seen the power of action potentials triggered by touch in other 'ordinary' plants – the characean algae *Chara* and *Nitella*. So it's very tempting to believe that electrical signals play an important role in the tactile life of peas and other higher plants. There are other small nuggets of evidence for this. Jaffe (1976b) has found that the electrical resistance of bean stems drops rapidly within 6 seconds of rubbing them, and other reports confirm disturbed electrical activity after mechanical stimulation in a variety of plants.

There the mystery of plant touchiness rested until a recent breakthrough accidentally revealed a group of touch-sensitive genes in a plant. Janet Braam and Ronald Davis of Stanford University were studying how plant hormones affect the genes of the weed *Arabidopsis*, a member of the mustard family. But during their experiments they noticed that simply spraying their plants with water stunted plant growth – the classic symptoms of touching that Jaffe had found. The major discovery, however, was that spraying water onto *Arabidopsis* leaves activated five separate genes in the plant, and that these genes make a protein called calmodulin (Braam and Davis, 1990).

As we saw earlier in box 9.4, calmodulin controls many key regulators of the cell's vital processes in both animals and plants. In fact, receptors for calmodulin exist in mammal brains, where they are involved in responding to sensory input. Braam and Davis suggest that plants and mammals may be detecting environmental changes through a similar mechanism that evolved hundreds of millions of years ago, long before the plant and animal kingdoms diverged.

They also suggest that the calmodulin may be responsible for triggering the burst of ethylene, and also for rearranging the cell's skeleton so that it can grow out sideways instead of longways. Thus, calcium and calmodulin can translate signals from the environment into messages that enable the plants to sense and respond to changes in their environment.

But exactly how calcium and calmodulin relate to the electrical activity of plant touchiness is far less clear, although the behaviour of animal sensors gives us clues. We know that plants, like animals, have special stretch-sensitive gates in their plasma membrane envelopes which let ions leave the cell. Although the stretched gates generally aren't too choosy about which ions they let through, calcium ions would be expected to be part of the ionic passage. Alternatively, calcium ions could be released from stores inside the cell as a result of the activation of substances called inositol phospholipids – a well-known phenomenon in animals (see p. 136).

Either way, the passage of ions would upset the membrane voltage, setting up a receptor potential. Given a sufficiently powerful stimulus, the receptor potential could trigger an action potential, which can tell other cells to respond. Thus, electricity could play a co-ordinating role in touch responses in plants.

Whatever the sequence, the Stanford scientists have given us a valuable breakthrough in trying to find how electrical activity, hormones, genes, calcium and calmodulin might all be choreographed into making plants respond to touch.

Touchiness in Fungi

Fungi can be even touchier than plants. We saw earlier in the touchy lassoos of some carnivorous fungi (chapter 4) that touchiness plays a vital part in ambushing passing prey. But there is a good deal more interest in how fungi actually attack crop plants using a sense of touch. The bean rust fungus provides a good example of how the sense of touch comes to the aid of fungi attacking plants. When spores of the fungus drop onto a bean leaf they germinate into long thread-like cells, but these alone cannot infect the plant. The hyphae need to grow across the leaf surface in search of stomatal pores, through which they launch specially flattened infection probes which invade the inside of the leaf and parasitize it. The mystery is how the hyphae know exactly where they are going in their invasion plans. No matter where on the leaf the spores land, the hyphae seem to make a bee-line for the stomata. Plant pathologist Harvey Hoch and his colleagues at Cornell University have shown that the secret is in microscopic corrugations over the surface of the leaf, which the bean rust fungus uses for tactile navigation (Hoch et al., 1987).

Hoch and his colleagues tested the touch-sensitivity of the fungus by growing it on plastic wafers specially etched with a variety of microscopic ridges spaced at various intervals and heights ranging from 0.03 to 5 microns high. The fungi were able to tell the differences in spacing between the ridges, guiding their growth towards some widths and not others. But only when they touched ridges 0.05 microns high did they grow infection probes. This height proved to be crucial, because when Hoch and his colleagues re-examined the bean plant they discovered ridges of the same height around the entrance to the stomata. Evidently the fungus uses its sense of touch to search out these tell-tale ridges to launch its attack through the stomata. Armed with the secret of the fungal navigation system, the researchers now hope to outwit the fungus by developing varieties of bean plants with no stomatal ridges, or with ridges of different heights.

Plants themselves also use their sense of touch to defend against the fungi. In fact plants are so sensitive to touch that a single microscopic fungus trying to penetrate a plant triggers off wall thickening around the attacked cell, stalling the fungus at least temporarily (Aist, 1977). Ethylene involvement in this defence was first suspected in an unusual story. Japanese farmers noticed that their rice plants no longer suffered a fungal disease called rice blast during seasons when the wind blew strongly. The wind bends and knocks the rice plants, triggering off their ethylene, which in turn tells the plants to grow tougher cell walls and also to produce natural fungicides which both help fight the disease. The sense of touch is particularly important because laboratory tests show that by rubbing bean plants it stimulates their natural fungicides. Ethylene was again strongly suspected to play an important part in this response; applying an inhibitor of ethylene prevented the plant defending itself (Takahashi and Jaffe, 1984).

The touchiness of crop plants can even be exploited. Norman Biddington and A. Dearman of the National Vegetable Research Station, Wellesbourne, Warwick, found that lettuce and celery seedlings brushed lightly with sheets of paper before transplanting survived much better than unstimulated plants (Biddington, 1985). This may explain why Japanese farmers stroke their sugar beet plants with brooms before transplanting them to the field. For some species, like chrysanthemums, stunted growth is desirable. But touching could have unfortunate results for other plants. During their work with plants, scientists often touch the specimens and unwittingly stimulate them. This is bound to affect their results as plant physiologist Frank Salisbury of Utah State University found out. He measured the lengths of cocklebur leaves daily using a ruler, which in itself slowed their growth and caused the leaves to fall off prematurely (Salisbury and Ross, 1978). Touch sensitivity could upset other scientific experiments. Since wind plays an important part in shaping plants, by knocking them into one another, scientists experimenting in draught-proof laboratories could be obtaining misleading results.

We've already seen how plants use electrical signals to respond to touching and wounding. Over the past few years exciting breakthroughs have been made with an entirely different aspect of the electrical behaviour of plants: their plumbing. This work came about as the result of two different lines of investigation combining by some serendipity to give birth to a completely new idea. In one line of research, botanists have noticed that water pressure inside cells plays an important part in plant development.

Young plant cells start dividing and growing when they become inflated with water, and can't divide when they lose water (Kirkham et al., 1972). It's as if water pressure tells the cells whether to divide or not. Indeed, the strength of hydraulic pressure inside the cells of the water plant *Callitriche heterophylla* even selects which of two different shapes the leaves will develop into (Deschamp and Cooke, 1983).

In an entirely separate line of investigation, it has been found that when many plant cells expand with water or collapse on losing water they trigger action potentials. Some researchers believe that such signals help regulate the water content of the cells, by ejecting ions such as potassium together with any surplus water (Zimmerman and Steudle, 1978). This might explain the potassium that spurts out of the cell of the alga *Acetabularia*, apparently released by spontaneous action potentials (Mummert and Gradman, 1976). The idea is intriguing, in that too much water inside the cell puts pressure on the plasma membrane, triggering an action potential, which in turn releases potassium, draws out water by osmosis and then deflates the cell. This is a particularly clever strategy for plants living in salt water, which struggle to survive by keeping salts outside and water inside their cells (Zimmerman and Beckers, 1978).

This electrical scenario also fits the touch-sensitive *Chara* story neatly. When action potentials spread through the alga's cells, they lock up the cytoplasm and also eject small spurts of potassium and water from the cell. Just as water pressure on the inside of a cell provokes action potentials, so a tiny pulse of pressure on the *outside* of the *Chara* cell is enough to trigger its action potential. The important lesson here is that plant cells apparently sense pressure, whether it be due to touch or water, or from the outside or the inside. The pressure is probably turned into an electrical signal by stretching the plasma membrane and opening ion gates (Zimmerman, 1980). Certainly, various plant cells sense pressure using hydrogen ion gates in their plasma membranes (Rheinhold et al., 1984; Kinraide and Wyse, 1986). It's a very appealing explanation of how all plants sense touch stimuli (something that we'll be coming to in more detail shortly). So, for example, as a trigger hair on the Venus's flytrap is bent over it stretches the sensor cells and their plasma membranes, opens their ion gates, triggers a receptor potential and delivers the action potential, which tells the rest of the trap to move. This is a good strategy for reacting quickly to sudden changes in the outside world, as you'd expect with an insect in a flytrap or from plants attacked and injured by animals.

It's especially interesting that stretch-sensitive ion channels have recently been discovered in stomata guard cells (Schroeder and Hedrich, 1989). We don't know what role these channels have, but it's another indication that stomata are probably touchy: you could imagine that they might sense the buffeting of leaves in wind, or animals touching the leaves, mentioned in chapter 8. And there's another possibility: the stretch channels might also help stomata move by responding to the pressure of the guard cells as they swell or shrink.

A sense of touch is no recent jump in evolution. Recent work on the common bacterium *Escherichia coli* using patch clamp recorders revealed unexpected fluctuations in electricity whenever the membrane was pushed or pulled. These were the unmistakable signs of ion gates responding to pressure, and the researchers even believe that, in the same way as we have speculated for plants, these channels help the bacteria to balance the water/osmotic pressure inside their cells (Martinac et al., 1987). As if this wasn't convincing enough evidence that touchiness is an ancient sense, touch-sensitive channels have been discovered in yeasts too (Gustin et al., 1988).

Action potentials can also tell the plant how to look after its food supply. Most of a plant's food is carried in long narrow pipes called phloem sieve tubes, named after the sieve-like plates that connect the end of one tube with its neighbour. The bulk of the food is the sugar sucrose, which is carried from green leaves to the rest of the plant where it is needed. But exactly how the sucrose is loaded into the sieve tubes and then unloaded at its final destination is not certain.

When Walter Eschrich of Göttingen University discovered action potentials in the sieve tubes of *Mimosa* (chapter 7), he wondered whether ordinary plants might behave the same way. The results of his most recent work throw open an entirely new area of research on the way that electricity might control plant food supplies.

In the common ornamental *Zucchini*, rapid action potentials were detected travelling at the same sort of speeds as in *Mimosa* – up to 10 centimetres a second (Eschrich et al., 1988). This is far too fast to be explained away as some sort of side-effect of sugar transport. Instead, the signals are generated quite independently. Although phloem tubes are supposedly designed for carrying sugar, they are also ideal for sending electrical messages: long cells, relatively few obstacles and an extensive network throughout the plant are all ideal features of an express route for sending long-distance electrical signals. They are perhaps the nearest thing that the plant kingdom has ever come to evolving a nervous system.

But what are the phloem's electrical signals doing? They may be telling the phloem to load or unload sugar. When sucrose is loaded in the sieve tubes of a leaf it grows more positively charged and fires an action potential. On the other hand, when sucrose is unloaded into a growing fruit the response is the opposite: it grows more negative and fires another action potential. These action potentials could be telling phloem some way away to pump or dump sucrose, as we saw in the *Mimosa* leaf movement (p. 102).

Action potentials in the phloem could even be translating what is happening in the outside world. We know that lighting conditions help tell the phloem when to unload its cargo: flashes of far-red light trigger unloading, whilst white light prevents unloading. What's interesting here is that both these sorts of light trigger different sorts of action potentials, but whether the electrical signals could be telling the phloem how to respond to the light outside we just don't know. Yet if this idea seems far-fetched, there are reports of action potentials in the phloem of other plants too (Hejnowick, 1970; Jones et al., 1974; Opritov, 1978). The electrical activity also supports the idea that electricity is somehow involved in the movement of sugar through the phloem (Fensom and Spanner, 1969; Fensom, 1972). There is even a theory that electrical activity regulates phloem loading and unloading by controlling the hydraulic pressure in the phloem – a neat tie-in with the ideas we discussed earlier about hydraulic pressure in ordinary plants (Smith and Milburn, 1980).

Wounding Plants

Wounding can be thought of as almost an exaggerated form of touching. Indeed, plants initially respond in much the same way whether they are touched or wounded: ethylene is released, electrical signals are triggered, respiration rises, phloem tubes become blocked, stomata close and later on leaves and flowers are often shed.

Some of the fastest responses to wounding were outlined by botanist Herbert Roberts in his work on plant respiration at the turn of the century (Roberts, 1896, 1897), and in the 1920s Winthrop Osterhout of the University of California at Berkeley found that wounding a plant has a significant effect on its electrical resistance and voltage (Osterhout, 1922).

More recently, Jerome van Sambeek and Barbara Pickard of Washington University, St Louis, confirmed a rise in the rate of respiration and a fall in photosynthesis and transpiration, all corresponding to the release of Ricca's factor and an electrical signal, upon injuring tomato and other 'ordinary' plants (van Sambeek and Pickard, 1976a, b, c). Another piece of recent work, on a liverwort, has backed up the correlation between electrical signalling and increases in respiration in wounded plants. What makes this research interesting is that an artificial electric shock also triggered the electrical signal and spurt of respiration. Furthermore, if the electric shock was not strong enough to trigger an electrical signal, it could not trigger the respiration rise either (Dziubinska et al., 1989).

The question of exactly how the electric signals might work was tackled by Eric Davies of the University of Nebraska in a brave theory that tries to embrace a plant's response to several stresses, including wounding (Davies, 1987). He claims that an action potential triggers a flux of ions in its target cells. The ions then trigger a cascade of reactions, involving cell wall chemicals, enzymes and cell structures such as microtubules. Davies believes that calcium, and possibly other ions such as potassium, released by electrical signals could turn on enzymes needed for making proteins to repair the wound. Davies and Schuster (1981a, b) also showed that wounding plants triggered a sudden arrangement of ribosomes into bead-like chains known as polysomes. This assembly is a conveyor belt which constructs proteins from their constituent amino acids. Most interesting of all, they found that the message to form polysomes travelled up and down the plant with a similar speed, direction and distance to those of the action potentials.

Davies' argument that electrical signals help plants cope with wounding recently received another boost from a group with a different interest. A team of British botanists based at Norwich and Leeds were looking at another phenomenon of long-range signalling in wounded plants. For twenty years it has been known that injured potato and tomato plants make poisons which attack invading bugs, a feature that plant breeders would like to exploit for creating insect-resistant crops. The plant makes these so-called proteinase inhibitors in leaves far away from the wound, and the problem that has puzzled researchers is what tells uninjured leaves to turn on their genes and make the proteinase inhibitors. The new research shows that when tomato leaves are crushed or burnt they trigger both an electrical signal and the production of proteinase inhibitors, and that both responses stop if aspirin or related chemicals are given. It is only circumstantial evidence, but given all the other evidence it is possible that electrical signals are playing some sort of role here (Wildon et al., 1989).

Interestingly enough, a recent paper has just shown that methyl jasmonate also turns on the production of proteinase inhibitors in plants (Farmer and Ryan, 1990). If you recall from chapter 10 methyl jasmonate is a close relative of both the plant hormone jasmonic acid and our own prostaglandin hormones, also released from injured tissues. And it's a remarkable coincidence that salicylic acid (the active ingredient of aspirin) blocks both the plant and animal responses to injury, in the latter case by stopping prostaglandins.

(Incidentally, this latest research also pointed to signs of communication between plants. Methyl jasmonate easily evaporates from one plant and turns on the proteinase inhibitor response in its neighbours. It's too early to say whether this is important in plants growing in nature, but it could have profound significance for an understanding of how plant communities grow, and it's seldom been considered by botanists before.)

Lastly, there is a strong suspicion that calcium and calmodulin are involved in wounded plants. When Braam and Davis made their breakthrough on touch-sensitive genes they also found that the same genes were sensitive to wound stimulation. This means that wounding turns on the production of calmodulin, which as we saw earlier binds to calcium and then switches on a cascade of events vital to the plant's functioning.

All this new work shows that plants have express alarm systems for dealing with any injury to themselves. It remains now to see whether we can fit all the jigsaw pieces together: the triggering of Ricca's factor, action potentials, ethylene, jasmonic acid, proteinase inhibitors, calmodulin, calcium, respiration, photosynthesis and transpiration.

Temperature Stress

Wounding and touching are not the only stresses that trigger electrical signals: chilling and invasion by fungi cause very similar responses. The Dutch botanist A.L. Houwink, who discovered the importance of Ricca's factor in *Mimosa* leaf movements, also found that a drop of ice-cold water fired off action potentials in ordinary plants (Houwink, 1938) (figure 12.1). Even brief chilling makes electrical waves (Pickard, 1984). I've often seen *Mimosa* leaves fold instantly when their *roots* are watered with cold water, which carries the exciting implication that roots might talk to their shoots using electrical signals.

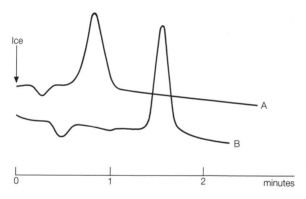

FIGURE 12.1 Action potentials occur in all plants, whether they show obvious signs of movement or not. This recording made in 1938 by A.L. Houwink on *Naravelia* shows an action potential passing from two places A and B on the stem after ice-cold water had been dropped on the plant at A.
(After Houwink, 1938)

Over the past ten years there has been a spate of new reports of temperature-triggered electrical behaviour in plants, confirming Houwink's earlier discoveries in algae and ordinary higher plants with or without motor organs. A sudden drop of only 1 or 2 °C is enough to trigger action potentials in cucumber plants or *Mimosa* (and makes its leaves move). A sudden drop in temperature will also trigger an action potential in the alga *Nitella* and stop its cytoplasm from streaming. In fact, the list of plant responses to rapid chilling are so varied – cytoplasmic streaming, transport of sugars in the phloem, hydraulic pressure and growth of cells, biochemistry and metabolism are all affected – that there is clearly a profound phenomenon at the heart of all of them. Calcium ions have been identified as playing a crucial role in these events, but whether electrical signals are the lynchpin remains to be seen, although as Minorvsky (1989) concludes in his recent review of the subject: 'stress-induced electrical responses are not solely a property of species showing motile responses, but are probably ubiquitous throughout the plant kingdom'.

The trouble with all these studies is trying to pin down exactly what is cause and what is effect. Is it really the electrical signal that triggers responses to stress, or is the signal the by-product of the repair process? The picture is much more complicated than I have described here because, apart from the possibility of electrical signals, wounded plants also use hormone signals. Once again, ethylene is involved in the early stages of wounding, followed by the action of other hormones such as abscisic acid and auxin. So far the evidence for the importance of electricity in wounded and other stressed plants is largely circumstantial, but there is so much that it is making the argument more and more credible. As the pattern of events takes place so quickly it's difficult to unravel the full story. What would really help the case would be to mimic the wound responses with artificial electric currents, using a technique known as voltage clamp (outlined in chapter 10). This technique allows the current to be increased slowly until an electrical signal is triggered, and with this sort of control other plant reactions can be more easily monitored simultaneously.

If electrical signals are important, they will give plants a rapid alarm system, warning them of an attack and preparing their defences. And from an evolutionary point of view these signals would also mark another way that electrical signals may have evolved in plants.

Touchy Sex

The sexual habits of ordinary flowers border on something between touch and injury. Pollination is a daunting task for any pollen grain. It has to travel from male to female sex organs (often involving long journeys carried in the air or on piggy-back on animals). Once landed on a fertile stigma, the pollen grain breaks out of its tough little coat and sprouts a finger – the

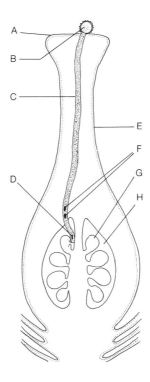

FIGURE 12.2 Pollination relies on senses of taste and touch, the hormone ethylene and perhaps even electrical signals. When a pollen grain B lands on a stigma A it sends out a pollen tube C, which grows down through the style E to one of the ovules G. There it delivers one of its sperm F, which slithers in amoeboid fashion to the female's egg cell G.

pollen tube – which elongates, ploughing through the style and the ovary to the egg cell buried inside (figure 12.2). Its journey through the female is guided by a sense of taste, growing away from air, towards moisture, and following a trail of sugars or sometimes calcium that leads to the egg. The ovules themselves even send out their own chemical signals to guide the pollen tube and its sperm as they draw near. In fact, it is easy to manipulate pollen tubes on an agar plate, making them grow towards extracts of stigmas or styles, or towards sugars (see the experiments in chapter 14). Without its sense of taste, the pollen tube would become lost.

Recent research by two Japanese botanists shows that the pollen tube may also feel its way to the egg by touch. Hirouchi Tokufumi and Shozo Suda of Kobe University grew pollen tubes from lily flowers on agar jelly and saw how, as expected, they followed a haphazard pattern. But when meshes of silk or synthetic material were laid on the agar, the growth pattern changed completely. The pollen tubes hugged the meshes, using them like railway lines. In fact they preferred meshes with openings of between 59 and 70 microns, and this size is significant. The inside of a lily style has cells arranged in rows of ridges and furrows 64 microns long and 24 microns wide, just the size to which a pollen tube is touch sensitive. But it must be said that when the pollen tubes were given the choice of meshes or chemicals from female sex organs, they found the female's perfume more seductive. So their sense of taste is stronger than their sense of touch (Tokufumi and Suda, 1975).

But that is still only half the story. The egg cells somehow anticipate the arrival of their suitors by respiring more heavily as the pollen grains germinate on the stigmas. Ethylene is involved – when the pollen tube grows through a style it actually wounds the tissue and releases ethylene. Furthermore, the pollen grains themselves come prepared by containing an ethylene precursor sufficient to trigger a burst of ethylene when they land on the stigma (Whitehead et al., 1984). This ethylene may well be the signal to boost the ovary's respiration, just as in a wounded plant.

Another idea is that electrical signals are triggered in the pollen tube as it grows through the style, and affect the ovary in the same way. The first signs of an electrical pollination signal were recorded in 1926 by botanists Robert Chodat and S.C. Guha of the Institute of Botany, Geneva (Chodat and Guha, 1926). Within 60 seconds of pollination they measured an electrical disturbance in the stigma, which then travelled down the style.

Only in 1967 did the Soviet scientists A.M. Sinyukhin and E.A. Britikov rediscover these signals. They reported an electrical impulse passing down a style at about 30 millimetres per second, and within tens of seconds respiration of the ovary increased (Sinyukhin and Britikov, 1967b). Their results have been backed up by other scientists, who have recorded a variety of electrical signals during pollination (Pickard, 1971). Spanjers (1981) at Toernooiveld, The Netherlands, found that the pollinating signal is remarkably similar to the signal produced by the wounded *Mimosa* leaf, and one wonders whether Ricca's factor is involved.

Another sure sign of touchiness in flowers is found in the common twayblade, a member of the orchid family. Charles Darwin observed that when he touched the centre of the flower, to mimic an insect pollinator, it oozed a drop of sticky fluid. Unfortunately this phenomenon has not been studied since. But Soviet plant physiologists G.P. Molotok, E.A. Britikov and A.M. Sinyukhin claim that the production of another flower secretion, nectar, from lime tree nectaries is triggered by an electrical signal in the glandular cells (Molotok et al., 1968). This has not been followed up either, but the whole idea of tactile glands is certainly interesting. After all, the digestive glands of the Venus's flytrap can be stimulated by repeated stroking, as mentioned in chapter 5. And if plant glands are electrically sensitive, we would have an impressive parallel with the way that animal glands are regulated by nerve impulses.

Touchiness in Climbing Plants

There is one sort of plant that has capitalized on a sense of touch for a special sort of growth. In general, flowering plants stand upright on their own, but some species have taken a rather lazier approach, wrapping themselves around nearby supports using their stems, leaf stalks or special thin fingery organs (modified from leaves) known as tendrils. More importantly from our point of view, all these climbers depend on a well-developed sense of touch.

Tendrils are the best studied of the climbing mechanisms. They sweep slowly and blindly through the air in their search for supports, and if they collide with a solid object, like a pole, they start coiling up and squeezing it. But the tendril is fussy about the sort of touch stimulation it needs: if the support is too smooth or is taken away soon after contact, the tendril uncoils. Some tendrils are more sensitive than human touch. If a thread of wool weighing no more than 0.000,25 grams is drawn along a tendril of *Sicyos*, coiling begins immediately. If a tendril completely fails to grasp a suitable support it coils up in the air and gradually dies. This ensures that the climbing plant only grasps a secure anchorage. Although the movements depend on the tendril's growth, its response can be remarkably fast. The world records are probably held by the passion flowers *Passiflora gracilis* and *Passiflora sicyoides* which you can actually see coiling up in about 20–30 seconds, and in *Cyclanthera pedata* coiling starts half a second after stimulation.

How then does a tendril judge the comfortable grip of a safe anchorage, not too smooth and just the right thickness? The sensitivity sometimes resides in special cells, but not always. The surface of the tendrils of *Eccremocarpus scaber* are lined with microscopic pimple-shaped cells which are extraordinarily touch sensitive. If just *one* of these pimples is stroked it triggers coiling along the *whole* tendril. French botanist Antonin Tronchet of Besançon claimed that the trigger to coil was fired as the pimples were stretched or pushed (Tronchet, 1962). Tronchet (1977) has also suggested that similar pimply cells covering the sensitive surfaces of touchy stamens, such as *Centaurea*, *Portulaca* and *Opuntia*, behave in the same way – converting touch stimulation into movement signals, as we saw in chapter 2. Indeed, pimples may be quite common in the tactile sensors of plants, because they have been discovered covering the sensitive parts of the touchy *Mimulus* stigma (Christalle, 1931). Yet microscope studies by Steffen Junker of the University of Aarhus, Denmark, failed to find anything special about these cells, apart from one property (Junker, 1977a). The cells were peppered with tiny pores (plasmodesmata) connecting one cell with another. These allow them to 'talk' to each other using chemical or electrical messages, just as the sensitive cells of the Venus's flytrap, leaf pulvini and other performing plants communicate. So once the touch sensors are excited, they quickly pass the message to the rest of the tendril.

The chemistry and physics behind the tendrils' sense of touch is obscure. The textbook explanation implicates auxin hormones, but auxins move too slowly and only in one direction. Instead, the evidence points more towards electrical signals and ethylene. We can speculate that the sequence might be as follows. When a tendril touches a support the plasma membrane of its sensitive cells stretches, just as we saw earlier for pressure and touch sensing in *Chara* and other plants. As the membrane stretches, it releases an electrical signal and ethylene, and this tells the other cells in the tendril to coil. Thereafter, hormones may take over control of the coiling.

The evidence for electrical signals in the tendrils is as follows. Jagadis Bose and others made tendrils coil by passing electric currents through them (Bose, 1927). Several years later, the German plant movement expert Karl Umrath measured action potentials travelling along the tendrils of melon plants within thousandths of a second of stroking them (Umrath, 1934), and others have found similar signals in other tendrils (Houwink, 1938; Pickard, 1971).

However, the orthodox explanation of the tendril's coiling involves ethylene, and on the face of it the case for a hormone is much stronger than that for any electrical signal. Within the first hour of stroking a tendril, the ethylene levels triple on the touched side. Because ethylene is already known to affect the activity of auxin and thereby slows the elongation growth of ordinary cells, there's an *a priori* case for its helping the tendril coil up. Certainly auxins have a very strong effect – if tendrils are dipped into solutions of IAA (indoleacetic acid) they begin coiling intensely within only a few seconds. Drugs that block the action of IAA also block coiling. The hormone probably makes one side of the tendril grow faster than the other, curling it round and round. But if auxin was the only thing involved, you would expect its activity in the tendril to change during stimulation. However, there is no evidence that auxin moves through the tendril any faster after touching stimulation, or that it moves from one side of the tendril to the other. Instead, Steffen Junker suggests that the tendril may become more sensitive to auxin on one side than on the other (Junker, 1977b).

Yet just to throw the cat among the pigeons, Mordechai Jaffe and Arthur Galston also suspected a 'muscle-like' contraction in the coiling. They believed that a protein with ATPase activity (fuelled by ATP (adenosine triphosphate)) contracts in the tendrils, somewhat like the proposed contractile ATPase fibres of the *Mimosa* pulvinus (Jaffe and Galston, 1967). The work of Jaffe and Galston remained a mere tantalizing suggestion until two Chinese researchers, Yong-Ze Ma and Lung-Fei Yen of Beijing Agricultural University, recently confirmed the presence of actin and myosin in pea tendrils (Ma and Yen, 1989). However, this evidence is circumstantial, and as yet there is no definite proof that actomyosin is the power behind tendril contraction. And turgor and growth are also involved.

During the first 30 minutes of tendril coiling, sucrose and other carbon compounds, ions (particularly hydrogen ions) and water are all lost from the side of the tendril touching the support. Therefore these cells lose their turgor and shorten whilst cells on the outside of the coil expand, and so the tendril curves (Jaffe and Galston, 1968). Yet quite how the action potentials, auxin, ethylene, contracting ATPase, ion movements, sugars and water all work together to make the tendril coil is unclear. Perhaps the reactions of the tendrils are like the putative spectacular growth movement of the Venus's flytrap (see chapter 5). Perhaps we are dealing with the meeting of two different evolutionary answers to the tendril phenomenon: the electrical trigger of movement of ions and water, and the hormone trigger of another cascade of reactions. The coiling tendril may have somehow neatly slotted the two systems together to achieve its grasp around a support.

Touchy Plants: Freak or Fact?

Judging by the scant attention it receives from most scientists and textbooks, you would think that plant touchiness is a freak of nature. I hope this chapter has laid that myth to rest: far from being unusual, touch-sensitivity is widespread in plants and fungi. And as we'll see in chapter 13 it is also a key piece in the jigsaw of the evolution of touchy plant movements. Yet it is astonishing that such a vital side of plant behaviour could have been overlooked, and I suspect it suffered the same prejudice that befell Burdon-Sanderson's historical discovery of an electrical signal in a touchy plant over a hundred years ago (Burdon-Sanderson, 1873) – it simply didn't fit the notion of what a plant *should* be.

The Evolution of the Plant Motor

And yet there seems to be a conundrum here: if plants are so excitable why don't they all behave with spectacular touchy movements like *Mimosa* leaves or *Sparmannia* flowers?

Part of the reason could be purely mechanical problems. Most plant cells wear a strait-jacket of cellulose which holds back outward cell movements. The important exception is stomata movements, which gives us an important clue to how all leaf movements may have evolved in the first place.

Stomata have specially weakened collapsible cell walls, giving them the flexibility to let their guard cells and the neighbouring cells accommodate movement. Collapsible thin cell walls also let motor cells move during leaf movements, but the leaf pulvini had to overcome an additional problem. Leaf stalks generally carry several rather inflexible bundles of internal plumbing, their vascular tissue. The mobile pulvini solved that problem by carrying all the vascular strands in one single flexible core, allowing them to behave as a hinge. So, with the mechanical problem solved, it's not stretching the imagination too far to see leaf movements evolving from stomata movements because they share the same water pump motor.

As we saw in chapter 8, there are striking parallels in the engines driving the motor cells of both stomata and pulvini. Both use osmotic water pumps driven by powerful floods of potassium ions across their plasma membrane envelopes. Water is irresistibly drawn with the potassium by osmosis, but the surge of electrical charge carried by the potassium ions has to be balanced by other ion movements, helped by special gates in the plasma membrane which are sensitive to the ions or to the voltage of the membrane, and fuelled by ATP and electrical power. None of this could be possible without the sensors to tell the motor cells when to move, and their signals trigger the movements which then have to be co-ordinated by the motor cells 'talking' chemically and electrically with each other, using plasmodesmata communication channels between their cells.

This scenario is the same for all stomata and pulvini movements, whether they are responding to light, touch, wounding, their own rhythms or any other stimulus. In other words, virtually all the components needed for leaves to move already existed long before in their stomata. Therefore it's tempting to explain the evolution of pulvini movements from stomata movements.

Why leaf stalks should evolve their own movements is another matter. One of the biggest demands of most plants is light. They need it for photosynthesis, for co-ordinating their daily and yearly clocks, and even for keeping warm. Small wonder that plants have evolved highly sophisticated light sensors which steer their leaf and stem growth towards sunlight. But there's a snag. As leaves mature they stop growing, so they lose their sun-tracking ability. But a motorized leaf overcomes that problem because it doesn't rely on growth for its movement, and so it can track the sun continuously. That means it achieves far better photosynthesis than its competitors, thereby improving the plant's chances of survival. And improved survival is the fundamental driving force behind evolution.

Inheriting the same light sensors as stomata, the pulvinus could also tell the difference between night and day. This allowed them to perform night-time 'sleep' movements which helped protect their younger leaves and buds from the cold at night, and perhaps preventing moonlight from upsetting their internal clock. Again, another survival advantage.

Therefore the evolution of leaf stalk movements from stomata movements seems a natural progression, and it's no surprise that light-guided leaf movements are fairly common. But now we come to the crunch question posed earlier: why are touchy leaf movements comparatively rare?

There is every evolutionary incentive for plants to develop touchy leaves. Forest plants need to drain their leaves of water, which reflects valuable sunlight, but their weak touch sensitivity ensures they only close during heavy storms. But plants growing in sunny open places can afford to close in light showers, and sensitive plants like *Mimosa* could have then turned this to defence against leaf-eating animals.

Given these strong survival motives, is some vast jump in anatomy or physiology needed to get from light-sensitive to touch-sensitive leaf movements? The answer is apparently no. Scientists have tried to find some sort of magical factor that turns slow-moving light-sensitive leaves into fast touch-sensitive movements, but there is frustratingly little difference between them. So there seems little evolutionary obstacle.

In fact, the evolution of touchy leaves is surprisingly easy to comprehend. Taken in sequence, touch firstly has to be felt, and because all plants can perceive touch that doesn't present much evolutionary obstacle to harnessing it for touchy movements. Then an express message has to be passed to the motor cells to tell them to move. The fastest messengers are action potentials, which we know all plants use for a variety of other purposes: wound signals, temperature signals and so on. The motor cells are turned on by the electrical signals, possibly using the voltage-sensitive gates in their plasma membranes which we know are abundant in stomata and pulvini.

The motor cells in the touchy pulvinus tell us few secrets. As we saw earlier, their ion and water pumps work so much like stomata and light-sensitive leaf movements that there seems a clear line of ancestry. The only minor clue is in the tannin vacuoles in the motor cells. The touchiest leaves have the largest tannin vacuoles, and the least touchy leaves have the smallest ones. Because we think that the tannin holds reservoirs of ions which are vital for movement, their size might limit the scope of the touch-sensitive movements.

But the size of tannin vacuoles is relatively trifling, so to speak. You could hardly imagine the whole evolution of touchy plant movements being stalled simply because they couldn't develop large enough tannin. But let's pursue this angle a little further. Why should a touchy leaf need more ions than a non-touchy one?

There are two possible answers. The touchy leaf movements are generally faster, so they need to marshall their ionic forces more quickly – tannin reservoirs might be the solution. The other possibility is far more outlandish, and I offer it up as speculation for others to test.

Touchy leaves might use another motor apart from osmosis. We saw in chapter 6 that *Mimosa* might use muscle-like fibres to help its contractions, controlled by calcium ions. If this is true – and we need a lot more evidence to be sure – it makes evolutionary sense. We know that actin and myosin contractile fibrils drive the streams of cytoplasm around inside plant cells. It's not inconceivable that similar strands of moving protein could help power the *Mimosa* movement. If so, then how many other touchy plant movements use them? And do they have any connection with the fibrils also seen in light-sensitive pulvini? These are tantalizing questions that urgently need answers.

Whatever the driving forces behind touchy leaves, we can expect similar systems behind the touch-sensitive flower movements. They too have evolved the 'off-the-shelf' stomata osmotic pump and plant 'nervousness' into a spectacular movement. They too follow a similar sequence of motor events (figure 12.3), a strong indication that they evolved along similar lines to touchy leaf movements.

But other plant movements are less easy to fathom from an evolutionary point of view. The putative growth movement of the Venus's flytrap is one such curiosity, particularly as it shows remarkable similarities to the coiling of a tendril. They both curve their growth in towards a touch stimulus, and it's conceivable that rapid tendril coiling was the template for the carnivorous trap. But what makes the Venus's flytrap very special is its sophisticated touch sensors – the trigger hairs. Although they follow the same basic architectural style as many other plant hairs, their sensor cells are full of endoplasmic reticulum, a manufacturing and storage organelle. These whorls of endoplasmic reticulum are so similar to the gravity sensors of roots that one wonders whether they are related. We believe that roots use their endoplasmic reticulum to hold reservoirs of calcium ions, and when grains of starch press down on them under the influence of gravity, some of these ions may be surrendered, triggering a cascade of reactions that eventually leads to growth down towards the pull of gravity. In the Venus's flytrap, the endoplasmic reticulum comes under pressure not from gravity but from the stretching of the sensor cell as the trigger hair is bent over by an insect.

There's also another way that the Venus's flytrap could make sense of touch stimuli. Stretch-sensitive ion gates in either the plasma membrane or endoplasmic reticulum membranes may be involved, since as we'll see later these have an ancient pedigree dating back to bacteria. Once a touch stimulus has opened the ion gates, a receptor potential could be created, ready to fire an action potential.

THE ORDINARY PLANT

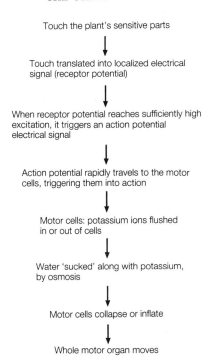

FIGURE 12.3 General scheme of events leading to plant movements.

However, some touchy movements seem to defy any logical evolutionary explanation. How, for instance, the bladderwort's trap originated is a mystery. Its hinged trapdoor and bladder is kept under vacuum by glands that pump salt and water, and is unlike any other plant movement. Not even other members of the same family, the Lentibulariaceae, give any hint of a bladder-like structure, and the nearest we can find to salt-pumping glands is in plants specially adapted to living in salty conditions.

But bladderwort aside, we can see that plant movements have incorporated many of the features of ordinary plants: touch-sensitivity, action potentials, water-pumping collapsible cells and so on. No single factor is extraordinary in itself – it's the combination of all the factors into one co-ordinated motor that makes plant movements so fascinating.

In the concluding chapter I want to sketch how touchy plant movements may have evolved, and where all plant excitability originated from. Along the way, you'll also see how our own nervous system evolved and why in so many respects animals and plants behave in the same way.

13

THE EVOLUTION OF THE NERVOUS PLANT

Looking back on this book we've made staggering progress to get from the wonders of touchy moving plants to the more humdrum world of common-or-garden plants. And yet we've made no extraordinary jumps of logic. We've seen that touchy plants have much in common with ordinary plants: mobile stomata (chapter 8), the same movements inside their cells (chapter 9), excitable chemistry (chapter 10) and they're all sensitive to touch and injury (chapter 12).

But then touchy plant movements are themselves no freak of nature: consider the sheer variety of touchy movements across such a wide range of families of plants (pp. 282–4), the range of touchy plant organs and their variety of different uses (pp. 279–82 and table 1.1). This all shows that these movements are no trifling accident of evolution.

But most of all, the similarities between plants and animals are astonishing. For a botanist like myself, brought up to believe that plants are somewhat apathetic organisms, this is a revelation. Their range of sensitivity is sometimes comparable with or even surpasses that of animals (pp. 275–6). Plants have virtually the full chemistry of an animal neuromotor system: nerve transmitters such as acetylcholine, norepinephrine, 5-hydroxytryptamine; messengers inside the cell such as calmodulin, inositol phospholipids and AMP (cyclic adenosine $3',5'$-monophosphate); cell motors such as actin, myosin, dynein and kinesin. They have much of the cell equipment of an animal non-muscle neuromotor system: mobile cilia and flagella, voltage- and touch-sensitive ion channels in their plasma membranes, sensors for chemicals, light, gravity and temperature. They can behave in the same way: receptor potentials, action potentials, memory systems and habituation learning. They even have hormones like insulin, prostaglandins and oestrogens. And so the list goes on.

THE EVOLUTION OF THE NERVOUS PLANT

With this weight of evidence we're bound to ask: how have plants come to behave like animals? With this single question we have finally arrived at the crux of the book, and in this concluding chapter I want to take you on a search for the origins of plant nervousness, in the course of which we'll see that plants have been poorly understood. Indeed, our notions of exactly what a plant is are highly questionable.

Given that the evolution of plants and animals parted company hundreds of millions of years ago, there is no way that plants could have 'borrowed' nerve-like behaviour from animals. On the other hand, you could argue that nerve-like activity in plants is a complete coincidence, and that it originated utterly independently long after plant and animal evolution split and went their separate ways. But that strikes me as carrying coincidence a little too far.

For one thing, there are so many similarities between plant and animal electrical signals. Can it really be coincidence that plants and animals can both translate information from the outside world into electrical codes – receptor potentials? That strong receptor potentials or a combination of weaker receptor potentials trigger an action potential? Is the fact that plant action potentials obey all the laws of animal action potentials – the rules of 'all-or-none', refractory periods and the facilitation of one impulse after another – stretching coincidence even further? And is it more coincidence that all these signals are driven by ion flows across excitable cell plasma membranes, and that excitable plants also obey the laws of animal narcosis? I think it is asking far too much of coincidence.

You might still be tempted to say that, because touchy movements are uncommon in plants, they must have arisen spontaneously and independently of each other in the course of evolution. After all, there are plenty of textbook cases of unrelated creatures independently evolving similar devices to solve similar problems. That might be the case if we were only dealing with a few families of plants, but as pages 282–4 show, there are seventeen widely related families of flowering plants with touchy movements: this is far too many cases to explain by independent evolution.

So instead of a list of extraordinary coincidences and freaks, let me give you this alternative evolutionary story: that the plant and animal kingdoms inherited the same basic neuromotors from the same mobile bacteria and protists, which formed the building blocks of the animal nervous system and its equivalent in plants. It's these simple creatures that hold the key to a number of mysteries: the evolution of the 'freaky' plants like *Mimosa* and Venus's flytrap, all the other excitable plant movements and even the touch-sensitivity of 'ordinary' plants. Almost every feature of both animal and plant excitability can be traced back to these ancient ancestors.

Over 3 billion years ago the bacteria were the main form of life. Two billion years later they gave rise to the eukaryotes – cells with a nucleus –

from which we and the rest of the animal, plant and fungus worlds eventually descended. But the inheritance handed down from our bacteria ancestors is still very much alive in all living eukaryote creatures. The clue is in the astonishing resemblance between certain of today's bacteria and the organelles in eukaryote cells. Let me illustrate this with two examples.

A chloroplast looks like a cyanobacterium (blue-green alga), with the same special photosynthesizing solar collecting plates wrapped inside membranes. Or take cilia and flagella in eukaryote cells, which look like the swimming bacteria called spirochaetes; they are even supported by the same microtubules. There are many other examples of the striking parallels between organelles and bacteria, and several decades ago they led to suggestions that organelles somehow evolved from bacteria. It was such a fantastic notion that few scientists dared voice their opinion in public. As the eminent cell biologist E.B. Wilson wrote in 1925 about the suggestion that mitochondria derived from respiring bacteria: 'Such speculations may appear too fantastic for present mention in polite biological society; nevertheless, it is in the range of possibility that they may some day call for more serious consideration' (Wilson, 1925).

The theory of the bacterial origins of organelles caused an uproar when it was formally proposed twenty years ago by Lynn Margulis, professor of biology at Boston University (Margulis, 1970). Only in the past few years has it gained respectability amongst mainstream scientists. Indeed, few biological ideas have swung so far from ridicule to respectability as Margulis's 'evolution by symbiosis'. Her thesis is that prokaryotic cells are the only true individual units of biology. She argues that the cells of higher organisms – the eukaryotes – actually began as communities of prokaryotes living in harmony. In a series of encounters, these formed composite organisms that were more evolutionarily viable than their ancestral donors. So they survived, and today's eukaryotic cell is the relic of those ancient communities. You could say that our bacterial ancestors are alive and well inside our cells.

Margulis's evidence turned on the presence of genetic information stored in DNA (deoxyribonucleic acid) in the mitochondria and chloroplasts, completely separate from the nucleus, the traditional home of DNA. This was a sure sign of a previous independent life, and it was backed up by years of electron microscope work which showed the detailed similarities between bacterial cells and their equivalent organelles.

What I want to stress here is the evolution of movement and sensitivity. The amoeboid slitherings of chloroplasts and mitochondria (and their precursors the proplastids), the beatings of cilia and flagella – all these once led free lives as swimming bacteria long ago in geological history. After they joined up into communities of bacteria, the power of movement was harnessed and passed on to their eukaryote cell descendants.

THE EVOLUTION OF THE NERVOUS PLANT

It's also no stretch of the imagination to trace our own nervous system and the nerve-like behaviour of plants back to the humble bacteria. After all, bacteria have all the credentials needed for excitability. Swimming bacteria such as *Halobacterium halobium* guide themselves using a rhodopsin light sensor, which may have become the forerunner of the eyespot of the single-cell alga *Chlamydomonas*, and maybe also of the eyes of higher animals. Bacteria talk to each other with chemical pheromones when they need sex. These same signals are the forerunners of the hormones which carry messages between cells in all multicellular creatures. The electrical signals believed by Julius Adler (1969) to drive bacteria cells (chapter 4) were probably the forerunners of nerve-like signals behind cilia- and flagella-swimming protozoa, fungi, plants and animals. It's quite conceivable that those electrical signals became the building blocks for the nervous systems of animals and the nerve-like behaviour of plants.

Even touch-sensitivity is no recent jump in evolution. Recent work on the common bacterium *Escherichia coli* has revealed pressure-sensitive ion gates and the researchers even believe that, in the same way as we have speculated for plants, these channels help the bacteria balance the water/ osmotic pressure inside their cells (Martinac et al., 1987). As if this wasn't convincing enough evidence that pressure/touchiness is an ancient sense, touch-sensitive channels have been discovered in yeasts too (Gustin et al., 1988). And so our sensitivity is probably descended in a continuous line through billions of years of evolution back to primeval bacteria.

The beauty of natural selection is that a device that works well once in nature often becomes something of a priceless family heirloom, preserved in pristine condition and handed down from generation to generation and, in the case of evolution, from species to species. Actomyosin, electrical signals, rhodopsin and all the other building blocks of sophisticated animal life are all living heirlooms that have survived intact over billions of years because they served a useful purpose. Sometimes they are adapted to different tasks. The microtubules, microscopic threads that drive the beatings of cilia and flagella, provide a good example. Their success in the early protists led to their surviving vitually unchanged in *every* cilia and flagella of the eukaryote world: protist, fungus, plant and animal – from sperm to the hairs that move mammalian eggs through oviducts or filter out debris in our bronchi. But the same microtubules are also essential for shuffling and tearing chromosomes apart during the division of eukaryote cells, they act as guidelines for small vacuoles, they structurally reinforce the cell in a cage-like skeleton and they even (in modified form) help give us sight in our retinas and hearing in our inner ears. Versatility is also a great virtue in nature.

The arrival of the single-celled eukaryote protists was a milestone in the evolution of neuromotors. They are often described as 'primitive', but they have all the attributes of a nerve-like system – sensors, signals and motors all rolled into one cell. They drive electrical signals like nerves, passing ions through their membranes using voltage-sensitive pores. Actomyosin powers the movement of their cytoplasm, just as it propels the contents of plant, fungus and animal muscle and non-muscle cells. They are even sophisticated enough to behave 'psychologically' like simple animals, as Philip Applewhite (1975) showed in his habituation experiments.

Comparing 'primitive' cells with animals is no wild flight of fancy. Biologists have been astonished by the recent discoveries in protists of insulin, oestrogens, acetylcholine, rhodopsin and other chemicals so sophisticated that many were previously considered exclusive to higher animals. After all, these are creatures so distantly related to us that they are now classified in their own separate kingdom. However, the astonishment is misplaced. Protist chemistry is simply a successful strategy that has survived intact, just as their cell organelles, motors, sensors and electrical signals have survived.

What we have here, then, are the building blocks of a nervous system laid down well before animals or nerves ever existed. The greatest milestone in the evolution of excitable creatures was the conversion of single-celled creatures into multicellular ones. Multicellular creatures have the luxury of appointing different jobs to specialized cells. Instead of sensors, nerves and muscles all rolled into one, their functions have become separated and specialized: nerve cells conduct signals, glands secrete hormones, retinas detect light and so on. All of that activity had to be co-ordinated, and the breakthrough came when cells started talking to each other with the chemical and electrical messages they'd inherited from their single-celled ancestors. This raised the level of organization to unparalleled heights of sophistication. It's the difference between a one-man band and a symphony orchestra – both use similar instruments but the musicians playing in an orchestra are vastly more sophisticated.

But let me add a sobering caveat. Our bodies still house refugees from our protist past. White blood cells behave like primeval amoebae, roaming lymph glands and blood vessels in search of foreign invaders to engulf and eat. Even though they are part of the human body, they still lead a semi-autonomous life. And the amoeba-like behaviour of cancer cells and embryonic cells, crucial to their development, is another tell-tale sign of a primeval heritage. Humans even have swimming sperm – a single cell that swims with a flagella, a direct descendant of the ancient swimming protists.

So, having glimpsed the origins of ourselves and our own nervous system, it's a little easier to grasp how plants evolved nervousness. Nearly all their nerve-like apparatus was inherited from our common bacteria and protozoa ancestors, before the plant and animal kingdoms diverged. From then on, from their earliest beginnings in the sea to the flowering plants on land, we

THE EVOLUTION OF THE NERVOUS PLANT

TABLE 13.1 Main groups of the terrestrial plant world (kingdom Plantae)

Division Bryophyta (mosses, liverworts, hornworts)
 Class Hepaticeae (liverworts)
 Class Anthocerotae (hornworts)
 Class Musci (mosses)

Division Tracheophyta (vascular plants)
 Subdivision Lycophytina (club mosses)
 Subdivision Psilophytina (whisk ferns)
 Subdivision Sphenophytina (horsetails)
 Subdivision Filicophytina (ferns)
 Subdivision Spermatophytina (seed plants)
  Class Cycadinae (cycads)
  Class Coniferinae (conifers)
  Class Gnetinae
  Class Angiospermae (flowering plants)
   Subclass Dicotyledoneae (dicotyledons)
   Subclass Monocotyledoneae (monocotyledons)

can trace a continuous thread of 'nervous' behaviour in the plant kingdom. The success of the swimming protists was repeated endlessly in the swimming sex and spore cells of the water and early land plants: algae, liverworts, hornworts, mosses, psilopsids, lycophytes, horsetails, ferns, cycads and gingkos (see table 13.1). In fact, cilia and flagella survived the evolutionary development of land plants until the appearance of conifers and flowering plants. But that profound change from plant life in water to life on land spelt doom for the swimming cells. They rely on water to carry them, and as plants became better adapted to dry habitats they dispensed with cilia and flagella as a bad risk. But they didn't junk their excitability; rather, the electrical signalling became refined into a communication system for different parts of the plant to talk to each other.

Even the earliest plants used their electrical signals for more than just driving their swimming movements. A beautiful example can be found in the sea where the churning of water at night by boats and animals generates an eery ghost-like glow. 'Bioluminescence' as it's known was first discovered in the dinoflagellates – protists which give off a flash of light when they're touched. No one is really sure why they do it; perhaps it frightens off predators, but that's only speculation.

The dinoflagellates come in two types – the plant-like (photosynthetic) and animal-like – but they both behave electrically. The plant-like dinoflagellate *Pyrocystis fusiformis* controls its flashes of light by triggering action potentials (Widder and Case, 1982). The animal-like dinoflagellate *Noctiluca miliaris* also controls both its swimming and bioluminescence with two different sorts of action potentials (Eckert and Sibaoka, 1967). There is even some evidence that it turns mechanical stimuli into receptor potentials before triggering an action potential (Eckert, 1965b). This sort of sophisticated electrical behaviour in such 'simple' creatures is another measure of the common inheritance of plants and animals. The touchy dinoflagellates weren't the only likely ancestors of 'nervous' plants. In plants and animals, egg cells burst with electrical energy when they're fertilized (p. 184). And another clue is from the algae *Chara* and *Nitella*, which as we saw earlier (chapter 8) use touch-sensitive electrical signals as a brake on the streaming movements of the cytoplasm inside their cells. All of which convincingly shows that primitive plants, with no outward sign of movement, none the less use electrical signals.

The same is becoming evident for ordinary flowering plants. We saw in the last chapter that a spate of recent discoveries has linked action potentials with feeling touch, wound and temperature stresses, with water balance and sugar transport, pollination and fertilization. We don't know the full significance of these signals yet, but such a rich variety of behaviour highlights how much electrical signals are a common feature of plant life.

It's almost certain that these common electrical signals are also the bedrock on which touch-sensitive movements evolved. With this excitability at their disposal, plants were free to evolve their own motors, mostly water pumps and contractile fibrils, which depend on rapid messages to sense touch stimuli, tell them when to move and co-ordinate the movement of their motor cells.

Yet there's a strange paradox, that as plants have become more sophisticated they've tended to become more sedentary. When they turned from an aquatic life to one on land they swapped swimming cells for a largely passive means of dispersing their sex cells and offspring. Instead, they used their latent excitability to evolve a completely new type of mobility. It all started with the opening and closing of the stomata leaf pores, developed into leaf movements, and then – in one of the most exciting twists of evolution – has turned into the spectacular touch-sensitive movements of flowers and leaves.

You could say that plants are once again becoming liberated, free to move and express their touchiness. The success of their touch-sensitive movements is self-evident from the variety of different plant families practising them, and the variety of uses they put them to: to help flowers cross-pollinate, to fold leaves to avoid damage and drain off rainwater, or to catch animals for food. 'Irritability', as the Victorian botanists called it, is part and parcel of plant life, and given the natural course of events we can imagine that plants are continuing to evolve into ever more animal-like creatures. Who knows, the day of the triffids might be closer than you think.

Nervous Plants?

And yet we come back to a nagging doubt that dogged the study of electrical signals in plants since the earliest days. When Sir John Burdon-Sanderson discovered action potentials in the Venus's flytrap in 1873, he was asked how plants could have nerve-like signals when they don't have nerves. The answer, quite simply, is that they don't need nerves. Plant cells can talk to their neighbours through tiny pores called plasmodesmata using electrical and chemical signals. Over longer distances they've made their phloem cells dual-purpose channels, for carrying sugar and sending electrical signals.

In contrast, nerves are much faster routes for channelling electrical signals (see p. 279) and they pass from cell to cell across tiny gaps called synapses which they bridge using chemical nerve transmitters. But what's interesting from an evolutionary point of view is that animals also appear to have something akin to plasmodesmata. Some animal cells join up with others using tiny pores called 'gap junctions' and pass both electrical and chemical messages to one another in much the same way as plant cells do. Recent evidence shows that plasmodesmata and gap junctions are lined with a similar substance (Meiners et al., 1988). This might be a case of parallel evolution – the same evolution to a problem arising independently – or it might have evolved from those distant ancestors, like *Volvox*, where single cells came together as communities and the need for cell-to-cell communication became paramount. Whatever the evolution, there is no anatomical reason why plants can't send electrical messages.

But before we get carried away on the idea that plants are evolving into some sort of green animal, it's worth trying to find a direct comparison to them in the animal world. To the disappointment of the more far-fetched fringes of the popular press (see chapter 12), plant neural systems are extremely primitive compared with those of animals. Not even the sophistication of a *Mimosa* or Venus's flytrap comes anywhere near the complexity of an earthworm, with its well co-ordinated movements, feeding and sexual behaviour. The most flattering comparison we could make is with jellyfish or sea anemones, with their simple networks of nerves lacking any central command centre.

So why is it that plants failed to become more sophisticated, even intelligent, creatures? The answer is simply one of necessity. The plant and animal kingdoms evolved far apart because their life styles became so very different. It boiled down to food, sex and protection: animals needed to find food and sometimes kill it, find and reproduce with a mate, and take shelter from danger. Plants, though, largely stayed rooted to one spot where they could make their own food by photosynthesis. The problem was that they were sitting ducks facing all the threats and hazards of the world, anything ranging from storms to the biting of caterpillars. So instead of developing locomotion to escape from trouble, they survived by adapting to the changes and stresses of their environment. For that they evolved a battery of sensors tuned to a great variety of environmental stimuli, their sensitivity even rivalling that of animals (see pp. 275-6).

The sensors in turn send hormone or electrical signals to change the plant's behaviour, growth or development. Hormones generally pass messages through the plant much more slowly than electrical signals, and are more likely to be involved in growth (compare the speeds of growth responses with those of touch-sensitive movements (p. 274). But whether hormones work together with electrical signals or even with gradients of electrical voltage is controversial and mostly beyond the realms of this book, although as we saw in chapter 10 hormones do upset the voltage of the plant membranes they latch onto (intriguing arguments for and against the electrical effects of hormones can be found in Pharis and Reid (1985)). Yet the details of how plants sense and respond to changes in the environment are still far from clear. Chemical responses may often be fine for long-term survival, but electrical signals may be better at coping with the most urgent problems that plants have to face.

And don't forget that electricity in plants works ar all sorts of levels, apart from long distance signalling. There's electrical power in photosynthesis, in shuttling ions in and out and through cells, in passing messages between neighbouring cells and through tissues and organs.

To sum up, then, it's a grave mistake to assume that because most plants don't show much sign of movement they lack nervousness. We have only scratched the surface of plant sensitivity and excitability, but clearly electrical behaviour plays an important part. To ignore it, as so many modern scientists have done, is a failure to understand what plants are and how they work. These are not the 'vegetable' organisms that textbooks would have us believe. They have much of the sensitivity of animals. They are excitable like animals. They can even move like animals. They are truly creatures of a nervous disposition.

The Future of Plant Electrophysiology

Of course it is all very well talking about mobile plants, tactile sensations and a pot pourri of other phenomena, but why should we bother taking oddball creatures like *Mimosa* and Venus's flytrap even remotely seriously?

History shows that plenty of other oddballs have spearheaded revolutionary solutions to mysteries in much more humble organisms. For centuries the ability of the Mediterranean electric fish (*Torpedo electricus*) to deliver shocks or numbness had been a curiosity. Then in 1758 a Dutch physicist,

Jean Allamand, realized that the eel's invisible punch felt like the shock from a Leyden jar (an early electric battery). He suggested that the electric fish might be discharging its own electricity – at that time an astonishing idea because it was believed that animal nerves worked on a 'nervous fluid'. The later confirmation that *Torpedo* indeed generated electric shocks led Luigi Galvani in 1791 to the idea that animal nerves might also work by an electrical force, which was the major stimulus for Alessandro Volta's invention of the electric battery that in time led to the age of electric power.

By the same token, the discovery of electric signals in Venus's flytrap over a hundred years ago has, belatedly, inspired the search for electrical excitement in ordinary plants. But the major stumbling block here is the paucity of outward signs of excitability in most plants. This has fostered the orthodox view that plants have no *need* for electrical signals. Even today, many botanists feel that if there are no nerves there cannot be any nerve-like activity!

That kind of prejudice goes back to the nineteenth-century German plant physiologists. They were fighting Aristotle's idea that plants had a soul in their pith, and this was one of the reasons why they attacked Burdon-Sanderson's work on the Venus's flytrap. More recently, sensationalist books have rekindled that prejudice.

Unfortunately this work has clouded the water for many bona fide scientists studying electrical signals and plant behaviour. Research money is notoriously difficult to find in this field. Barbara Pickard's pioneering work on cracking the formula of Ricca's factor has ground to a halt for lack of funds. Her joint discovery with Jerome van Sambeek of Ricca's factor in ordinary crop plants has gone virtually unnoticed. But consider the benefits that her work could bring. Because Ricca's factor may tell ordinary plants to close their leaf pores, there may be agricultural applications in helping crop plants cope with drought and other stresses. In a perverse way, some plants need to be stressed to be of economic use – consider the latex flow from a wounded rubber tree. Since ethylene encourages this flow, there might be scope for finding out whether Ricca's factor or even an artificial electric current increases the latex yield.

I have touched on plenty of other possible applications throughout the book. Animal hormones, or drugs that influence animal sensory systems, might usefully affect crop plants. The evidence that anaesthetics break dormancy in seeds is encouraging. Stroking seedlings before transplanting them could lead to better survival in the field. And with the 'magic' of genetic engineering, plant breeders have the tools to transfer the genes for sun-tracking or sun-avoiding movements into crop plants that need more or less sunlight respectively.

Academic Work

We've now come full circle, returning to the opening sentiments of the book. The study of reversible plant movements has been largely forgotten in mainstream plant science. It's considered an irrelevance, mostly a backwater for a few specialists to indulge in, but otherwise not much use.

Yet we now have new techniques for unlocking even more of the secrets of mobile plants. Nuclear magnetic resonance allows us to peer inside a plant's waterworks during movement without having to cut it up. Voltage clamping can highlight the role that electricity plays in a plant, and patch voltage clamping pinpoints the membrane gates that drive the electrical activity. Each year new chemical markers are being discovered which latch onto and highlight calcium ions as they move around living cells. The combination of antibodies with fluorescent tags can pinpoint individual molecules in a cell, and hopefully may reveal the mysteries of the function of actomyosin in leaf movements.

Most plant scientists have been brought up to believe that hormones are *the* language for plant cells to talk to each other and to co-ordinate the activities of the whole plant. Yet studies of the seemingly bizarre habits of a few mobile plants have already given us valuable clues as to how ordinary plants respond to touching and wounding, and how they use electrical signals. Furthermore, since it's now been shown unequivocally that all plants use tactile senses, this subject has been given a new lease of life. Yet its full implications still seem lost on most scientists.

But all is not lost. Animal physiologists and even psychologists have taken a great deal of interest in our botanical curiosities, although for slightly different reasons. They are searching for chemical or electrical clues as to how behaviour evolved and works in animals. To them the plant is a simple and easily observable model. For example, the Venus's flytrap is a good model for studying memory. The rhodopsin light receptor in *Chlamydomonas* is a useful tool for exploring vision in higher animals. Scientists interested in how muscles work continue to use the slime mould fungi. And the behaviour of single-celled plants and animals, and even bacteria, is constantly giving fresh insights into the origins of the nervous system and its control of behaviour. Perhaps all these scientists feel less inhibited because they don't see the protist, fungus, plant and animal kingdoms all rigidly divided from each other. Certainly, one of the great biological shocks of the past few years is the realization that simple organisms can have the sophisticated chemistry of a mammal and all the elements of a nervous system without nerves. Indeed, we and the plant world seem to have evolved from the same neuromotors: those in the animal kingdom evolved into nervous systems, and those in plants evolved into a simpler neural network.

THE EVOLUTION OF THE NERVOUS PLANT

Throughout this book we have seen how electrical signals control a whole circus of mobile plants: leaf movement, carnivorous plant traps, flower parts, flashing lights in algae, the beating of tiny propeller hairs in single-celled plants and animals and so on. In fact, if all cilia and flagella movements are controlled by electrical signals, then all algae, fungi, liverworts, mosses and ferns, and even some primitive conifers, use electricity, because all these plants have sperm driven by cilia or flagella. The higher flowering plants don't have cilia or flagella, but electrical signals may control the pollination of flowers and the secretion of glands, such as the digestive glands in the Venus's flytrap or the nectaries in lime trees. In all plants, any sort of touching or wounding could be translated into electrical information. The transport of food inside individual plant cells is partly driven by electrical power, and so is photosynthesis, but that's another story. . . . In fact plants seem to be alive with electrical activity of some sort; the trouble is finding out exactly what it is doing. Above all, we are still very unsure about much of this subject, and that is why it needs research – mobile plants are probably only the tip of the iceberg. The nervous plant is probably ubiquitous.

14

EXPERIMENTS

The *Mimosa pudica* Water Pump

Put the plants into a bright warm room. Cut a leaf stalk diagonally with a sharp razor-blade and let the plant rest under a damp jar for at least 30 minutes, until it has recovered from the wounding. When the stalk is upright again, touch the pulvinus with a brush and watch the cut end of the leaf stalk. A drop of liquid should ooze out from the bottom half of the cut surface. The liquid has been expelled from the bottom half of the pulvinus, as the motor cells lose their turgor.

Mimosa pudica Behaviour

Habituation

Lightly strike the lower side of the main pulvinus (between the main leaf stalk and stem) or a leaflet repeatedly ten times a minute for about 5 minutes (leaves can also be cut off and floated in water, where they recover and can then be tested). It is best to find a way of always delivering the same intensity of stimulus; for example, let a drop of water fall from a height of 4 centimetres. The stalk/leaflet moves at first but eventually regains its original starting position in spite of the continued blows. It then 'ignores' further stimulation for some time – an example of 'habituation'.

If several leaflets are cut off, floated in water and hit every 2 seconds, the leaflets close, but after about 13 minutes of this stimulation they become fully open and insensitive to stimulation. Increasing the test intervals to 15 seconds produces opening after about 30 minutes. When the intensity of stimulation is increased, with the interval between stimuli constant, the leaflets open after a much longer period of stimulation. As with many animal experiments, in *Mimosa* the longer the interval between stimuli, and the more intense the stimulation, the longer it takes for the organism to adapt. Electric shocks through the water have the same effect.

Discrimination

However, if the apparently 'numb' leaves are given another type of stimulus – burning, cutting, electric shock, ice-cold drops, drops of salt solution – they do respond. Thus the plant discriminates between different sorts of stimuli.

Conditioning

Try to get a 'conditioned response' by 'training' a plant with two different sorts of stimuli. After delivering a gentle stimulus (for example, a light stroke), wait 10 seconds for a reaction before giving a strong stimulus (for example, an electric shock) to provoke a full collapse of the leaf. Let the leaves rest and then repeat the same routine several times more. If the leaves become conditioned, eventually they will give a full response to the weak stimulation alone, as if they were 'expecting' the strong stimulus to follow.

Mimosa pudica and Anaesthetics

Mimosa is affected by anaesthetics. Stimulate a leaf so that the leaflets close up and then cut off the leaf and put it in a small bottle of water supported by a cottonwool plug. Place the bottle in a dish containing chloroform or ether, and cover the whole with a glass jar in good light. In a few minutes the leaflets expand, but they are completely insensitive to all stimulation. When the anaesthetic is taken away, the effect wears off and the leaves recover fully.

Mimosa pudica and Wounding

Place two *Mimosa* leaves in two small water-filled bottles, supported again by cottonwool plugs. Let the leaves recover in good light and warmth, and then burn one of the leaves. After the leaf has collapsed, take a drop of the water from its bottle and put it into the water of the unstimulated leaf. The second leaf should then react, because Ricca's factor released from the wounded leaf is a chemical signal for leaves to move. Now repeat the same experiment using a leaf from an 'ordinary' plant species with no touch-sensitive leaf movements. When this leaf is burnt, it too should release Ricca's factor, which will make a *Mimosa* leaf collapse. Indeed, Ricca's factor should make stomata close, make touch-sensitive stamens bend and probably affect many other movements which have yet to be tested.

Touch-sensitive Stamen Movements

Cut flowers and their stalks of *Berberis* or *Mahonia* and put the cut ends into water. Let the flowers recover so that the flowers have opened up and the stamens have spread out. With the help of a magnifying lens, gently touch the bottom of the stalk (filament) of a stamen with a needle. The stamen bends very quickly inwards towards the style. Try stimulating the upper part of the stalk, and there is no reaction. Notice also that touching one stamen does not affect the other stamens.

Repeat the exercise with *Sparmannia africana*. Only the base of the outside of the stamens is sensitive to touch, but the stimulation is signalled from one stamen filament to another.

Take the flowers of *Centaurea*. With the aid of a magnifying glass, cut off the petals to reveal the tube of stamens in the centre and place the cut flower stalk in water to recover for 10 minutes. Stroke the filaments of the stamens with a needle and watch the tube contract, squeezing out its pollen and revealing the stigma and style inside.

Take the flower of *Mimulus*. With the aid of a magnifying glass, cut off the petals to reveal the twin-lobed stigma and its stalk (style). Place the cut end of the flower stalk in water and brush the inside of the stigma lobes with a brush or needle. The lobes fold together. Repeat the exercise with pollen from another *Mimulus* flower and then with pollen from a completely different flower. After a few hours the lobes will reopen if pollen from a foreign plant has been used, but if pollination is successful they remain shut.

Venus's Flytrap

Flick over the trigger hair of a trap with a pencil or dissecting needle. Only use healthy traps, and try not to re-use a trap that has recovered from a previous closure as these tend to lose some of their vitality. One touch in warm conditions is not enough to make the trap shut, but another stroke of the trigger hair within 35 seconds usually makes it close. If the second flick is delayed, however, more flicks are needed. Record how long it takes before the trap closes when the flicks are delayed over longer intervals, for example one flick every 3, 4, 5, 10, 15 and 20 minutes. This shows the Venus's flytrap 'memory' at work.

EXPERIMENTS

How quickly will a trap reopen? That depends on the triggering stimulus. Darwin found that the trap would open after about 22 hours if there was nothing edible inside. Traps with food take much longer to reopen and may never do so, depending on the size of the trap and its meal.

Experiment with how different 'foods' affect the closure. Try placing one of the following in a trap: a fly-sized piece of meat, cheese, glass; a drop of nitrogen-containing fertilizer solution; a drop of salt solution; a drop of sugar solution. Any substance containing nitrogen will stimulate immediate trap closure and slower 'tightening' of the trap around the stimulus. However, salt solutions will also trigger action potentials in the trap and cause the trap lobes to close. Sugar should have little effect, and glass either will have no effect or, if it hits the trigger hairs, will be rejected after a few hours.

You can feel the strength of the trap by attaching a piece of wire to a specimen of meat and letting the trap close around it. Then try tugging the meat out − not as easy as you might think.

Sundew

How sensitive is the sundew tentacle's sense of taste and touch? Darwin experimented on the sundew's sensitivity with a range of stimuli: small bits of paper and hair; grains of sand; drops of water; the tips of needles; small pieces of meat or cheese. He also tested a wide variety of solutions on the leaves, including milk, urine, egg white, saliva, meat juices, blood, chloroform, cyanide, cobra venom, tea and sherry. With the end of a fine needle he placed small drops of the substances on the tentacles and found that solutions were effective when they contained nitrogen.

How dilute a solution can the sundew detect? Darwin took solutions known to trigger tentacle movement and diluted them until there was no longer any response. He found that the tentacles were amazingly sensitive; for example, only 0.000,002,16 grams of ammonium phosphate caused a response!

Similarly, their sense of touch is acute. Darwin used pieces of cotton, hair and strips of paper, which he weighed accurately and then cut into small pieces. He gently placed them onto the sticky fluid of a single tentacle and waited for a curling response. The tentacles responded to a mass as small as 0.000,822 grams.

How fast does a sundew tentacle or leaf respond? Darwin discovered that speed of response depended on the type of stimulus. Between 20 and 70 minutes after stimulating a tentacle with a fine hair or needle, a bending movement was seen. The response to a grain of sand was much faster after it˙

had been dipped in a solution containing nitrogen. Does the texture of a material also matter? Test cotton fibres and smooth nylon ones.

How do the tentacles digest meat? Place small cubes of egg white onto tentacles. They have usually disappeared after several hours. By adding drops of dilute alkali such as bicarbonate of soda to the albumin cubes, Darwin showed that the digestive enzymes in the tentacles' secretions work in acidic conditions. Digestion stopped until a drop of dilute acid (such as vinegar) was added. The acidity of digestive juices can be detected using blue litmus paper, which turns red in acid solutions. If no litmus is available, soak a strip of blotting paper in the blue extract left over from boiling purple-coloured cabbage.

How to Record Electrical Activity from Plants

There are three pieces of apparatus needed for measuring electrical signals: two electrodes, measuring equipment and a recorder. Electrodes are simply probes that conduct electric current from the organism or its surroundings to the monitoring equipment. They can be made from a piece of copper, silver or brass wire, but these sorts of electrodes have a drawback – they generate their own unknown voltages when they touch an organism or its surroundings. This is most important if you are trying to measure the exact voltage of a specimen. In these cases you need to make the electrodes 'non-polarizing', by matching their material with something in common with the chemistry of the organism. A common sort of non-polarizing electrode is a silver wire coated with silver chloride, made either by dunking the wire into molten silver chloride or by electroplating it. (*Note*: Plate it anodically from a concentrated lithium chloride solution in series with a concentrated solution of hydrochloric acid and a platinum cathode, bridged by a lithium chloride–agar salt bridge, using a constant-current power source.) But if you only want to measure relative changes in voltage then simple metal electrodes are adequate.

An electrical circuit for processing the signals from recording electrodes is shown in figure 14.1. The steady component of the potentials of the organism and the electrode is filtered out through a resistance–capacitance circuit comprising C_1 and R_1. This allows only the varying part of the signal through. The signal needs to be amplified, and this circuit uses a single MOSFET integrated circuit connected as a non-inverting amplifier with a gain of about 100, so that a signal of 50 millivolts from the plant would register as 5 volts at the output of the amplifier. The integrated circuit

FIGURE 14.1 Circuit board design for recording electrical signals from plants: (a) circuit diagram; (b) layout from top.

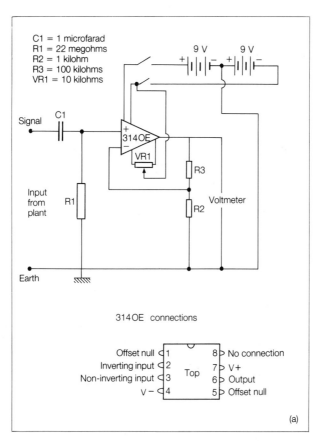

(a)

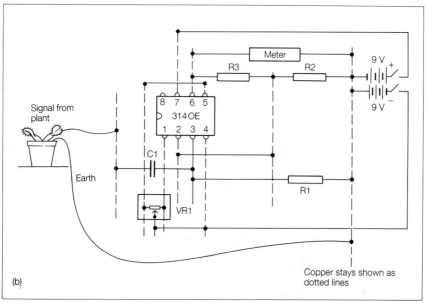

(b)

(3140E in the circuit diagram) may be difficult to purchase, but there are suppliers of amplifiers specifically designed for biological recordings.

The voltmeter can be a centre-zero meter, which allows negative and positive voltages to be measured, or a simple moving-coil voltmeter. Alternatively, you can use an oscilloscope or moving-wire galvanometer to display the electrical activity as it occurs. However, none of these instruments makes a permanent recording of electrical signals, and for that you need a chart recorder. Commercial types might be too expensive, but it is possible to make your own using a hifi loudspeaker (Stong, 1962).

The circuit can be built on stripboard. The components and wire connections are fitted on the opposite side to the copper strips and the leads are passed through the holes and soldered to the strips. It is important that you cut the strips where they pass beneath the integrated circuit, or its opposite terminals will be short-circuited. Single-stranded bell wire can be used for connections between strips and to the meter and batteries. If available, light-weight screened cable should be used for the connection to the plant, with the screen connected to earth. If this is not available, use bell wire but the input and earth leads to the plant should be twisted tightly together and kept away from any mains wiring and from the output of the amplifier. The equipment needs a dual power supply of ±9 volts, which can be provided by two 9 volt batteries connected in series with the centre point earthed. A double-pole on/off switch should be used to break both supplies when the equipment is switched off.

When the circuit is complete, it should be possible to vary the reading on the meter over about ±4 volts by adjusting VR1, which is used to zero the readings. A quick check to see whether the equipment is working is to hold the bare end of the input lead between finger and thumb and briefly touch the ±9 volt power rails in turn with a finger of the other hand. The needle should swing as a minute current flows through your body.

For making measurements from potted plants, place the measuring electrode on a small drop of tap water or dilute salt solution on the plant, and clamp it in position without injuring the plant. Place the earth electrode either in the soil or preferably on a distant part of the plant shoot. Allow the equipment to stabilize before making any readings (Young, 1975; Goldsworthy, 1984).

With this apparatus the world of electrical excitability in plants opens up! Good specimens to test are Venus's flytrap and *Mimosa*, because they are both relatively large and give strong signals. Also try investigating the effects of touching and wounding on ordinary plants.

Apparatus for Measuring Leaf Movements

Attach a leaf stalk by a cotton thread to an electronic arm (Harris), supplied

EXPERIMENTS

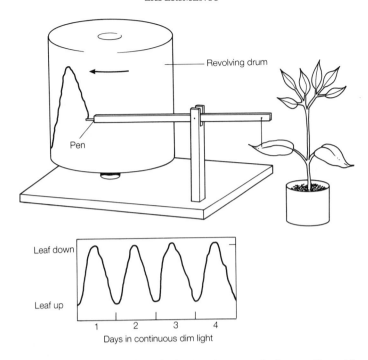

FIGURE 14.2 Recording leaf movements. The classic method, recording with a pen mark on a revolving drum, using a delicately balanced pen-and-lever system attached to a leaf by a fine thread. Many plants such as the common bean continue their leaf movements for several days, even when kept in continual light or dark.

with 9 volts power and monitored with a voltmeter (figure 14.2). The electronic arm produces a voltage proportional to the position of its movable arm. A small piece of putty is used as a counterbalance on the opposite end of the arm to the plant. Care must be taken not to exceed the limits of the output voltage of the arm. Apart from this, recordings are not difficult to interpret. The apparatus can be connected to a personal computer such as the BBC microcomputer using an analogue socket (for further details see Marsden and Brylewski, 1985).

A much more laborious method is to take individual readings with the naked eye, using a transparent protractor. You merely hold the protractor between your eye and the leaf, centre the apex of the protractor on the axis of the stalk of the leaf (or the stem if a whole leaf is being measured), and turn the protractor so that the base line is aligned with the leaflets on one side of the leaf (or the stem). The angle of the opposing leaflets (or leaf stalk) is then recorded (Stong, 1961).

Aspirin

According to folklore, cut flowers can be kept fresh by feeding them an aspirin in their water. One way of testing this idea is to place flowering shoots of your choice into tap water and compare them with flowering shoots kept in water containing half, one, five or ten soluble aspirins. This range allows you to compare the effects of concentrations of apirin.

Count the number of days before the flowers in the treatments turn colour, wilt and die. The flowers in ordinary water are the 'control' against which the aspirin treatments are compared. From these results a graph can be plotted with the number of flowers surviving drawn on the vertical axis and the number of days of flower survival on the horizontal axis. Differences between aspirin treatments and the control indicate that aspirin does indeed have an effect on flowers, and this probably depends on the concentration of aspirin used.

Light-sensitive Flowers

Pot plants of *Gazania nigens* are placed under a light as bright as possible (neon tubes or incandescent lamps with a water filter) in an otherwise dark room. To avoid differences in temperature between light and dark periods (which would trigger temperature-sensitive movements) the plants must not stand too close to the light. Furthermore, it is best to have the room ventilated. When the flowerheads have opened completely, the light is switched off. The following 'dark' movement can only be seen with a green safety light. According to the kind of plant and the experimental conditions, the petals of the marginal flowers close together within 10–30 minutes. Now the light is switched on, and the flowers open, although slightly more slowly. The opening and closing can be repeated several times. The movement works by differences in the rate of growth of the inner and outer side of the petals.

Rhythmic Autonomic Movements of Leaves

Bean plants are a good specimen because they are easy to grow and make strong leaf movements. The movements are recorded on a cylinder rotating about once every 7 days. You can use the recording drums used in hygrometers (for measuring air moisture). The recording lever of the cylinder is attached to the joint of the leaf stalk with the leaf blade by a silk thread 30 centimetres long. It is important to mark off the days and middays on the cylinder so that the period of each leaf movement can be measured. The plants are then kept in continual darkness or continual light. When recordings have been taken for several days, the ink trace left on the recording drum shows the pattern of leaf movements.

Motorized Seeds

Cover a strip of paper thickly with portions of the twisting awn of feather grass (*Stipa*), gum these to the paper and to each other and leave the whole to dry; the structure will exhibit a close left-hand twining with the paper on the outside (from Josts' *Plant Physiology*, 1907). Also try Stamp's research paper and Peart's research paper. Also Kerner (1904): use *Erodium* as a hygrometer. Suspend *Erodium* seed over a circle drawn on paper with the pointer (sharp) end of the seed downwards. Marks are made on the circumference of the circle corresponding to the positions of the seed in extremes of damp and dry weather. Anything in between relates to these two extremes. (Robert Hooke also made a hygrometer from grass seed awns.) (Anon., 1937.)

Experiments Needing a Microscope

Stomatal Behaviour

The leaf (epidermis) of many species can easily be peeled off and looked at under a simple microscope. Two favourite specimens are the ornamental plant *Commelina communis* and the skin of onion bulbs. Remove the epidermis from the underside of the youngest fully expanded leaves of *Commelina*, or the epidermis of an onion, by bending them back until the epidermis snaps. Peel off a slither of epidermis with tweezers or forceps and place *immediately* in one of the following solutions: distilled water (the 'control', used for comparing the effects of the other treatments); tap water; dilute and concentrated salt solutions (a few grains and a pinch of salt respectively dissolved in a cup of distilled water); dilute and concentrated sugar solutions (measures as for salt solutions); a solution of one soluble aspirin in a cup of distilled water.

Leave the specimens for an hour under good light, and then mount them in their respective solutions on a microscope slide with coverslip under a light microscope at $\times 100$, $\times 200$ and $\times 400$ magnification. The different treatments will affect the stomatal pores. If the cells are too heavily pigmented green from chlorophyll, try taking specimens from the chlorophyll-free parts of variegated *Tradescantia* leaves or purple-pigmented *Tradescantia* leaves (Godsell, 1985).

Note: To make temporary slide preparations such as these take the following precautions.

The slides and coverslips should be clean.
A drop of the mounting solution should be placed in the centre of the slide.
Only a small specimen should be used.
Put the coverslip on by resting one of its edges on the mounting liquid and gently lowering the slip down – this helps to avoid trapping air bubbles.
Take care not to get any liquid on the lens of the microscope.

Stomatal Replicas

This method gives a permanent replica of stomata, but unfortunately it doesn't let you see movement in action. Paint a thin coat of clear nail varnish on the underside of a leaf. Make sure the coat is continuous or it will be difficult to peel off afterwards. Allow the varnish to dry for about 15 minutes. Then, using a scalpel or other sharp knife, lift up an edge of the coat of varnish (use a low powered microscope to help, if necessary). Take hold of the open edge with a pair of fine tweezers or forceps and peel the replica off the leaf. If the varnish stretches as you peel it, it isn't dry enough and you'll have to wait a few minutes more.

Float the peel off in a drop of water on a slide, with the replica face upwards. Drain off the surplus water. Don't use a coverslip. The contrast of the replica will improve as the water dries out, and the replica will stick to the slide. At ×400 magnification you should see the guard cells and stomatal pores in detail.

Cytoplasmic Streaming

Use flowers of spiderplant (*Tradescantia*) or *Zebrina pendula* or, if no flowers are available, young pumpkin shoots. With a pair of fine tweezers or forceps, peel off the hairs running along the flower stalks or pumpkin shoots, place onto a drop of tap water on a microscope slide, and gently slide on a coverslip. Observe at ×500 magnification. The hairs of the flower stalks of *Tradescantia* or *Zebrina* consist of a single line of barrel-shaped cells. The cytoplasm is a thin layer clinging to the inside of the cell wall, and straddles the large vacuole in the middle of the cell. The cytoplasm constantly flows around the cell at about 2–5 microns per second. For further experiments see Howells and Fell (1979) and Perley and Glass (1979).

Cytoplasmic streaming is also easily seen in the leaves of the Canadian pondweed (*Elodea canadensis*) or *Vallisneria spiralis*, both common aquarium plants. Peel off the epidermis of the leaves and observe at ×500 magnification. Streaming is particularly noticeable in the elongated cells of the midrib vein of the leaves of *Elodea* and in the bottom parts of *Vallisneria* leaves. Green chloroplasts can be seen moving with the cytoplasm.

Chloroplast Movements

Leaflets of the moss *Funaria* are good specimens because they are only one cell thick, except for their midribs, and easy to observe under the microscope. The duskweed (*Lemna*) is another good specimen. After 2 hours in darkness, chloroplasts lie along a cell's side-walls which adjoin neighbouring cells. When the microscope light is switched on the chloroplasts migrate to the cell walls lying parallel to the light.

Cell Division

Another good specimen for studying how the parts of a cell move is the alga *Spirogyra*. Its tubular cells grow in long filaments and are common in stagnant ponds. The filaments are mounted on a slide in a drop of pond water with a coverslip. Mitosis cell division is clearly visible in unstained living cells, but usually between midnight and 2 a.m.! At ×300 magnification the division first becomes apparent when the nucleus disappears, leaving a network of fine fibres (the spindle) in its place. At the same time, the prospective cross walls, which will eventually separate the cell into two halves, begin to appear in the middle of the cell. The spindle then elongates and the two new daughter nuclei begin to appear. The new cell wall appears at the centre, separating the two cells to which the two daughter nuclei migrate. To see mitosis, specimens need to be kept out of strong light as much as possible, and as it takes over 2 hours for cell division to progress, the specimens need to be kept in the dark between observations (Hartree, 1988).

Phototaxis

These experiments are much easier if a culture of *Chlamydomonas*, *Volvox*, *Euglena* or some other single species of swimming alga is used. One of the simplest ways of showing phototaxis is simply to shine a beam of light onto a bottle of the algae. The millions of congregating cells turn the water green, and the effect can be made more distinct by covering a bottle of the culture with black paper containing a single opening ¼ inch in diameter. When the culture is exposed to light, the algae swim to the opening in less than an hour, and when the mask is completely removed the green 'window' remains behind.

To show what influence the colour of light has on the swimming response, a black paper sleeve is made with windows covered by transparent plastic of various colours, behaving as crude light filters. A series of ¼ inch holes ⅜ inch apart are punched in the paper with a paper punch. A small strip of clear or coloured cellophane is placed over each hole and secured with tape. As different filters also let through different *intensities* of light, this must be taken into account by measuring each filter with a photographic lightmeter. Light is then shone on the black-paper-covered culture bottle. The results should show that white light gives the quickest response, followed by blue, green, yellow and red. The number of algae congregating at each window also declines in the same sequence (Stong, 1964).

Chemotropism

Different species of pollen need different amounts of sugar to germinate, so to find the best level for the pollen you have chosen, sow pollen grains in various concentrations of sugar solution on a cover glass in a moist chamber in a warm dark place. The tubes appear within a few hours. Observe under a microscope, and note the marked streaming of the cytoplasm.

Place a drop of liquid containing pollen tubes onto an agar jelly covered with a film of tap water in an enclosed dish. Place a drop of sugar solution onto one part of the agar, and observe in which direction the pollen tubes grow. Pollen tubes are attracted to sugar. Try experimenting with different test substances to see which ones the pollen tubes are attracted to or repelled from: extracts made from crushed stigma-styles, calcium chloride, acids, alkalis.

Slime Moulds

The plasmodial slime moulds are fairly easy to collect and preserve. They are predominantly autumnal woodland organisms, most often found on rotten wood or leaf litter. Some species prefer wood or coniferous trees and are becoming more common as new forestry plantations mature. Others are found in deep litter under holly bushes, bramble clumps, ferns or stands of rosebay willows.

Slime moulds can be kept on a porridge medium, made with 8 grams of rolled oats, 1 gram of agar jelly and 100 millilitres of water. The ingredients are best autoclaved (subjected to steam and pressure treatment to prevent contamination). Sterile techniques are ideal, again to prevent the medium from being spoiled by other fungi or bacteria contaminants (Ashworth and Dee, 1975).

A culture chamber can be built from an inverted saucer or Petri dish within a larger covered container, such as a casserole dish or a pair of facing

EXPERIMENTS

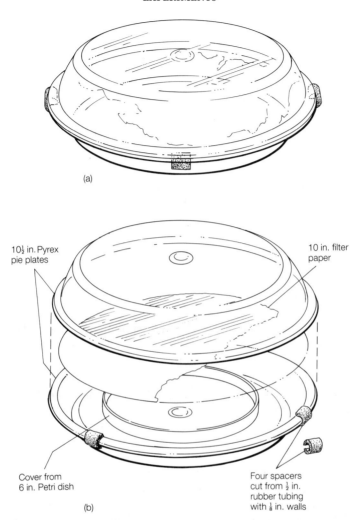

FIGURE 14.3 Slime moulds can be cultivated using a petri dish covered by a larger container and kept moist with sheets of filter paper, blotting paper or towelling paper. The slime plasmodia can be fed moistened flakes of uncooked traditional porridge oats. (a) Arrangement of an improvised chamber for cultivating slime mould; (b) details of the culture chamber.

Pyrex pie plates, so that the smaller dishes serve as a platform (figure 14.3). Lay a sheet of filter paper, blotting paper or paper towelling across the platform, allowing it to droop to the bottom of the dish. Wet the paper thoroughly with distilled water, pouring off any excess. From time to time add enough water to prevent drying.

Place a small fragment of slime mould sclerotium (the dried stage of the plasmodium life cycle) on the platform. Wet the specimen in a drop of water, and within a few hours the organism will awake from its 'hibernation'. When the plasmodium has emerged and begun to seek food, give it a flake of uncooked traditional porridge oats moistened in a drop of distilled water. As the plasmodium grows, feed it more moist oat flakes along its growing front. Keep the dish covered at room temperature without direct sunlight. To keep the culture clean, transfer the organisms to a fresh sheet of paper weekly, avoid overfeeding and remove abandoned food.

To see the pulsating cytoplasmic streaming inside the plasmodium, transfer several oat flakes that are being eaten by the slime mould to a Petri dish containing a thin layer of non-nutrient 1.5 per cent agar jelly, maintained like the paper culture but without adding water. Mount it on a miscoscope at ×50 magnification. Observe the effects of cutting one of the veins when the circulation is active; scrape a piece of plasmodium onto a fresh medium and watch it reconstitute itself. Place a drop of vinegar near the plasmodium and note the reaction.

Usually cultures can be kept in the plasmodial stage for some weeks by giving them sufficient food and water. If the food is removed, the organism will transform itself into the fruiting (sporangial) stage. The entire plasmodium becomes more orange than yellow, and makes blobs on stalks. Each of these airborne bodies develops thousands of microscopic amoeboid and then flagellated swarm cells (which can be seen at ×500 magnification). The whole transformation takes about 3 days, but starts suddenly, usually at night (Stong, 1966).

Carnivorous Fungi

Carnivorous fungi are particularly common in soils rich in organic matter. To isolate them, about 0.5–1 gram of farmyard soil or compost is sprinkled onto an agar or jelly surface. Eelworms (nematodes) are usually present. After about 5–7 days, the plates should be looked at under a microscope for captured nematodes (Barron, 1977; Nordbring-Hertz, 1984).

How to Grow Some Mobile Plants

Aeschynomene

Form: Herbs and shrubs. Not often cultivated in temperate countries.
Availability: Unknown.
Conditions: Needs greenhouse for warmth.
Soil: Compost of good rich sandy loam.
Propagation: Raise from seed in good heat, or from cuttings in sand in a closed frame with brisk heat.

A. aspersa Herbaceous perennial, 6–8 ft (1.8–2.4 m) tall. Native of East Indies.

A. indica Native of East Indies.

A. sensitiva Shrub 3–6 ft (0.9–1.8 m) tall. Native of Jamaica.

Aldrovanda vesiculosa

Distribution: Grows in still water in Southern Europe through to Japan, Australia, Africa.
Form: Rootless stem.
Availability: Specialist nurseries (see Specialist carnivorous plant nurseries).
Conditions: Will not tolerate direct sunlight. Grow in an aquarium on a 4 cm layer of peat with water containing water fleas. Tolerates quite low temperatures during winter, but grows best at 60–70 °F (16–21 °C).
Propagation: Cut plants longer than 1 inch (2.5 cm) into two pieces.

Apocynum androsemifolium

Distribution: Common in many parts of North America, but not cultivated.
Form: Perennial, growing 1–2 ft (0.3–0.6 m) tall.
Conditions: Easily grown in peaty soil, as for azaleas.
Propagation: By seed or suckers or by division in spring as new growth begins.

Arctotis

Distribution: Mostly South African species.
Availability: Seeds fairly easy to obtain.
Cultivation: Sunny dry position outside during summer, but greenhouse in winter. Good for growing outdoors when the summers are not too bad – good flower garden annual. Flowers close up, though, at night or on cloudy days.
Compost: Loam and leaf-soil kept open by sand.
Propagation: Easily grown from seed. Sowing indoors at 60–70 °F (16–21 °C) is satisfactory. Sow outdoors in spring after the threat of frost has gone. Thin to 8–12 inches (20–30 cm) apart. Greenhouse sowing: keep night temperature 50–60 °F (10–16 °C) and day temperature a few degrees higher. Then transplant to a garden cold frame or sheltered area. Avoid overwatering. Young plants tolerate dryish, rather than excessively wet, conditions. Cuttings of perennials is easily done at any time of the year; the shoots should be placed in a very sandy compost in warmth but uncovered (otherwise they damp-off).

Bauhinia

Distribution: Tropics and subtropics.
Form: Evergreen shrubs, trees, many of them climbers. Explosive seed capsules.
Availability: Specialist nurseries, e.g. Thompson & Morgan (Ipswich) Ltd, London Road, Ipswich, Suffolk IP2 0BA.
Cultivation: Most species need a heated greenhouse.
Soil: Compost of sand, loam, peat. Good drainage and moderately firm potting.
Propagation: Cuttings of wood – not too ripe or too young.
Plant in sand after removal of leaves and place under bell-glass in moist heat – will usually root.

Berberis

Distribution: Temperate.
Form: Deciduous and quite hardy.
Availability: Bushes/trees easily obtainable.
Propagation: By layering or cuttings. Seeds germinate freely.

Bignonia

Distribution: Mainly sub-tropical or tropical trees, shrubs, herbs.
Availability: Specialist nurseries, e.g. Sherrards Garden Centre Ltd, Wantage Road, Donnington, Newbury, Berks RG16 9BE.
Conditions: Moist, warm; not difficult to grow. Grow in the border of a greenhouse.
Soil: Compost of 2 parts fibrous loam, 1 part rough peat, 1 part leaf mould plus some sharp sand. Good drainage.
Propagation: Easily by seed or by cuttings of two or three joints from good strong shoots in spring. Place cuttings in well-drained pots of sandy soil under a glass, pot when rooted, and grow until large enough to be planted out.

Biophytum sensitivum

Distribution: Tropical perennial herbaceous plants.
Availability: Common in tropics; seeds not easily obtainable for cultivation.
Form: Squat rosette of leaves.
Conditions: Needs greenhouse. Minimum temperature 60 °F (16 °C) and humid. Needs shading from strong sun. Grow in single 3 inch (8 cm) pots or three or four plants in 6 inch (15 cm) trays.
Compost: Loam and peat. Plenty of organic material in soil. Keep fairly moist, but drained to avoid stagnation.
Propagation: By seed sown in spring in sandy peaty soil at 60–70 °F (16–21 °C). Young plants are easily transplanted. Plants self-sow their seed thanks to their explosive seed dispersal.

Cassia nictitans

Distribution: North America from Maine to Georgia.
Form: Hardy annual up to 15 inches (38 cm) tall.

Availability: Unknown.
Cultivation: Loves plenty of sun.
Soil: Open sandy loam, well drained.
Propagation: By seed.

Catalpa

Distribution: North America, West Indies, eastern Asia.
Form: Deciduous and evergreen trees, usually not more than 45 ft (13.5 m) tall.

C. bignoides

Distribution: Georgia to Florida and Mississippi.
Availability: Widely available at specialist nurseries, e.g. Hillier Nurseries (Winchester) Ltd, Ampfield House, Ampfield, near Romsey, Hants SO51 9PA.
Conditions: Sunny not windy locations.
Soil: Any ordinary reasonably moist soil. Best in fertile soils, though.
Propagation: Leaf cuttings under mist in summer. Root cuttings. Graft onto seedling understocks in greenhouse in later winter.

Catasetum

Distribution: Tropical America, Brazil to West Indies.
Form: Epiphytic orchids, many producing aerial roots which need considerable humidity.
Availability: Specialist orchid nurseries.
Conditions: A good orchid species for beginners to grow. Needs heat and humidity. Expose to light in autumn. Winter temperature must not fall below 60 °F (16 °C). Thrives in well-drained pots that are not too large or hanging baskets. Minimum winter night temperature 60–65 °F (16–18 °C), plus an extra 5–15 °F (3–8 °C) during the day. During active growth must not give excessive water, and water must not lodge in the leaf bases of young shoots, otherwise they will rot. Mild fertilization helpful during the growing season.
Compost: 3 parts Osmunda fibre, 1 part sphagnum. Repotting must be done before the rooting medium becomes stale or sour.
Propagation: New pseudobulbs at full size and maturity are broken and rested for 3–4 weeks before potting.

Centaurea

Many species occur in temperate zones and the tropics. Temperate cultivation is described here.
Availability: Easily obtainable as seeds from most garden shops.
Conditions: Easily grown. Water in moderation during dry weather. Flowers in July.
Soil: Ordinary soil.
Propagation: Sow in early spring in an open border, and thin out to about 9 inches (23 cm) apart. *C. cyanus* can be sown in September for flowering in early summer.

Cereus

Distribution: Tropical America, particularly northern and eastern South America.
Form: Tree-like, shrubby or sometimes sprawling cacti.
Availability: Specialist nurseries.
Cultivation: As for most cacti. Flowers open at night.

Cyclanthera explodens (C. brachystachya)

Distribution: South America.
Form: Gourd-like climbing plant growing to 10 ft (3 m) or more.
Availability: Unknown.
Conditions: Grown like cucumbers, gourds and melons. If the growing season is long, can sow directly outdoors, but in northern latitudes sow indoors some 8 weeks before it is safe to transplant without the danger of frost. Young plants should be grown indoors at 65–70 °F (18–21 °C) in fairly humid conditions. Potted singly. Harden off before completing transplanting. Do not allow to dry out. Grow outside in warm situations from May to autumn. Climbing supports needed – trellis, canes, etc.
Soil: Rich soil; benefits from manure beneath roots. Needs plenty of water during summer, but drainage must be good.

Desmodium motorium (D. gyrans)

Distribution: Tropical Asia.
Form: Perennial shrub growing to 4 ft (1.2 m).

Availability: Specialist nurseries, e.g. M. Holtzhausen, 14 High Cross Street, St Austell, Cornwall; Thompson & Morgan (Ipswich) Ltd, London Road, Ipswich IP2 0BA.
Conditions: High temperatures and humidity. In greenhouses, minimum night temperature 70 °F (21 °C) with an extra 5–15 °F (3–8 °C) during the day. Water to keep moderately moist. When the temperature exceeds 72 °F (22 °C), and especially in sunshine, the two small side leaflets of each leaf move continuously.
Soil: Ordinary fertile potting soil.
Propagation: Treat as an annual; raise seeds sown in late winter at 70–75 °F (21–24 °C). Cuttings in spring root readily in greenhouse propagating bed with a little bottom heat. When well established, the tips of the young plants are pinched out to encourage branching.

Dionaea muscipula

Distribution: Confined to coastal areas of North and South Carolina.
Availability: Good plants obtainable from specialist nurseries (see Specialist carnivorous plant nurseries).
Conditions: Keep permanently wet in summer by standing in acidic or distilled water, and keep moist in winter. Not hardy to frost, so keep indoors or in a greenhouse in winter. Prefers heated greenhouse or south-facing window-sill.
Soil: Mixture of fine sand and peat with living sphagnum around the plant, or moss peat and sphagnum moss, either separately or as a mixture with sharp sand added.
Propagation: Seed. Leaf cuttings – leaf plus stalk – in compost. Division of corms from mature plants.

Drosera

Distribution: All over the world, with the greatest number of species in Australia.
Availability: Specialist nurseries (see Specialist carnivorous plant nurseries).
Conditions: Plenty of light. Pots standing in distilled or acidic water. Northern hemisphere species must be kept cold during winter. Most other species need well-ventilated greenhouse, but tropical species need warmth and high humidity.
Soil: Peat or mixture with sphagnum moss; good drainage.
Propagation: Seeds sown on compost surface. *D. binata* and *D. hamiltonii* by root cuttings. Leaf cuttings placed on level compost and partly covered

by peat. Pygmy sundews propagated by gemmae (tiny reproductive scales) sown like seed on compost surface.

Dryas octopetala

Distribution: America, Europe, Asia. Alpine plant.
Availability: Widely available at specialist nurseries, e.g. Potterton & Martin, The Cottage Nursery, Moortown Road, Nettleton, Nr Caistor, Lincs LN7 6HX.
Conditions: Among the easiest alpines to grow; particularly suitable for rock gardens. Succeeds well even in low-lying areas, but prefers cool summers. Full sun preferred.
Soil: Well drained. Some limestone helpful. Each spring, a top dressing of sandy soil and some crushed limestone plus peat moss mixed in May can be worked in around the plants.
Propagation: Easily done by seed, summer cuttings, layering.

Ecballium elaterium

Distribution: Middle and eastern Mediterranean.
Form: Climbing perennial. Related to cucumbers, gourds, melons.
Availability: Unknown.
Conditions: As for *Cyclanthera explodens*.

Echinocereus

Distribution: Southwest United States and Mexico.
Form: Cacti.
Availability: Specialist nurseries, e.g. Holly Gate Nurseries Ltd, Billingshurst Lane, Ashington, West Sussex RH20 3BA.
Conditions: Basically the same as desert cacti. Water moderately spring to autumn, but not in winter. Keep cool in winter with night temperatures 45–50 °F (7–10 °C) ample and an extra 5–15 °F (3–9 °C) during the day. Summer conditions – sunny and temperatures as high as in habitat.
Soil: Well drained, preferably highly fertile.
Propagation: Seeds, offsets, cuttings, graftings.

Incarvillea

I. delavayi

Distribution: Asia.
Form: Perennial; reasonably hardy. Very showy flowers; 2–2.5 inch (5–6 cm) long corolla.
Availability: Obtainable from many garden shops.
Conditions: Warm sheltered sunny positions, with shade from fiercest sun. Must not allow excessive drying.
Soil: Porous, well drained, nourishing.
Propagation: Sowing seed best in spring. Division in spring. Flowering at 3 years old in greenhouse.

Malvastrum

Form: Shrubs.
Availability: Unknown.
Conditions: Not demanding. Sun and hot weather.
Soils: Variety of soils, well drained – not excessively dry or wet. Helped by reasonable organic content.
Propagation: Planting in spring or early autumn (in cold climates plant in summer).

Martynia

Distribution: Warm parts of North America.
Availability: Specialist nurseries.
Conditions: Sheltered positions.
Soil: Well drained porous.
Propagation: Seeds sown in warm moist conditions in greenhouse in early spring. Seedlings put out in open border in June, or grown on in pots in greenhouse.

M. annua

Distribution: Mexico, Central America, West Indies, tropical Asia.
Conditions: Sun and hot weather. Needs adequate watering. Growing conditions very much like those for cucumbers, gourds and melons.
Soil: Fertile well drained.
Propagation: By seed. Often takes 3–4 weeks for seedling to break soil

surface. Indoor growing temperature 70 °F (21 °C). Transplant into coarse fertile soil in 4 inch (10 cm) pots, and grow in sunny warm greenhouse for 7–10 days before transplanting outdoors. Then stand in cold frame or sheltered spot to harden. Space out between 1 and 1.5 ft (0.3–0.45 m) outdoors.

Masdevallia muscosa

Distribution: Cool wooded mountainous regions of Columbia.
Availability: Specialist nurseries.
Cultivation: Has a reputation for being a difficult orchid to grow, but basically needs cool greenhouse without draughts. Winter night temperature of 50 °F (10 °C), with a few extra degrees during the day. Needs high humidity and good aeration. Water generously throughout the year.
Soil: Very porous rooting medium, e.g. Osmunda fibre plus peat moss plus sphagnum moss, plus charcoal, with very sharp drainage.
Propagation: Division of plants.

Mazus

Distribution: Japan, China, Himalayas to Australasia.
Availability: Specialist nurseries, e.g. Mansfield Nurseries, Eastwood Rise, Eastwood, Essex SS9 5DA.
Cultivation: Good for rock gardens. Full sun with part shadow.
Soil: Gritty fairly nourishing soil, not overly dry.
Propagation: Easily done by seed, or division, or by cuttings.

Mesembryanthemum

Distribution: South Africa, Mediterranean, Sahara Desert, Southwest Asia, coastal South California and Baja California Mexico.
Form: Annual and non-hardy biennial, succulent herbaceous plants.

M. crystallinum

Availability: Widely available from garden shops.
Cultivation: Easy to grow. Good plant for dry sunny locations. Water only when obviously dry.
Soil: Thrives in poor soils – coarse, not excessively fertile. Good drainage.
Propagation: Seeds sown outdoors in spring or earlier indoors at 60–65 °F

(15–18 °C) about 8 weeks before young plants are to be transplanted. Sow seeds with shallow covering.

Mimosa

M. pudica

Distribution: Brazil.
Form: Sub-shrubby perennial, usually grown as an annual, growing to a few inches high.
Availability: Specialist nurseries, including Thomas Butcher Ltd, 60 Wickham Road, Shirley, Croydon, Surrey CR9 8AG.
Cultivation: Greenhouse, window-sill (but better in terrarium, covered glass, transparent plastic bell-jar or bowl). In sun or light but shade from strong sunlight in pots or beds. Indoors winter temperature 50–60 °F (10–15 °C) satisfactory. Easily damaged by draughts of cold air or too dry an atmosphere. Main pest is mealybugs. Movement occurs between 60 and 104 °F (15–40 °C) and is most rapid between 75 and 85 °F (24–29 °C).
Soil: Well-drained ordinary garden soil. Must be moist at all times, with weekly dilute liquid fertilizer application after roots fill containers.
Propagation: Easily grown from seed at 65–70 °F (18–21 °C) – cover with plastic/glass to raise humidity if necessary. Transfer seedlings to 2.5 inch (6 cm) pots, and then later to 4–5 inch (10–12 cm) pots.

M. speggazinii (M. polycarpa)

Distribution: South America.
Form: Shrub 4–10 ft (1.2–3 m) tall. Less sensitive than M. pudica.
Availability: Unknown.

Mimulus

Distribution: Widely distributed over extra-tropical regions, especially western North America.
From: Herbaceous annual, perennial.
Availability: Widely available from garden shops.
Conditions: Very easily cultivated. Similar growing conditions to cucumbers and melons.
Soil: Fertile well drained.
Propagation: Very easily raised from seed, shallow sowing. Sow in spring in greenhouse in trays or small pots, and plant out when no frost.

M. lutea

Distribution: Brazil.
Form: 2 inches (5 cm) tall; flowers in August.

Molinia coerula

Distribution: Europe, wet moorlands.
Form: Perennial tufted grass, 1–4 ft (0.3–1.2 m) tall.
Availability: Specialist nurseries, e.g. F.G. Barcock & Co., Garden House Farm, Drinkstone, Suffolk IP30 9TN.
Soil: Moist acid soil or waterside.
Propagation: Easily propagated by division in spring or early autumn by seed.

Mormodes buccinator

Distribution: Mexico and Venezuela.
Cultivation: As for *Catasetum*.
Availability: Unknown.

Neptunia

Distribution: Chiefly tropics, subtropics of Americas, Asia, Australia, Africa.

N. prostrata (N. oleracea)

Form: Erect, sprawling or prostrate stems.
Availability: Unknown.
Cultivation: Not very easy to grow. Needs a humid tropical environment. Grows best in waterside soil where its stems can float out over the water surface in a sunny location. Can grow in heated aquarium.
Propagation: By seed.

N. plena

Cultivation and availability: As for *N. prostrata*.

Notocactus

Distribution: North America.
Form: Large cacti.
Availability: Specialist cacti nurseries.
Cultivation: Good for rock gardens. Warm dry conditions. Good for greenhouses or window-sills.
Soil: Well-drained slightly acid soils.
Propagation: Readily propagated by offsets, or seeds.

Opuntia

Distribution: Americas, from Utah and Nebraska to Patagonia.
Form: Large upright cacti.
Availability: Specialist cacti nurseries.
Cultivation: Easily grown. Few species need heat.
Soil: Well-drained soil with plenty of lime.
Propagation: Easy from stems.
Warning: The tiny hairs on the stem easily detach and lodge in your skin – they are difficult to remove and, although not poisonous, cause irritation.

Oxalis

Cultivation: Easy to grow. Shade from hot sun and heat.
Soil: Gritty, well drained. High proportion of leaf mould or other organic matter.

O. acetosella

Distribution: North temperate woods.
Form: Perennial, about 3 inches (7 cm) high. Flowers in spring.
Availability: Specialist nurseries, e.g. Sifelle Nursery, The Walled Garden, Newick Park, Newick, Sussex.
Cultivation: Dappled shade or overhead shade needed in hot sun.
Soil: Needs abundant organic matter and moderate moisture.
Propagation: Easily from offset bulbs/tubers.

Portulaca

P. grandiflora

Distribution: Brazil.
Form: Annual, 6 inches (15 cm) tall.
Availability: Widely available from most garden shops.
Cultivation: Easy to grow in borders, banks, rock gardens, window boxes. Full sun.
Soil: Porous well-drained soil, not necessarily fertile.

Pterostylis

Distribution: Mostly Australasian.
Availability: Unknown.
Form: Perennial.
Cultivation: Most species seem to like partly shaded positions.
Soil: Pots should be a third filled with broken crocks for good drainage.

Rehmannia

Distribution: China, Japan.
Form: Perennial herbs with very showy flowers, about 3–4 ft (0.9–1.2 m) tall.
Availability: Widely available from specialist nurseries, e.g. The Old Manor Nursery, Twyning, Gloucestershire GL20 GDB.
Cultivation: Can withstand considerable cold, but not hard freezing. Good for growing in pots in greenhouses.
Soil: Fertile, well drained, but not dry.
Propagation: From seed sown about midsummer. Root cuttings. Blooms freely the following year.

Schrankia

Distribution: Warmer parts of America.
Form: Perennial herbs, sub-shrubs.
Availability: Unknown.
Soil: Mixture of loam, peat, sand.
Propagation: Young cuttings in sand under bell-glass in heat. Separation of root tubers.

HOW TO GROW SOME MOBILE PLANTS

S. uncinata (sensitive briar)

Distribution: United States, from Illinois, South Dakota to Arkansas, Texas.
Form: Prostrate/arching branched stems, about 2 ft (0.6 m) tall.
Availability: Unknown.
Cultivation: Full sun.
Soil: Well drained, not excessively fertile.

Smithia

Distribution: Tropical Asia, Africa.
Form: Herbs, sub-shrubs, shrubs.
Availability and Cultivation: Unknown.

Sparmannia africana

Distribution: South Africa.
Form: Herbaceous, 10–20 ft (3–6 m) tall.
Availability: Widely available from specialist nurseries, e.g. Oak Cottage Herb Farm, Nesscliffe, near Shrewsbury, Shropshire SY4 1DB.
Cultivation: Needs full sun and sufficient water to stop leaves wilting. Greenhouses – winter night temperature 50 °F (10 °C) is adequate (too high is detrimental), with day temperatures 5–10 °F (3–6 °C) higher. In summer, keep greenhouses cool and airy. Regular applications of dilute liquid fertilizer throughout growing season.
Soil: Ordinary well-drained garden soil. If growing in containers, coarse loamy nourishing earth, as for chrysanthemums and geraniums.
Propagation: From seed. Cuttings root readily – take in spring in 7 or 8 inch (17 or 20 cm) pots, for blooming the following winter.

Spathodea

Distribution: Tropical Africa.
Form: Trees.
Cultivation and Availability: Unknown.

Stylidium

Distribution: Australasia, Southeast Asia.
Form: Small annuals, herbaceous perennials and sub-shrubs.

Availability: Specialist or Australian nurseries, e.g. Nindethana Seed Service, Narrikup 6326, Western Australia.
Cultivation: Occasionally cultivated in greenhouses and outdoors in Californian-type climates. Sunny locations. Greenhouse winter night temperature about 50 °F (10 °C), and extra 5–15 °F (3–8 °C) during day.
Soil: Well-drained sandy peaty soil. Keep moderately moist.
Propagation: Readily from seed, less readily from cuttings.

Tecoma

T. stans

Distribution: South Florida, West Indies, South America. Grows on sandy high pine land.
Form: Shrubs, small trees up to 25 ft (7.5 m) tall.
Availability: Specialist nurseries, e.g. Thompson & Morgan (Ipswich) Ltd, London Road, Ipswich, Suffolk IP2 0BA.
Conditions: Sunny positions.
Soil: Fertile well drained.

Utricularia

Distribution: Worldwide. Boggy ponds, peat and sphagnum slurries.
Form: Most species are water plants; others are terrestrial; some species are epiphytes growing on trees in tropical rainforests.
Availability: Specialist nurseries (see Specialist carnivorous plant nurseries).
Conditions: Terrestrial bladderworts need sunlight with shading from direct sunlight in summer. Winter minimum temperature of 40 °F (4 °C). Moderate light, with water either just below or just above the soil surface. Aquatic bladderworts – some tolerate freezing conditions. Grow in rainwater with a bed of moss peat covered by lime-free gravel. Epiphytic bladderworts can grow in an unheated terrarium/greenhouse with high humidity and must be well shaded.
Soil: Terrestrial bladderworts need acid soil – equal parts moss peat and sand kept very wet all year. Epiphytic bladderworts need bark, moss peat.
Propagation: Cuttings taken in summer. Seed.

GLOSSARY

Action potential: electric signal which causes an action to happen. The signal is distinguished by its obeying the 'all-or-nothing' law – no matter how strong or weak the stimulus, the action potential is the same. This means that an action potential can only be fired after a critical level of excitation has been reached. The waveform of an action potential usually has a sharp spike followed by a more gradual return to the original 'resting' potential. During and immediately after the passage of an action potential a cell cannot fire another potential – this is called the 'refractory' or rest period.

Electrical current: as referred to in this book, a flow of ions carrying either positive or negative charges. Current can also be carried by electrons, but that is not dealt with here.

Excitation: responsiveness to changes in the environment by complex, usually adaptive, activity.

Hormone: any chemical that is made in one part of a cell/tissue/plant and is targeted on another part, where it has a marked effect on plant behaviour, growth or development at very low concentrations.

Ion: any electrically charged molecule or atom formed by the loss or gain of an electron(s).

Move: in this book refers to *independent* movement: 'to cause to change place or posture; to set in motion; to impel; to excite to action'.

Receptor potential: changes in electricity that convert stimuli into electric code. Given sufficient stimulation, receptor potentials trigger action potentials. They do not obey the 'all-or-nothing' law, and do not spread beyond the stimulated cell.

Taste is the detection and identification by the sensory system of dissolved chemicals in contact with the organism.

Touch is the ability to perceive mechanical stimulation of the organism.

Transpiration stream: the flow of water through a plant from the root tips to the stomatal pores of the shoot.

Vision is the reception by means of light-sensitive tissue of information conveyed by light rays.

Parts of the Plant

Anther: the part of the stamen that holds the pollen, in sacs.

Carpel: the female sex organ of flowers. It usually consists of an ovary, style and stigma.

Epidermis: the surface 'skin', usually consisting of a single layer of cells.

Flagella cilia: threadlike hairs on the surface of various bacteria, fungi, plants and animal cells, that beat in a rhythmic pulsation. In many cases this moves the cell along.

Gamete: a male or female sex cell.

Motor tissue: tissue capable of moving repeatedly.

Ovary: the part of the carpel that contains the ovules. The ovary is usually swollen and hollow and may hold one or more ovules in its cavities.

Ovule: the female sex cell (gamete) and its protective and nutritious tissue, which develops into the seed after fertilization. The ovule is held by a placenta to the ovary.

Parenchyma: relatively unspecialized tissue. Parenchyma cells are often present in great numbers forming a 'packing tissue' in which the other tissues are embedded. However, the xylem and phloem each have their own particular sort of parenchyma wrapped around them, and these cells often help transport chemicals and electrical signals.

Phloem: the tissue that mainly carries sugars, most hormones, and other organic material.

Pulvinus: a swollen motor organ making a hinge between a leaf blade and its stalk, or a leaf stalk and its stem. When a pulvinus moves it carries the leaf/stalk with it.

GLOSSARY

Stamen: the male sex organ of flowers.

Stamen filament: the stalk that holds the anthers.

Stomata: pores in the epidermis of the plant shoot, particularly in the leaves, which open and close using a pair of mobile guard cells and associated companion cells and allow gases to be exchanged and water to evaporate.

Stigma: the reception area of the carpel where pollen is collected before sprouting a pollen tube which grows down through the style into the ovary.

Style: the conduit for carrying pollen tubes from stigma to ovary.

Vascular tissue: the phloem and xylem.

Xylem: the tissue that mainly carries water and dissolved minerals.

Specialist Carnivorous Plant Nurseries and Societies

Societies

The Carnivorous Plant Society
Dudley and Margaret Watts
174 Baldwins Lane
Croxley Green
Herts WD3 3LQ

Specialist Carnivorous Plant Nurseries

British

Cyril Brown
65 Highfield Crescent
Hornchurch
Essex RM12 6PX

Heldon Nurseries
Asbourne Road
Spath
Uttoxeter ST14 5AD

Chiltern Seeds
Bortree Stile
Ulverston
Cumbria LA12 7PB

W.T. Neale & Co. Ltd
B.M. & S. Lamb
16/18 Franklin Road
Worthing
Sussex BN13 2PQ

SOCIETIES AND SPECIALIST NURSERIES

Potterton & Martin
The Nursery
Moortown Road
Nettleton
Nr Caistor
Lincs LN7 6HX

Sarracenia Nurseries
Links Side
Courtland Avenue
Mill Hill
London NW7

Southwest Seeds
200 Spring Road
Kempston
Bedford MK42 8ND

(Also specialists in cacti
and succulent plant seeds)

Rest of Europe

Fleischfressendepflanzen
Harald Weiner
Kaiserstrasse 74
3250 Hameln 1
Germany

Thysanotus Seed Malorder
Postfach 44–8109
2800 Bremen 44
Germany

Marcel Lecoufle
5 rue de Paris
94470 Boisse St. Leger
France

Australia

Carnivor and Insectivor Plants
78–80 Deering Street
P.O. Box 78
Diamond Creek
Victoria 3089

Carnivorous Gardens
P.O. Box 318
Acacia Ridge 4110
Brisbane
Queensland

Carnivorous Supplies
P.O. Box 179
Albion Park
N.S.W. 2527

Allen Lowrie
6 Glenn Place
Duncraig 6023
Western Australia

SOCIETIES AND SPECIALIST NURSERIES

Millingimbi Nursery
World of C.P.
P.O. Box 5
Seaforth
N.S.W. 2092

Japan

Hinode-Kadan Nursery
2735 Nakanogo
Hacijyot
Hacijyot Island
Tokyo 100–16

USA

Carolina Exotic Gardens
P.O. Box 1492
Greenville
NC 27834

Country Hills Greenhouse
Route 2
Corning
OH 43730

Orgel's Orchids
Route 2, Box 90
Miami
FL 33187

Plant Shops Botanical
18007 Topham
Reseda
CA 91335

Chatham Botanical
P.O. Box 691
Carrboro
NC 27510

Lee's Botanical Gardens
12731 Southwest 14th St.
Miami
FL 33184

Peter Pauls Nursery
Canandaigua
NY 14424

Mobile Plant Statistics

The Speed of Various Touch-sensitive Plants

(Modified from Jaffe, 1985)

Phenomenon	Plant	Response	Recovery
Leaf movements	*Mimosa pudica* (Sensitive Plant)	3 s	16 min
	Biophytum sensitivum	1 s	3 min
	Neptunia olearacea	3 min	1 h
	Dionaea muscipula (Venus's flytrap)	1 s	>24 h
	Aldrovanda (waterwheel plant)	0.04 s	1 h
	Drosera (sundew)	3 min	?
Stamens	*Portulaca grandiflora*	5 s	6 h
Tendrils	*Pisum sativum* (pea)	32 min	4 h
	Cucumis sativum	1.5 h	5 h
Carnivorous fungus	*Arthrobtrys oligospora*	0.1 s	–
Bending growth	*Avena sativa* (oat)	<5 min	–
	Bean rust fungus	< 24 h	–
Thickening of stems	*Bryonia dioica* (white bryony)	15 h	3 days
	Phaseolus vulgaris (bean)	6 h	3 days
	Lycopersicon esculentum (tomato)	15 h	3 days

Range of Response Thresholds

(From Shropshire, 1979)

Sense	Man	Receptor	Plants and micro-organisms
Vision	100 quanta to the eye or 10 quanta absorbed by rods (McBurney and Collings, 1977) 1 quantum, dark-adapted retinal rod (Hecht et al., 1942)	Photo-receptor	0.19 erg/cm^2 (458 nm) *Avena* coleoptiles 4.4 × 10^{10} quanta/cm^2 (phototropism) (Blaauw and Blaauw-Jansen, 1970) 500 quanta absorbed (440 nm) *Phycomyces* sporangiophore (phototropism) (Bergman et al., 1969) 7 quanta absorbed (660 nm) seed of *Lactuca sativa* (seed germination) (Blaauw et al., 1976)
Hearing	0.0002 dyn/cm^2 (McBurney and Collings, 1977) An order of magnitude above Brownian movements in fluid molecules of endolymph bathing the inner ear (Davis, 1959)	Audio-receptor	None demonstrated at present
Taste	0.0001 M NaCl in H$_2$O (McBurney and Collings, 1977)	Chemo-receptor	6 × 10^{-8} M L-aspartate (*E. coli*) 2.5 × 10^{-4} M valine-chemotaxis (Adler and Wung-Wai, 1974)
Smell	1 × 10^{-13} M 2-furaldehyde (Beidler, 1953) 1 × 10^{-12} M ethyl mercaptan (McBurney and Collings, 1977)		4.4 × 10^{-7} methylene (*Phycomyces*) growth response (Russo, 1977) 2 × 10^{-12} M antheridiol (*Achlya*) initiator of sexual reproduction (Barksdale, 1969)

MOBILE PLANT STATISTICS

Smell			0.4%/cm (*Lupinus*) gradient for hydrotropism of roots (Ball, 1969)
Touch	10 mg force (facial) (McBurney and Collings, 1977)	Mechano-receptor	0.5 mg load per sporangiophore (*Phycomyces*) growth response (Dennison and Roth, 1967)
			< 1 mg per tendril for a few seconds (pea tendrils) (Meyer and Anderson, 1952)
			0.32 mg (*Drosera*) curving of tentacles (Darwin, 1896)
Temperature	0.001 °C/s cutaneous temperature change 1.5×10^{-4} g cal/s cm^2 (Hardy and Oppel, 1937)	Thermo-receptor	0.04 °C/cm half-maximal thermotactic response to gradient (*Dictyostelium discoideum*)
			0.0004 °C across average pseudoplasmodium (Poff and Skokut, 1977)
			Δ 0.5 °C (*Crocus*) flower opening and closing (Bale, 1969)
Gravity and acceleration (motion)	0.5°/s^2 angular acceleration 0.04 cm/s^2 tangential acceleration at the labyrinth (Gualtierotti, 1971)	Gravi-receptor	240 gravity/s (oat coleoptiles) (Johnson, 1965)
			0.0014 gravity (68 h) $\equiv$ 340 gravity/s (oat coleoptiles) (Shen-Miller, 1970)
			2–10 dyn/cm^2 force of amyloplast on a membrane in a cell (*Helianthus*) (Audus, 1975)

Swimming Speed of some Algae

(From Sommer, 1988)

Species	Swimming velocity (microns per second)	Diameter (microns)
Ceratium tripos	82	250
Gonyaulax polyedra	58	250
Ceratium fusus	50	250
Prorocentrum mariae-lebouriae	42	83
Gyrodinium	7.8	230
Ochromonas mimima	3.6	75

Swimming Speed of some Bacteria

	Velocity (microns per second)	Cell lengths (per second)
Thiospirillium jenemie	87	2
Pseudomonas aeruginosa	56	37
Escherichia coli	17	8
Scurainia urae	28	7

The fastest animal (cheetah) moves at about three body-lengths per second.

Rates of Chloroplast Movements, Reacting to Strong Light

(From Zurzycki, 1959)

Species	Light intensity (lux)	Average speed for several cells (microns per second)	Maximum speed (microns per second)
Arabis arenosa	15,000	0.417	1.6
Funaria hygrometrica	25,000	0.246	1.5
Lemna triscula	10,000	5.41	26.4
Elodea densa	800	25.02	36.0

Action Potentials of Various Plants and Animals

Species	Lag time before start of action potential (seconds)	Speed of action (centimetres per second)
Nitella	1.2	4.0
Mimosa pudica leaf stalk	1.0	4.4
Berberis stamen	0.05	
Dionaea flytrap	0.4	20.0
Animals:		
Anodonta junction nerve	0.1	4.6
Octopus mantle nerve	0.01	300
Eledone moschata mantle nerve	0.003	452
Vertebrate nerve	0.000,2	100,000

Species	Rest period between firings of action potentials (seconds)
Nitella	4–40
Sparmannia stamen	30–60
Dionaea flytrap	0.6
Rana esculenta (frog) rectum	0.05
Vertebrate nerve	0.000,5

Taxonomy of Mobile Plants

Reversible Touch-sensitive Plant Movements

Leaf Movements (Protection from Animals or Rain)
Acacia lopantha
Aeschynomene indica
Biophytum sensitivum, B. reinwardtii, B. apodiscias, B. dendroides
Cassia nictitans
Machaerium arboreum
Mimosa pudica, M. invisa, M. speggazinii, M. casta, M. dormiens, M. humilis
Neptunia plena, N. olearacea
Oxalis acetosella, O. sensitiva, O. dendroides, O. hedera
Smithia sensitiva
Schrankia microphylla

Stamen Movements (Cross-pollination)
Apocynum androsemifolium
Berberis vulgaris and other Berberis species
Cereus speciosissimus and other Cereus species
Centaurea arten and other Centaurea species
Cichorium intybus
Cynara scolymus and other Cynara species
Echinocereus species
Hemianthemum vulgare
Mahonia aquifolium and other Mahonia species

Mesembryanthemum species
Notocactus species
Opuntia vulgaris and other *Opuntia* species
Portulaca grandiflora and other *Portulaca* species
Sparmannia africana
Telekia speciosa

Style Movements (Cross-pollination)

Arctotis aspera and other *Arctotis* species
Glossostigma elatinoides
Goldfussia anisophylla

Stigma Movements (Cross-pollination)

Bignonia species
Catalpa species
Crescentia cujete
Diplacus glutinosus
Incarvillea delavayi and other *Incarvillea* species
Martynia lutea and other *Martynia* species
Mimulus luteus and other *Mimulus* species
Rehmannia species
Tecoma stans and other *Tecoma* species
Torrenia fournieri
Utricularia vulgaris and other *Utricularia* species

Combined Stamen–Style (Trigger) Movements (Cross-pollination)

Stylidium species

Orchid Corolla Movements (Cross-pollination)

Caleana
Drakaea
Masdevallaia muscosa
Pterostylis species
Porroglossum echidnum and other *Porroglossum* species

Other Corolla Movements (Purpose Unknown)

Gentiana asclepiadea, G. bavarica, G. nivalis, G. pneumonanthe,
G. quadrifaria, G. utriculosa, G. verna

Carnivorous Traps (Catching Animal Prey)

Aldrovanda vesiculosa and other Aldrovanda species
Drosera rotundifolia and many other Drosera species
Dionaea muscipula
Utricularia vulgaris and other Utricularia species

Examples of Irreversible Touch-Sensitive Movements in Plants

Dehiscence of Fruits

Cyclanthera explodens
Ecballium elaterium
Cardamine hirsuta
Impatiens noli-me-tangere and other Impatiens species
Ulex species

Stamen Movements

Kalmia latifolia
Parietaria officinalis
Urtica dioica

Opening of Flowers

Genista tinctaria
Sarothamnus species
Stanhopea oculata

Pollinia Disposal

Catasetum macrocarpum and many other Catasetum species
Mormodes buccinor

Fast Autonomic Leaf Movements

Desmodium gyrans
Eleiotis sororia
Hedysarum gyrans

Relationships of Touch-sensitive Species

DICOTYLEDONS
  MAGNOLIIDAE
    Ranunculales
      Berberidaceae
        Berberis, Mahonia stamen movements

  CARYOPHYLLIDAE
    Caryophyllales
      Cactaceae
        Cereus, Echinocereus, Mamillaria stamen movements

    Aizoaceae
      Mesembryanthemum stamen movements

    Portulaceae
      Portulaca stamen movements

  DILLENIIDAE
    Violales
      Cistaceae
        Cistus stamen movements

  ROSIDAE
    Rosales
      Droseraceae
        Drosera, Dionaea, Aldrovanda carnivorous traps

    Fabales
      Leguminosae
        Acacia, Mimosa, Neptunia, Cassia leaf movements

    Geraniales
      Oxalidaceae
        Oxalis, Biophytum leaf movements

TAXONOMY OF MOBILE PLANTS

Evolutionary tree diagram showing the relationships between the orders of the flowering plants and their relative degrees of specialization. Plant orders that are underlined contain species showing touch-sensitive movement. Note how widespread they are throughout the flowering plants, including primitive orders (nearest the centre) and advanced orders (furthest from the centre). Thus touch-sensitive movements are truly widespread.
(After Heywood, 1974; courtesy of Oxford University Press)

ASTERIDAE
  Gentianales
    Gentianaceae
      Gentiana corolla movements
    Apocynaceae
      Apocynum stamen movements

283

Scrophulariales
 Scrophulariaceae
 Bignoniaceae
  Bignonia, *Catalpa* stigma movements
 Martyniaceae
  Martynia stigma movements
 Lentibulariaceae
  Utricularia stigma and carnivorous trap movements
 Stylidiaceae
  Stylidium, *Leevenhookia* trigger movements
Asterales
 Compositae
  style movements, stigma movements, stamen movements

MONOCOTYLEDONS
 LILIIDAE
  ORCHIDALES
   Orchidaceae
    Pterostylis, labellum movements

REFERENCES

Achenbach, F. and Weisenseel, M.H. (1981) Ionic currents traverse the slime mould *Physarum. Cell Biology International Reports*, 5, 375–9.
Adams, A.E.M. and Pringle, J.R. (1984) Relationship of actin and tubulin distribution to bud growth in wild-type and morphogenetic-mutant *Saccharomyces cerevisiae. Journal of Cell Biology*, 98, 934–45.
Adler, J. (1969) Chemotaxis in bacteria. *Science*, 166, 1588–97.
Adler, J. and Wung-Wai, T. (1974) 'Decision'-making in bacteria: chemotactic response of *Escherichia coli* to conflicting simuli. *Science*, 184, 1292–4.
Aist, J.R. (1977) Mechanically induced wall appositions of plant cells can prevent penetration of a parasitic fungus. *Science*, 197, 568–71.
Aist, J.R. and Williams, P.H. (1971) The cytology and kinetics of cabbage root hair penetration by *Plasmodiophora brassicae. Canadian Journal of Botany*, 49, 2023–34.
Allen, R.D. (1987) The microtubule as an intracellular engine. *Scientific American*, 256, 26–33.
Anon. (1887) *The Gardeners' Chronicle*, 25 June, 836.
Anon. (1937) List of plants with self-burying seeds: Die verbreitungsokologischen Verhaltnisse der Pflanzen Palastinas. I. Die antitelechorischen Erscheinungen. *Beiheft zum Botanischen Centralblatt A*, 56, 1–155.
Applewhite, P.B. (1973) Serotonin and norepinephrine in plant tissues. *Phytochemistry*, 12, 191–2.
Applewhite, P.B. (1975) Learning in bacteria, fungi, and plants. In W.C. Corning, V.A. Dual and A.O.D. Williams (eds), *Invertebrate Learning*, vol. 3, New York: Plenum, 179–86.
Applewhite, P.B. and Gardner, F.T. (1971) Rapid leaf closure of *Mimosa* in response to light. *Nature*, 233, 279–80.
Ashworth, J.M. and Dee, J. (1975) *The Biology of Slime Moulds*, Institute of Biology's Studies in Biology no. 56, London: Edward Arnold, 63–7.

REFERENCES

Audus, L.J. (1975) Geotropism in roots. In J.G. Torrey and D.T. Clarkson (eds), *Development and Function of Roots*, London: Academic Press, ch. 16, 327–63.
Backster, C. (1968) Evidence of a primary perception in plant life. *International Journal of Parapsychology*, 10, 329–48.
Balan, J. and Gerber, N.N. (1972) Attraction and killing of the nematode *Panagrellis redivivus* by the predacious fungus *Arthrobotrys dactyloides*. *Nematologica*, 18, 163–7.
Ball, N.G. (1969) Tropic, nastic, and tactic responses, plant physiology, a treatise. In F.C. Steward (ed.), *Analysis of Growth: Behaviour of Plants and Their Organs*, vol. Va, New York: Academic Press, ch. 3.
Barksdale, A.W. (1969) Sexual hormones of *Achlya* and other fungi. *Science*, 166, 831–7.
Barnard, C. (1875) *Leçons sur les Anesthesiques et sur L'asphyxie*, Paris: Baillère.
Barnard, C. (1878) *Leçons sur les Phénomènes de la Vie Communs aux Animaux et Végétaux*, 278.
Barron, G.L. (1977) *The Nematode-Destroying Fungi*, Topics in Mycobiology, Guelph, Ontario: Canadian Biological Publications, 117–32.
Barron, G.L. (1980) A new *Haptoglossa* attacking rotifers by rapid injection of an infective sporidium. *Mycologia*, 72, 1186–94.
Barron, G.L. (1987) The gun cell of *Haptoglossa mirablilis*, *Mycologia*, 79, 877–83.
Bates, G.W. and Goldsmith, M.H.M. (1983) Rapid response of the plasma-membrane potential in oat coleoptiles to auxin and other weak acids. *Planta*, 159, 231–7.
Begg, D. and Rehbun, L.I. (1979) pH regulates the polymerization of actin in the sea urchin egg cortex. *Journal of Cell Biology*, 83, 241–8.
Beidler, L.M. (1953) Physiological problems in odour research. *Annals of the New York Academy of Sciences*, 58, 52–7.
Bergman, K., Burke, P.V., Cerdá-Olmedo, E., David, C.N., Delbrück, M., Foster, K.W., Goodell, E.W., Heisenberg, M., Meissner, G., Zalokar, M., Dennison, D.S. and Shropshire, W., Jr (1969) Phycomyces. *Bacteriology Review*, 33, 99–157.
Biddington, N.L. (1985) A review of mechanically induced stress in plants. *Scientific Horticulture*, 36, 12–20.
Biedermann, W. (1898) *Electro-physiology*, vol. II, translated by F.A. Welby, London: Macmillan.
Biswas, S. and Bose, D.M. (1972) An ATPase in sensitive plant *Mimosa pudica*. I. Purification and characterization. *Archives of Biochemistry and Biophysics*, 148, 199–207.
Blaauw, O.H. and Blaauw-Jensen, G. (1970) The phototropic responses of *Avena* coleoptiles. *Acta Botaniques Néerlandais*, 19, 755–63.
Blaauw, O.H., Blaauw-Jensen, G. and Elgersma, O. (1976) Determination of hit numbers from dose–response curves for phytochrome control of seed germination (*Lactuca sativa* cv. NORAN). *Acta Botaniques Néerlandais*, 25, 341–8.
Blackman, V.H. and Paine, S.G. (1918) Studies in the permeability of the pulvinus of *Mimosa pudica*. *Annals of Botany*, 32, 69–85.
Blatt, M.R. (1990) Potassium channel currents in intact stomatal guard cells: rapid enhancement by abscisic acid. *Planta*, 180, 445–55.

REFERENCES

Blatt, M.R., Thiel, G. and Trentham, D.R. (1990) Reversible inactivation of K^+ channels of *Vicia* stomatal guard cells following the photolysis of caged inostil 1,4,5-triphosphate. *Nature*, 346, 766–9.
Bonner, J.T. (1971) Aggregation and differentiation in the cellular slime molds. *Annual Review of Microbiology*, 25, 75–92.
Bopp, M. and Weber, W. (1981) Hormonal regulation of the leaf blade movement of *Drosera capensis*. *Physiologia Plantarum*, 53, 491–6.
Bose, J.C. (1906) *Plant Response as a Means of Physiological Investigation*, London: Longman.
Bose, J.C. (1907) *Comparative Electro-physiology*, London: Longman, Green.
Bose, J.C. (1927) *Plant Autographs and their Revelations*, New York: Macmillan.
Braam, J. and Davis, R.W. (1990) Rain-, wind- and touch-induced expression of calmodulin and calmodulin-related genes in *Arabidopsis*. *Cell*, 60, 357–69.
Bremekamp, C.E.B. (1915) Stossreizbakeit der Blumenkrone bei *Gentiana quadrifaria*. *Recueil des Travaux Botaniques Néerlandais*, 12, 26–30.
Brittain, R.T. and Collier, H.O.J. (1956) Antagonism of 5-hydroxytryptamine by dock leaf extracts. *Proceedings of the Physiological Society*, 135, 58–9.
Britz, S.J. (1979) Chloroplast and nuclear migration. In W. Haupt and M.E. Feinleib (eds), *Physiology of Movements, Encyclopedia of Plant Physiology, New Series*, vol. 7, Berlin: Springer-Verlag, 170–205.
Britz, S.J. and Briggs, W.R. (1976) Circadian rhythms of chloroplast orientation and photosynthetic ability in *Ulva*. *Plant Physiology*, 58, 22–7.
Britz, S.J. and Briggs, W.R. (1983) Rhythmic chloroplast migration in the green alga *Ulva*: dissection of movement mechanism by differential inhibitor effects. *European Journal of Cell Biology*, 31, 1–8.
Brown, R. (1874) *A Manual of Botany*, Edinburgh: Blackwood, 572.
Brown, W.H. (1916) The mechanism of movement and the duration of the effect of stimulation in the leaves of *Dionaea*. *American Journal of Botany*, 3, 68–90.
Buchen, B. and Schroder, W.H. (1986) Localization of calcium in the sensory cells of the *Dionaea* trigger hair by laser micro-mass analysis (LAMMA). In A.J. Trewavas (ed.), *Molecular and Cellular Aspects of Calcium in Plant Development*, NATO Series A: Life Sciences, vol. 104, New York and London: Plenum, 233–40.
Buchen, B., Hensel, D. and Sievers, A. (1983) Polarity in mechanoreceptor cells of trigger hairs of *Dionaea muscipula* Ellis. *Planta*, 158, 458–68.
Bünning, E. (1930) Die Reizbewegungen der Staubblatter von *Sparmannia africana*. *Protoplasma*, 11, 49–84.
Bünning, E. (1934) Elektrische Potentialänderungen an seismonastisch gereizten Staubfäden. *Planta* 22, 251–68.
Bünning, E. (1977) Fifty years of research in the wake of Wilhelm Pfeffer. *Annual Review of Plant Physiology*, 28, 1–22.
Bünning, E. and Moser, I. (1969) Interference of moonlight with the photoperiodic measurement of time by plants, and their adaptive reaction. *Proceedings of the National Academy of Sciences, USA*, 62, 1018–22.
Burdon-Sanderson, J. (1873) Note on the electrical phenomena which accompany stimulation of leaf of *Dionaea muscipula*. *Proceedings of the Royal Society, London*, 21, 495–6.

REFERENCES

Burdon-Sanderson, J. (1899) On the relation of motion in animals and plants to the electrical phenomena which are associated with it. Croonian lecture, *Proceedings of the Royal Society, London*, 65, 37–64.

Burdon-Sanderson, J. (1903) Referee's report, filed in the Linnean Society of London.

Burdon-Sanderson, J. (1911) Excitability of plants. In *Sir John Burdon-Sanderson, a Memoir . . . with a Selection from his Papers and Addresses*, Oxford: Clarendon Press, 187.

Campbell, N.A., Sitka, K.M. and Morrison, G.H. (1979) Calcium and potassium in the motor organ of the sensitive plant: localization by ion microscopy. *Science*, 204, 185–7.

Cande, W.Z. and McDonald, K.L. (1985) In vitro reactivation of anaphase spindle elongation using isolated diatom spindles. *Nature*, 316, 168–70.

de Candolle, A.P. (1832) *Physiologie Végétale*, Bechet jeune, Librairie Faculté de Médecine, Paris.

Casser, M., Hodick, D., Buchen, B. and Sievers, A. (1985) Correlation of excitability and bipolar arrangement of endoplasmic reticulum during the development of sensory cells in trigger hairs of *Dionaea muscipula* Ellis. *European Journal of Cell Biology*, 36, Supplement 7, 12.

Checcucci, A. (1976) Molecular sensory physiology of *Euglena*. *Naturwissenschaften*, 63, 412–17.

Chodat, R. and Guha, S.C. (1926) La pollinisation et les réponses électriques du pistil. *Compte Rendu des Séances de la Société des Physique et d'histoire Naturelle de Genève*, 43, 105–11.

Christalle, W. (1931) Uber die Reizerscheinungen der Narben von Mimulus luteus. *Botanisches Archiv*, 34, 115–45.

Clifford, H.T. and Monteith, G.B. (1989) A three phase seed dispersal mechanism in Australian quine bush (*Pelastostigma pubescens* Domin.). *Biotropica*, 21, 284–6.

Collier, E., Watkinson, A., Cleland, C.F. and Roth, J. (1987) Partial purification and characterisation of an insulin-like material from spinach and *Lemna gibba* G3. *Journal of Biological Chemistry*, 262, 6238–47.

Cooksey, B. and Cooksey, K.E. (1980) Calcium is necessary for motility in the diatom *Amphora coffeaeformis*. *Plant Physiology*, 65, 129–31.

Corti, B. (1774) *Osservazioni microscopiche sulla tremellae sulla circolazione del fluido in una pianta aquajuola*, Lucca: Apresso Rogghi, 66.

Couch, J.N. (1937) The formation and operation of the traps in the nematode-catching fungus, *Dactylella bembicoides* Drechsler. *Journal of the Elisha Mitchell Science Society*, 53, 301–9.

Csaba, G. (1980) Phylogeny and ontogeny of hormone receptors: the selection theory of receptor formation and hormonal imprinting. *Biological Review*, 55, 47–63.

Csaba, G. and Pal, K. (1982) Effects of insulin, triiodothyronine, and serotonin on plant seed development. *Protoplasma*, 110, 20–2.

Darwin, C. (1862) *The Various Contrivances by Which Orchids are Fertilised by Insects*, London: John Murray, ch. 7.

Darwin, C. (1865) *The Movements and Habits of Climbing Plants*, London: John Murray.

REFERENCES

Darwin, C. (1875) *Insectivorous Plants*, London: John Murray.
Darwin, C. (1896) *The Power of Movement in Plants*, New York: D. Appleton, 194.
Darwin, C. and Darwin, F. (1880) *The Power of Movement in Plants*, London: John Murray.
Davidson, J.G.N. and Barron, G.L. (1973) Nematophagous fungi: *Haptoglossa*. *Canadian Journal of Botany*, 51, 1317–23.
Davies, E. (1987) Action potentials as multifunctional signals in plants: a unifying hypothesis to explain apparently disparate wound responses. *Plant, Cell and Environment*, 10, 623–31.
Davies, E. and Schuster, A. (1981a) Intercellular communication in plants: evidence for a rapidly generated, bidirectionally transmitted wound signal. *Proceedings of the National Academy of Sciences, USA*, 78, 2422–6.
Davies, E. and Schuster, A. (1981b) Wounding, action potentials and polysome formation. *Plant Physiology*, 67, 538.
Davis, A.B. (1982) The development of anaesthesia. *American Scientist*, 70, 522–8.
Davis, H. (1959) Excitation of audiotry receptors. In J.E. Hall and H.W. Maoun (eds), *Handbook of Physiology, Neurophysiology Field*, Section 1, vol. 1, Washington: American Physiological Society, 565–84.
Dean, J.M. and Smith, A.P. (1978) Behavioral and morphological adaptation of a tropical plant to high rainfall. *Biotropica*, 10, 152–4.
De Mairan, J. (1729) Observation botanique. *Histoire de l'Academie Royale des Sciences, Paris*, 35.
Dennison, D.S. and Roth, C.C. (1967) *Phycomyces* sporangiophores: fungal stretch receptors. *Science*, 156, 1386–8.
Desbiez, M.O., Kergosien, Y., Champagnat, P. and Thellier, M. (1984) Memorization and delayed expression of regulatory messages in plants. *Planta*, 160, 392–9.
Deschamp, P.A. and Cooke, T.J. (1983) Leaf dimorphism in aquatic angiosperms: significance of turgor pressure and cell expansion. *Science*, 219, 505–7.
Dietz, E. and Uhlenbruck, G. (1989) An insulin receptor in microorganisms: fact or fiction? *Naturwissenschaften*, 76, 269–70.
Dixon, R. (1914) *Human Side of Plants*, New York: Stokes.
Dorscheid, T. and Wartenberg, A. (1966) Chlorophyll als photoreceptor bei der Schwachlichtbewegung des *Mesotaenium*-chloroplasten. *Planta*, 70, 187–92.
Duddington, C.L. (1957) *The Friendly Fungi*, London: Faber & Faber.
Dunny, G.M., Brown, B.L. and Clewell, D. (1978) Induced cell aggregation and mating in *Streptococcus faecilis*: evidence for a bacterial sex pheromone. *Proceedings of the National Academy of Sciences, USA*, 75, 3479–83.
Dziubinska, H., Trebacz, K. and Zawadzki, T. (1989) The effect of excitation on the rate of respiration in the liverwort *Conocephalum conicum*. *Physiologia Plantarum*, 75, 417–23.
Earnest, D.J. and Turek, F.W. (1983) Role for acetylcholine in mediating effects of light on reproduction. *Science*, 219, 77–9.
Eckert, R. (1965a) Bioelectric control of bioluminescence in the dinoflagellate *Noctiluca*. I. Specific nature of triggering events. *Science*, 147, 1140–2.
Eckert, R. (1965b) Bioelectric control of bioluminescence in the dinoflagellate *Noctiluca*. II. Asynchronous flash initiation by a propagated triggering potential. *Science*, 147, 1142–5.

REFERENCES

Eckert, R. (1972) Bioelectric control of ciliary activity. *Science*, 176, 473–81.
Eckert, R. and Machemer, H. (1975) Regulation of ciliary beating frequency by the surface membrane. In R. Stephens and S. Inoue (eds), *Symposium of the Society for General Physiology*, New York: Raven Press, 151–63.
Eckert, R. and Sibaoka, T. (1967) Bioelectric regulation of tentacle movement in a dinoflagellate. *Journal of Experimental Biology*, 47, 433–46.
Eckert, R., Naitoh, Y. and Machemer, H. (1976) Calcium in the bioelectric and motor functions of *Paramecium*. *Symposium for Experimental Biology*, 30, 233–55.
Edgar, L.A. and Pickett-Heaps, J.D. (1983) The mechanism of diatom locomotion. I: An ultrastructural study of the motility apparatus; II: The identification of actin. *Proceedings of the Royal Society of London B*, 218, 331–43; 345–8.
Edwards, S.E. (1980) Spore discharge in *Calymperes*. *Journal of Bryology*, 11, 95–7.
Ehleringer, J. and Forseth, I. (1980) Solar tracking by plants. *Science*, 210, 1094–8.
Eisner, T. (1981) Leaf folding in a sensitive plant: a defensive thorn-exposure mechanism? *Proceedings of the National Academy of Sciences, USA*, 78, 402–4.
Ellis, H.W. and Turner, E.R. (1978) The effect of electricity on plant growth. *Science Progress, Oxford*, 65, 395–407.
Emmelin, N. and Feldberg, W. (1947) The mechanism of the sting of the common nettle (*Urtica urens*). *Journal of Physiology*, 106, 440–55.
Emmelin, N. and Feldberg, W. (1949) Distribution of acetylcholine and histamine in nettle plants. *New Phytologist*, 48, 143–8.
Enright, J.T. (1982) Sleep movements of leaves: in defense of Darwin's interpretation. *Oecologia*, 54, 253–9.
Erickson, R. (1958) *Trigger Plants*, Perth, Australia: Paterson Brokensha.
Eschrich, W., Fromm, J. and Evert, R.F. (1988) Transmission of electric signals in sieve tubes of *Zucchini* plants. *Botanica Acta*, 101, 327–31.
Ewart, J. (1903) *On the Physics and Physiology of Protoplasmic Streaming in Plants*, Oxford: Clarendon Press.
Farmer, E. and Ryan, C.A. (1990) Interplant communication: airborne methyl jasmonate induces synthesis of proteinase inhibitors in plant leaves. *Proceedings of the National Academy of Sciences, USA*, 87, 7713–16.
Feldman, D., Do, Y., Burshell, A., Stathis, P.A. and Loose, D.S. (1982) An estrogen-binding protein and endogenous ligand in *Saccharomyces cerevisiae*: possible hormone receptor system. *Science*, 218, 297–8.
Feldman, D., Tökés, L.G., Stathis, P.A., Miller, S.C., Kurz, W. and Harvey, D. (1984) Identification of 17β-estradiol as the estrogenic substance in *Saccharomyces cerevisiae*. *Proceedings of the National Academy of Sciences, USA*, 81, 4722–6.
Fensom, D.S. (1972) A theory of translocation in phloem of *Heracleum* by contractile protein microfibrillar material. *Canadian Journal of Botany*, 50, 479–97.
Fensom, D.S. and Spanner, D.C. (1969) Electro-osmotic and biopotential measurements on phloem strands of *Nymphoides*. *Planta*, 88, 321–31.
Findlay, G.P. (1975) Anatomy and movement of the column in *Stylidium*. *Australian Journal of Plant Physiology*, 2, 597–621.

REFERENCES

Findlay, G.P. and Pallaghy, C.K. (1978) Potassium chloride in the motor tissue of *Stylidium*. *Australian Journal of Plant Physiology*, 5, 219–29.
Fineran, B.A. (1985) Glandular trichomes in *Utricularia*: a review of their structure and function. *Israel Journal of Botany*, 34, 295–330.
Fitzgerald, R.D. (1882) *Australian Orchids*, vol. 1, Sydney.
Fleurat-Lessard, P.M. (1988) Structural and ultrastructural features of cortical cells in motor organs of sensitive plants. *Biological Reviews*, 63, 1–22.
Fleurat-Lessard, P.M. and Bonnemain, J.-L. (1978) Structural and ultrastructural characteristics of the vascular apparatus of the Sensitive Plant (*Mimosa pudica* L.). *Protoplasma*, 94, 127–43.
Fleurat-Lessard, P.M. and Roblin, G. (1982) Comparative histocytology of the petiole and the main pulvinus in *Mimosa pudica* L. *Annals of Botany*, 50, 83–92.
Fleurat-Lessard, P.M. and Satter, R. (1985) Relationships between structure and motility of *Albizzia* motor organs: changes in ultrastructure of cortical cells during dark-induced closure. *Protoplasma*, 128, 72–9.
Fleurat-Lessard, P.M., Roblin, G., Bonmort, J. and Beese, C. (1988) Effects of colchicine, vinblastine, cytochalasin B and phalloidin on the seismonastic movements of *Mimosa pudica* leaf and on motor cell ultrastructure. *Journal of Experimental Botany*, 39, 209–21.
Forseth, I.N. and Ehleringer, J.R. (1983) Ecophysiology of two solar tracking desert winter annuals. IV: Effects of leaf orientation on calculated daily carbon gain and water use efficiency. *Oecologia*, 58, 10–18.
Foster, K.W. and Smyth, R.D. (1980a) The visual system of green algae: a forerunner of human vision. *Federation of American Societies for Experimental Biology, Federation Proceedings*, 39, 2137.
Foster, K.W. and Smyth, R.D. (1980b) Light antennas in phototactic algae. *Microbiological Reviews*, 44, 572–630.
Foster, K.W., Saranak, J., Patel, N., Zarilli, G., Okabe, M., Kline, T. and Nakanishi, K. (1984) A rhodopsin is the functional photoreceptor for phototaxis in the unicellular eukaryote *Chlamydomonas*. *Nature*, 311, 756–9.
Frachisse, J.-M., Desbiez, M.-O., Champagnat, P. and Thellier, M. (1985) Transmission of a traumatic signal via a wave of electric depolarization, and induction of correlations between the cotyledonary buds in *Bidens pilosus*. *Physiologia Plantarum*, 64, 48–52.
Franks, N.P. and Loeb, W.R. (1984) Do anaesthetics act by competitive binding to specific receptors? *Nature*, 310, 599–601.
Fromm, J. and Eschrich, W. (1988) Transport processes in stimulated and non-stimulated leaves of *Mimosa pudica*: I, II. *Trees*, 2, 7–17; 18–24.
Gaillochet, J., Poussel, H. and Gavaudan, P. (1975) Recherche de conditions 'standard' de narcose chez *Mimosa pudica* L. en vue de l'application à divers problèmes du mouvement foliaire. *Comptes Rendus des Séances de la Société de Biologie*, 169, 652–9.
Galston, A.W. and Slayman, C.L. (1979) The not-so-secret life of plants. *American Scientist*, 67, 337–44.
Gilroy, S., Read, N.D. and Trewavas, A.J. (1990) Elevation of cytoplasmic calcium by caged calcium or caged inositol triphosphate initiates stomatal closure. *Nature*, 346, 769–71.

REFERENCES

Giridhar, G. and Jaffe, M.J. (1988) Thigmomorphogenesis. XXIII: Promotion of foliar senescence by mechanical perturbation of *Avena sativa* and four other species. *Physiologia Plantarum*, 74, 473–80.

Godsell, P.M. (1985) Stomata. 1: Measuring stomatal opening; 2: The mechanism of stomatal opening. *The School Science Review*, 67, 74–7; 298–301.

Goebel, K. (1920) *Die Entfaltungsbewegungen der Pflanzen und deren teleogische Deutung*, Jena: Fischer.

Goeschl, J.D., Rappaport, L. and Pratt, H.K. (1966) Ethylene as a factor regulating the growth of pea epicotyls subjected to physical stress. *Plant Physiology*, 41, 877–84.

Goldsworthy, A. (1984) Measuring electric signals in plants. *Carnivorous Plant Society Journal*, 8, Spring, 8–11; Autumn, 11.

Gotto, Y. and Ueda, A.K. (1988) Microfilament bundles of F-actin in *Spirogyra* by fluorescence microscopy. *Planta*, 173, 442–6.

Gould, M. and Stephano, J.L. (1987) Electrical responses of eggs to acrosomal protein similar to those induced by sperm. *Science*, 235, 1654–6.

Groenewald, E.G. and Visser, J.H. (1978) The effect of arachidonic acid, prostaglandins and inhibitors of prostaglandin synthetase, on the flowering of excised *Pharbitis nil* shoot apices under different photoperiods. *Zeitschrift für Pflanzenphysiologie*, 88, 423–9.

Groenewald, E.G., Visser, J.H. and Grobbelaar, N. (1983) The occurrence of prostaglandin $(PG)F_2\alpha$ in *Pharbitis nil* seedlings grown under short days or long days. *South African Journal of Botany*, 2, 82.

Grove, M.D., Spencer, G.F., Rohwedder, W.K., Mandava, N.B., Worley, J.F., Warthen, J.D., Jr, Steffens, G.L., Flippen-Anderson, J.L. and Cook, J.C. (1979) Brassinolide, a plant growth-promoting steroid isolated from *Brassica napus* pollen. *Nature*, 281, 216–17.

Gualtierotti, T. (1971) The gravity sensing mechanism of the inner ear. In *Gravity and the Organism*, Chicago: University of Chicago Press, 263–81.

Gustin, M.C., Zhou, X.-L., Martinac, B. and Kung, C. (1988) A mechanosensitive ion channel in the yeast plasma membrane. *Science*, 242, 762–5.

Haberlandt, G. (1884) *Physiological Plant Anatomy*, translated by M. Drummond, London: Macmillan.

Haberlandt, G. (1914) *Physiological Plant Anatomy*, 4th edn, translated by M. Drummond, London: Macmillan, 613.

Häder, D.-P. (1977) Influence of electric fields on photophobic reactions in blue-green algae. *Archives of Microbiology*, 114, 83–6.

Häder, D.-P. (1978) Extracellular and intracellular determination of light-induced potential changes during photophobic reactions in blue-green algae. *Archives of Microbiology*, 119, 75–9.

Häder, D.-P. and Poff, K.L. (1979) Light-induced accumulations of *Dictyostelium discoideum* amoebae. *Photochemistry and Photobiology*; cited by D.P. Häder and K.L. Poff (1979) Photodispersal from light traps by amoebas of *Dictyostelium discoideum*, *Experimental Mycology*, 3, 121–31.

Hagiwara, S. and Jaffe, L.A. (1979) Electrical properties of egg cell membranes. *Annual Review of Biophysics and Bioengineering*, 8, 385–416.

REFERENCES

Hardy, J.D. and Oppel, T.W. (1937) Studies in temperature sensation. III: The sensitivity of the body to heat and the spatial summation of the end organ responses. *Journal of Clinical Investigation*, 16, 533–40.
Hartree, D. (1988) Studies of *Spirogyra* sp. *The School Science Review*, 69, 71–4.
Hasezawa, S., Nagata, T. and Syono, K. (1988) The presence of ring formed actin filaments in plant cells. *Protoplasma*, 146, 61–3.
Hatano, S. and Oosawa, F. (1966) Isolation and characterization of plasmodial actin. *Biochimica et Biophysica Acta*, 127, 488–98.
Haupt, W. (1973) Role of light in chloroplast movement. *Bioscience*, 23, 289–96.
Hazum, E., Sabatka, J.J., Kwen-Jen, C., Brent, D.A., Findlay, J.W.A. and Cuatrecasas, P. (1981) Morphine in cow and human milk: could dietary morphine constitute a ligand for specific (μ) receptors? *Science*, 213, 1010–12.
Heath, O.V.S. and Russell, J. (1954) Studies in stomatal behaviour. VI: An investigation of the light responses of wheat stomata with the attempted elimination of control by the mesophyll; I: Effects of light independent of carbon dioxide and their transmission from one part of the leaf to another. *Journal of Experimental Botany*, 5, 1–15.
Hecht, S., Shlaer, S. and Pirenne, M.H. (1942) Energy, quanta and vision. *Journal of General Physiology*, 25, 819–40.
Hedrich, R. and Neher, E. (1987) Cytoplasmic calcium regulates voltage-dependent ion channels in plant vacuoles. *Nature*, 329, 833–6.
Hejnowick, Z. (1970) Propagated disturbances of transverse potential gradient in intracellular fibrils as the source of motive forces for longitudinal transport in cells. *Protoplasma*, 71, 343–64.
Hepler, P.K. (1985) Calcium restriction prolongs metaphase in dividing *Tradescantia* stamen hairs. *Journal of Cell Biology*, 100, 1363–8.
Hepler, P.K. and Palevitz, B.A. (1974) Microtubules and microfilaments. *Annual Review of Plant Physiology*, 25, 309–62.
Hesketh, T.R., Moore, J.P., Morris, J.O., Taylor, M.V., Rogers, J., Smith, G.A. and Metcalfe, J.C. (1985) A common sequence of calcium and pH signals in the mitogenic stimulation of eukaryotic cells. *Nature*, 313, 481–4.
Heslop-Harrison, J. and Heslop-Harrison, Y. (1989) Conformation and movement of the vegetative nucleus of the angiosperm pollen tube: association with the action cytoskeleton. *Journal of Cell Science*, 93, 299–308.
Heywood, V.H. (1978) *Flowering Plants of the World*, Oxford: Oxford University Press.
Hoch, H.C., Staples, R.C., Whitehead, B., Comeau, J. and Wolf, E.D. (1987) Signaling for growth orientation and cell differentiation by surface topography in *Uromyces*. *Science*, 235, 1659–62.
Hodick, D. and Sievers, A. (1986) The influence of Ca^{2+} on the action potential in mesophyll cells of *Dionaea muscipula* Ellis. *Protoplasma*, 133, 83–4.
Hodick, D. and Sievers, A. (1988) The action potential of *Dionaea muscipula* Ellis. *Planta*, 174, 8–18.
Hodick, D. and Sievers, A. (1989) On the mechanism of trap closure of Venus flytrap (*Dionaea muscipula* Ellis). *Planta*, 179, 32–42.
Holmes, E. and Gruenberg, G. (1965) Learning in plants. *Worm Runners Digest*, 7, 9–12.

REFERENCES

Hooker, J. (1860) *Flora of Tasmania*, London: Lovell Reeve.

Hope, A.B. and Walker, N.A. (1975) *The Physiology of Giant Algal Cells*, Cambridge: Cambridge University Press.

Hormann, G. (1898) *Studien über die Protoplasmastromung bei den Characeen*, Jena: Fischer, 65.

Horowitz, A.L., Lewis, D.C. and Gasteiger, E.L. (1975) Plant 'primary perception': electrophysiological unresponsiveness to brine shrimp killing. *Science*, 189, 478–80.

Houwink, A.L. (1935) The conduction of excitation in *Mimosa pudica*. *Recueil des Travaux Botaniques Néerlandais*, 32, 51–91.

Houwink, A.L. (1938) The conduction of excitation in *Clematis zeylanica* and in *Mimosa pudica*. *Annales du Jardin Botanique de Buitenzorg*, 48, 10–16.

Howells, K.F. and Fell, D.A. (1979) Quantitative aspects of cyclosis in plant cells. *Journal of Biological Education*, 13, 88–90.

Iijima, T. and Sibaoka, T. (1982) Propagation of action potential over the trap-lobes of *Aldrovanda vesiculosa*. *Plant and Cell Physiology*, 223, 679–88.

Iijima, T. and Sibaoka, T. (1985) Membrane potentials on excitable cells of *Aldrovanda vesiculosa*. *Plant and Cell Physiology*, 26, 1–13.

Ilker, R. (1975) Complexing of P-Protein filaments in the phloem of *Vicia faba* and *Xylosoma congestum* with heavy meromyosin. *Plant Physiology, Supplement*, 56, 17.

Insell, J.P. and Zacharaiah, K. (1978) The mechanism of the ring trap of the predacious hyphomycete *Dactylella brochopaga* Drechsler. *Protoplasma*, 95, 175–91.

Izant, J.G. (1983) The role of calcium ions during mitosis. Calcium participates in the anaphase trigger. *Chromosoma*, 88, 1–10.

Jacobshagen, S., Altmuller, D., Grolig, F. and Wagner, G. (1986) Calcium pools, calmodulin and light-regulated chloroplast movements in *Mougeotia* and *Mesotaenium*. In A.J. Trewavas (ed.), *Molecular and Cellular Aspects of Calcium in Plant Development*, NATO Series A: Life Sciences, vol. 104, New York: Plenum.

Jacobson, S.L. (1965) Receptor response in the Venus's flytrap. *Journal of General Physiology*, 49, 117–29.

Jacobson, S.L. (1974) Effect of ionic environment on the response of the sensory hair of Venus's flytrap. *Canadian Journal of Botany*, 52, 1293–302.

Jaffe, M.J. (1968) Phytochrome-mediated bioelectric potentials in mung bean seedlings. *Science*, 162, 1016–17.

Jaffe, M.J. (1976a) Thigmomorphogenesis: a detailed characterization of the response of beans (*Phaseolus vulgaris* L.) to mechanical stimulation. *Zeitschrifte für Pflanzenphysiologie*, 77, 437–53.

Jaffe, M.J. (1976b) Thigmomorphogenesis: electrical resistance and mechanical correlates of the early events of growth retardation due to mechanical stimulation in beans. *Zeitschrifte für Pflanzenphysiologie*, 78, 24–32.

Jaffe, M.J. (1977) Experimental separation of sensory and motor functions in pea tendrils. *Science*, 195, 191–2.

Jaffe, M.J. (1985) Wind and other mechanical effects in the development and behavior of plants, with special emphasis on the role of hormones. In R.P. Pharis and D.M. Reid (eds), *Encyclopedia of Plant Physiology, New Series. Hormonal Regulation of Development. II: Role of Environmental Factors*, Berlin: Springer-Verlag, 444–84.

Jaffe, M.J. and Biro, R. (1979) Thigmomorphogenesis: the effect of mechanical perturbation on the growth of plants, with special reference to anatomical changes, the role of ethylene, and interaction with other environmental stresses. In H. Mussell and R. Staples (eds), *Stress Physiology in Crop Plants*, New York: Wiley, 25–59.

Jaffe, M.J. and Galston, A.W. (1967) Physiological studies on pea tendrils. III: ATPase activity and contractility associated with coiling. *Plant Physiology*, 42, 845–7.

Jaffe, M.J. and Galston, A.W. (1968) Physiological studies on pea tendrils. V: Membrane changes and water movement associated with contact coiling. *Plant Physiology*, 43, 537–42.

Jaffe, M.J., Gibson, C. and Biro, R. (1977) Physiological studies of mechanically stimulated motor responses of flower parts. I: Characterization of the thigmotropic stamens of *Portulaca grandiflora* Hook. *Botanical Gazette*, 138, 438–47.

Joel, D.M., Rea, P.A. and Juniper, B.E. (1983) The cuticle of *Dionaea muscipula* Ellis (Venus's flytrap) in relation to stimulation, secretion and absorption. *Protoplasma*, 114, 44–51.

Johnson, A. (1965) Investigations of the reciprocity rule by means of geotropic and geoelectric measurements. *Physiologia Plantarum*, 18, 945–67.

Jonas, H. (1976) *Mimosa pudica* L., responses to electrical and mechanical stimuli, cardenolides and light. *Zeitschrift für Pflanzenphysiologie*, 80, 395–406.

Jones, C. and Wilson, J.M. (1982) The effects of temperature and action potentials on the chill sensitive seismonastic plant *Biophytum sensitivum*. *Journal of Experimental Botany*, 33, 313–20.

Jones, M.G.K., Novacky, A. and Dropkin, V.H. (1974) 'Action potentials' in nematode induced plant transfer cells. *Protoplasma*, 80, 401–5.

Jones, R.S. and Stutte, C.A. (1986) Acetylcholine and red-light influence on ethylene evolution from soyabean leaf tissues. *Annals of Botany*, 57, 897–900.

Josephson, J.O. (1966) Some bioelectric properties of *Amoeba proteus*. *Acta Physiologica Scandinavica*, 66, 395–405.

Jost, L. (1907) *Lectures on Plant Physiology*, translated by R.J.H. Gibson, Oxford: Clarendon Press, 414.

Junker, S. (1977a) Ultrastructure of tactile papillae on tendrils of *Eccremocarpus scaber* R. et P. *New Phytologist*, 78, 607–10.

Junker, S. (1977b) Thigmonastic coiling of tendrils of *Passiflora quadrangularis* is not caused by lateral redistribution of auxin. *Physiologia Plantarum*, 41, 51–4.

Kabsch, W. (1861) Anatomische und physiologische Untersuchungen über einige Bewegungserscheinungen im Pflanzenreiche. *Botanische Zeitung*, 19, 369–75.

Kamiya, N. (1959) *Protoplasmatologia*, vol. 8: *Physiologie der Protoplasmas*, No. 3: *Motilitat. Protoplasmic streaming*, Vienna: Springer-Verlag, 1011.

Kamiya, N. and Abe, S. (1950) Bioelectric phenomena in the myxomycete plasmodium and their relation to protoplasmic flow. *Journal of Colloid Science*, 5, 149–63.

REFERENCES

Kato, T. and Tonomura, Y. (1975) *Physarum* tropomyosin-troponin complex isolation and properties. *Journal of Biochemistry*, 78, 583–8.
Kerner von Marilaun, A. (1904) *The Natural History of Plants*, English version translated by F.W. Oliver, London: Gresham.
Kersey, Y.M., Hepler, P.K., Palevitz, B.A. and Wessells, N.K. (1976) Polarity of atin filaments in characean algae. *Proceedings of the National Academy of Sciences, USA*, 73, 165–7.
Keskin, B. and Fuchs, W.H. (1969) Der Infektions-vorgang bei *Polymyxa betae*. *Archive Mikrobiologie*, 68, 218–26.
Kilmartin, J.V. and Adams, A.E.M. (1984) Structural rearrangements of tubulin and actin during the cell cycle of the yeast *Saccharomyces*. *Journal of Cell Biology*, 98, 922–33.
Kinraide, T.B. and Wyse, R.W. (1986) Electrical evidence for turgor inhibition of proton extrusion in sugar beet taproot. *Plant Physiology*, 82, 1148–50.
Kirkham, M.B., Gardner, W.R. and Gerloff, G.C. (1972) Regulation of cell division and cell enlargement by turgor pressure. *Plant Physiology*, 49, 961–2.
Kjellberg, B., Karlsson, S. and Kerstensson, I. (1982) Effects of heliotropic movements of *Dryas octopetala* L. on gynoecium temperature and seed development. *Oecologia*, 54, 10–13.
Klein, K., Wagner, G. and Blatt, M.R. (1980) Heavy-meromycin-decoration of microfilaments from *Mougeotia* protoplasts. *Planta*, 150, 354–6.
Kline, D., Simoncini, L., Mandel, G., Maue, R.A., Kado, T. and Jaffe, L.A. (1988) Fertilization events induced by neurotransmitters after injection of mRNA in *Xenopus* eggs. *Science*, 241, 464–7.
Knapp, J.L. (1838) *The Journal of a Naturalist*, London: Murray.
Kohno, T. and Shimen, T. (1988) Mechanism of Ca^{2+} inhibition of cytoplasmic streaming in lily pollen tubes. *Journal of Cell Science*, 91, 501–9.
Kolata, G. (1984) Steroid hormone systems found in yeast. *Science*, 225, 913–14.
Kordan, H.A. (1977) Amytal-induced elongation in light-germinated rice seedlings. *Annals of Botany*, 41, 257–9.
Kordan, H.A. (1978) Effect of barbiturates on rice seedling germination and developmental behaviour. *Annals of Botany*, 42, 73–81.
Kordan, H.A. (1980) Largactil and hydroxylamine induced geotropic disorientation in lettuce (*Lactuca sativa* L.). *Zeitschrift für Pflanzenphysiologie*, 100, 273–8.
Kordan, H.A. (1984) *In vivo* effects of barbiturates on seed germination and seedling growth. *Journal of Biological Education*, 18, 266–8.
Kordan, H.A. and Mumford, P.M. (1978) Effect of barbiturates on *Impatiens holstii* pollen germination. *Annals of Botany*, 38, 997–9.
Kordan, H.A. and Rengel, Z. (1988) Impaired anthocyanin production in barbiturate-treated *Zea mays* seedlings. *Annals of Botany*, 61, 221–3.
Koshland, D.E., Mitchison, T.J. and Kirschner, M.W. (1988) Polewards chromosome movement driven by microtubule depolymerization *in vitro*. *Nature*, 331, 499–504.
Kunkel, A.J. (1878) Uber elektromotorische Wirkungen an unverletzten lebenden Pflanzen theilen. *Arbeiten aus dem botanischen Institut in Würzburge*, 2, 1–17.
Kursanov, A.L., Kulikova, A.L. and Turkina, M.V. (1983) Actin-like protein from the phloem of *Heracleum sosnowskyi*. *Physiologie Vegetale*, 21, 353–60.

REFERENCES

La Claire, J.W., II (1989) Actin cytoskeleton in intact and wounded coenocytic green algae. *Planta*, 177, 47–57.

Lambeck, F. and Skofitsch, G. (1984) Distribution of serotonin in *Juglans regia* seeds during ontogenic development and germination. *Zeitschrift für Pflanzenphysiologie*, 114, 349–53.

Larqé-Saavedra, A. (1979) Studies on the effect of prostaglandins on four plant bioassay systems. *Zeitschrift für Pflanzenphysiologie*, 92, 263–70.

Leach, C.M. (1976) Electrostatic theory to explain violent spore liberation by *Drechslera turcica* and other fungi. *Mycologia*, 68, 63–86.

Leach, C.M. and Apple, J.D. (1984) Leaf surface electrostatics: behavior of detached leaves of beans, maize, and other plants under natural conditions. *Phytopathology*, 74, 704–9.

Legros, P., Uytdenhoef, P., Dumont, I., Hanson, B., Jeanmart, J., Massart, B. and Conard, V. (1975) Specific binding of insulin to the unicellular alga *Acetabularia mediterranea*. *Protoplasma*, 86, 119–34.

Lehmann, M., Vorbrodt, H.-M., Dam, G. and Koolman, J. (1988) Antiecdysteroid activity of brassinosteroids. *Experientia*, 44, 355–7.

Le Page-Degrivy, M.-Th., Bidard, J.-N., Ronvier, E., Bulard, C. and Ladzuski, M. (1986) Presence of abscisic acid, a phytohormone, in the mammalian brain. *Proceedings of the National Academy of Sciences, USA*, 83, 1155–8.

Le Roith, D., Shiloach, J., Roth, J. and Lesniak, M.A. (1980) Evolutionary origins of vertebrate hormones: substances similar to mammalian insulins are native to unicellular eukaryotes. *Proceedings of the National Academy of Sciences, USA*, 77, 6184–8.

Le Roith, D., Shiloach, J., Roth, J. and Lasniak, M.A. (1981) Insulin or a closely related molecule is native to *Escherichia coli*. *Journal of Biological Chemistry*, 256, 6533–6.

Lichtner, F.T. and Williams, S.E. (1977) Prey capture and factors controlling trap narrowing in *Dionaea* (Droseraceae). *American Journal of Botany*, 64, 881–6.

Litvin, F.F., Sineshchekov, O.A. and Sineshchekov, V.A. (1978) Photoreceptor electric potential in the phototaxis of the alga *Haematococcus pluvialis*. *Nature*, 271, 476–8.

Lloyd, C. (1988) Actin in plants. *Journal of Cell Science*, 90, 185–8.

Loewy, A.G. (1952) An actomyosin-like substance from the plasmodium of a myxomycete. *Journal of Cellular and Comparative Physiology*, 40, 127–56.

Lonsdale, M. and Braithwaite, R. (1988) The shrub that conquered the shrub. *New Scientist*, 1634 (15 October), 52–5.

Loose, D.S., Stover, E.P., Restrepo, A., Stevens, D.A. and Feldman, D. (1983) Estradiol binds to a receptor-like cytosol binding protein and initiates a biological response in *Paracoccidiodes brasiliensis*. *Proceedings of the National Academy of Sciences, USA*, 80, 7659–63.

Loumaye, E., Thorner, J. and Catt, K.J. (1982) Yeast mating pheromone activates mammalian gonadotrophs: evolutionary conservation of a reproductive hormone? *Science*, 218, 1323–5.

Lunsden, P.J. and Vince-Prue, D. (1981) The perception of dusk signals in photoperiodic time-measurements. *Physiologia Plantarum*, 60, 427–32.

Ma, Y.-Z. and Yen, L.-F. (1989) Actin and myosin in pea tendrils. *Plant Physiology*, 89, 586–9.

REFERENCES

MacRobbie, E.A.C. (1971) Fluxes and compartmentation in plant cells. *Annual Review of Plant Physiology*, 22, 75–96.

Malamy, J., Carr, J.P., Klessig, D.F. and Raskin, I. (1990) Salicylic acid: a likely endogenous signal in the resistance response of tobacco to viral infection. *Science*, 250, 1002–4.

Margulis, L. (1970) *Origin of Eukaryotic Cells*, New Haven, CT: Yale University Press.

Marsden, J. and Brylewski, A.J. (1985) Use of the BBC computer to monitor movement in *Mimosa pudica*. *The School Science Review*, 67, 77–80.

Martin, E.V. and Clements, F.E. (1935) Studies on the effects of artificial wind on growth and transpiration in *Helianthus annuus*. *Plant Physiology*, 10, 613–36.

Martinac, B., Buechner, M., Delcour, A.H., Adler, J. and Kung, C. (1987) Pressure-sensitive ion channel in *Escherichia coli*. *Proceedings of the National Academy of Sciences, USA*, 84, 2297–301.

Maugh, T.H. (1982) The scent makes sense. *Science*, 215, 1224.

McAinsh, M.R., Brownlee, C. and Hetherington, A.M. (1990) Abscisic acid-induced elevation of guard cell cytosolic Ca^{2+} precedes stomatal closure. *Nature*, 343, 186–8.

McBurney, D.H. and Collings, V.B. (1977) *Introduction to Sensation/Perception*, New Jersey: Prentice Hall.

McCann, F.V., Cole, J.J., Guyre, P.M. and Russell, J.A.G. (1983) Action potentials in macrophages derived from human monocytes. *Science*, 219, 991–3.

Meeuse, B.J.D. and Raskin, I. (1988) Sexual reproduction in the arum lily family. *Sexual Plant Reproduction*, 1, 3–5.

Meiners, S., Baron-Epel, O. and Schindler, M. (1988) Intercellular communication – filling in the gaps. *Plant Physiology*, 88, 791–3.

Menzel, D. (1989) Chromosomes moved by actin? *Botanica Acta*, 102, 5–6.

Métraux, J.P., Signer, H., Ryals, J., Ward, E., Wyss-Benz, M., Gaudin, J., Raschdorf, K., Schmid, E., Blum, W. and Inverardi, B. (1990) Increase in salicylic acid at the onset of systemic acquired resistance in cucumber. *Science*, 250, 1004–6.

Meyer, B.S. and Anderson, D.B. (1952) *Plant Physiology*, Toronto: Van Nostrand, 747.

Miller, A.L., Raven, J.A., Sprent, J.I. and Weisenseel, M.H. (1986) Endogenous ion currents traverse growing roots and root hairs of *Trifolium repens*. *Plant, Cell and Environment*, 9, 79–83.

Miller, A.L., Shand, E. and Gow, N.A.R. (1988) Ion currents associated with root tips, emerging laterals and induced wound sites in *Nicotiana tabacum*: spatial relationship proposed between resulting electrical fields and phytophthoran zoospore infection. *Plant, Cell and Environment*, 11, 21–5.

Millet, B. and Thibert, P. (1976) Enregistrement de la réponse mécanique des étamines de *Berberis* stimulées électriquement. *Comptes rendus des séances de la Société de Biologie*, 170, 87–93.

Minorvsky, P.V. (1989) Temperature sensing by plants: a review and hypothesis. *Plant, Cell and Environment*, 12, 119–35.

Miura, G.A. and Shih, T.-M. (1984) Cholinergic constituents in plants: characteriza-

tion and distribution of acetylcholine and choline. *Physiologia Plantarum*, 61, 417–21.
Miyares, C. and Menendez, C. (1976) In B. Samuelson and R. Paoletti (eds) *Advances in Prostaglandin and Thromboxane Research*, vol. 2, New York: Raven Press, 877.
Mockrin, S.C. and Spudich, J.A. (1976) Calcium control of actin-activated myosine adenosine triphosphatase from *Dictyostelium discoideum*. *Proceedings of the National Academy of Sciences, USA*, 73, 2321–5.
Mohl, H. von (1856) Welche Ursachen bewirken Erweiterung und Verengung der Spaltoffnungen? *Botanisches Zeitung*, 14, 697–704, 713–21.
Mohri, H., Mohri, T., Mabuchi, I., Yazaki, I., Sakai, H. and Ogawa, K. (1976) Localization of dynein in sea urchin eggs during cleavage. *Development, Growth and Differentiation*, 18, 391–8.
Molotok, G.P., Britikov, E.A. and Sinyukhin, A.M. (1968) The electrophysiological and functional activity of nectary glands in the lime tree subjected to mechanical stimulation. *Doklady Akademiya Nauk SSSR*, 181, 750–3.
Moody, W. and Zeiger, E. (1978) Electrophysiological properties of onion guard cells. *Planta*, 139, 159–65.
Moran, N., Ehrenstein, G., Iwasa, K., Bore, C. and Mischke, C. (1984) Ion channels in plasmalemma of wheat protoplasts. *Science*, 226, 835–8.
Morren, C. (1839) Recherches sur le mouvement et l'anatomie du style du *Goldfussia anisophylla*. *Nouveaux Mémoires de l'Académie royale de Bruxelles*, 12, 3–34.
Morris, D.A. (1980) The influence of small direct electric currents on the transport of auxin in intact plants. *Planta*, 150, 431–4.
Morse, M.J. and Satter, R.L. (1987) Effects of external K^+ on K^+ channels in *Samanea* protoplasts. *Plant Physiology (Supplement)*, 83, 112.
Morse, M.J., Coté, G.G., Crain, R.C. and Satter, R.L. (1987) Light-stimulated inositol phospholipid turnover in *Samanea saman* leaf pulvini. *Proceedings of the National Academy of Sciences, USA*, 84, 7075–8.
Morse, M.J., Satter, R.L., Crain, R.C. and Coté, G.G. (1989) Signal transduction and phosphatidylinositol turnover in plants. *Physiologia Plantarum*, 76, 118–21.
Muller, H. (1883) *The Mechanism of Flowers*, translated by D'Arcy Thompson, London: Macmillan.
Mummert, E. and Gradmann, D. (1976) Voltage dependent potassium fluxes and the significance of action potentials in *Acetabularia*. *Biochimica et Biophysica Acta*, 443, 443–50.
Naitoh, Y. and Eckert, R. (1974) The control of ciliary activity in Protozoa. In M. Slegh (ed.), *Cilia and Flagella*, London: Academic Press, 305–52.
Neel, P.L. and Harris, R.W. (1971) Motion-induced inhibition of elongation and induction of dormancy in *Liquidambar*. *Science*, 173, 58–9.
Newcombe, F.C. (1920) Response of sensitive stigmas to unusual stimuli. *Report of the Michigan Academy of Sciences*, 22, 145–6.
Newcombe, F.C. (1922) Significance of the behavior of sensitive stigmas. *American Journal of Botany*, 9, 99–120.
Nordbring-Hertz, B. (1984) The biology of a nematode-trapping fungus. *Journal of Biological Education*, 18, 21–4.

REFERENCES

Oesterhelt, D. and Stoeckenius, W. (1971) Rhodopsin-like protein from the purple membrane of *Halobacterium halobium*. *Nature*, 233, 149–52.

Oliver, F.W. (1888) On the sensitive labellum of *Masdevallia muscosa*, Rehb. f. *Annals of Botany, Old Series*, 1, 237–52.

Opritov, V.A. (1978) Propagating excitation and assimilate transport in the phloem. *Soviet Plant Physiology*, 25, 828–37.

Ortega, J.K.E. and Gamow, R.I. (1970) *Phycomyces*: habituation of the light growth response. *Science*, 168, 1374–5.

Osterhout, W.J.V. (1922) *Injury, Recovery and Death in Relation to Conductivity and Permeability*, Philadelphia, PA: Lippincott, ch. 5.

Pallaghy, C.K. (1968) Electrophysiological studies in guard cells of tobacco. *Planta*, 80, 147–53.

Peart, M.H. (1979) Experiments on the biological significance of the morphology of seed-dispersal units in grasses. *Journal of Ecology*, 67, 843–63.

Peart, M.H. (1981) Further experiments on the biological significance of the morphology of seed-dispersal units in grasses. *Journal of Ecology*, 69, 425–36.

Percival, M.S. (1965) *Floral Biology*, Oxford: Pergamon.

Perley, J.E. and Glass, A.D.M. (1979) Cytoplasmic streaming and ion transport: a laboratory exercise which tests a longstanding botanical concept. *Journal of Biological Education*, 13, 199–203.

Petzelt, C. and Hafner, M. (1986) Visualization of the Ca^{2+}-transporting system of the mitotic apparatus of sea urchin eggs with a monoclonal antibody. *Proceedings of the National Academy of Sciences, USA*, 83, 1719–22.

Pfarr, C.M., Cove, M., Grissom, P.M., Hays, T.S., Porter, M.E. and McIntosh, J.R. (1990) Cytoplasmic dynein is localized to kinetochores during mitosis. *Nature*, 345, 263–5.

Pfeffer, W.F.P. (1875) *Physiologische Untersuchungen*, Leipzig: Engelmann, 56.

Pfeffer, W.F.P. (1877) *Osmotische Untersuchungen, Studien zur Zellmachanik*, Leipzig: Engelmann.

Pfeffer, W.F.P. (1888) Ueber chemotaktische Bewegungen von Bakterien, Flagellaten und Volvocineen. *Untersuchungen zur Botanisches des Institut de Tübingen*, 2, 582–663.

Pfeffer, W.F.P. (1905) *The Physiology of Plants*, translated by A.J. Ewart, vol. 3, Oxford: Clarendon Press.

Pharis, R.P. and Reid, D.M. (1985) *Encyclopedia of Plant Physiology. Hormonal Regulation of Development. III: Role of Environmental Factors*: B.G. Pickard, chapter 7, Roles of hormones, protons and calcium in geotropism; B.G. Pickard, chapter 10, Role of hormones in phototropism; D.S. Fensom, chapter 17, Electrical and magnetic stimuli. Berlin: Springer-Verlag.

Pickard, B.G. (1971) Action potentials resulting from mechanical stimulation of pea epicotyls. *Planta*, 97, 106–15.

Pickard, B.G. (1984) Voltage transients elicited by brief chilling. *Plant, Cell and Environment*, 7, 171–8.

Ping, Z., Mimura, T. and Tazawa, M. (1990) Jumping transmission of action potential between separately place internodal cells of *Chara corallina*. *Plant Cell Physiology*, 31, 299–302.

REFERENCES

Poenie, M., Alderton, J., Steinhardt, R. and Tsien, R. (1986) Calcium rises abruptly and briefly throughout the cell at the onset of anaphase. *Science*, 233, 886–9.

Poff, K.L. and Skokut, M. (1977) Thermotaxis by pseudoplasmodia of *Dictyostelium discoideum*. *Proceedings of the National Academy of Sciences, USA*, 74, 2007–10.

Poff, K.L. and Whitaker, B.D. (1979) Movement of slime molds. In W. Haupt and M.E. Feinleib (eds), *Encyclopedia of Plant Physiology, New Series*, vol. 7, Berlin: Springer-Verlag, 356–82.

Racusen, R.H. and Etherton, B. (1975) Role of membrane-bound fixed-charge changes in phytochrome-mediated mung bean root tip adherence phenomenon. *Plant Physiology*, 55, 491–5.

Racusen, R.H. and Satter, R.L. (1975) Rhythmic and phytochrome-regulated changes in transmembrane potential in *Samanea* pulvini. *Nature*, 255, 408–10.

Raschke, K. (1976) Transfer of ions and products of photosynthesis in guard cells. In I.F. Wardlaw and I.B. Passioura (eds), *Transport and Transfer Processes in Plants*, New York: Academic Press, 203–15.

Raskin, I., Ehmann, A., Melander, W.R. and Meeuse, B.J.D. (1987) Salicylic acid: a natural inducer of heat production in *Arum* lilies. *Science*, 237, 1601–2.

Raven, P.H., Evert, R.F. and Eichhorn, S.E. (1986) *Biology of Plants*, 4th edn, New York: Worth.

Ray, J. (1693) *Historia Plantarum*, London: S.J. Camello.

Rea, P.A. (1982) Fluid composition and factors that elicit secretion by the trap lobes of *Dionaea muscipula* Ellis. *Zeitschrift für Pflanzenphysiologie*, 108, 255–72.

Rea, P.A., Joel, D.M. and Juniper, B.E. (1983) Secretion and redistribution of chloride in the digestive glands of *Dionaea muscipula* (Venus's flytrap) upon secretion stimulation. *New Phytologist*, 94, 359–66.

Rheinhold, L., Seiden, A. and Volokita, M. (1984) Is modulation of the rate of proton pumping a key event in osmoregulation? *Plant Physiology*, 75, 846–9.

Ricca, U. (1908) I movimenti d'irritazione delle piante. *Malpighia*, 22, 173–98.

Ricca, U. (1916) Soluzione d'un problema di fisiologia. La propagazione di stimulo nella '*Mimosa*'. *Nuovo Giornale Botanico Italiano, Nuovo Serie*, 23, 51–170.

Ridgway, E.B. and Durham, A.C.H. (1976) Oscillations of calcium ion concentrations in *Physarum polycephalum*. *Journal of Cell Biology*, 69, 223–6.

Robb, E.J. and Barron, G.L. (1982) Nature's ballistic missile. *Science*, 218, 1221–2.

Roberts, H.M. (1896) The respiration of wounded plants. *Annals of Botany, Old Series*, 10, 531–82.

Roberts, H.M. (1897) The evolution of heat by wounded plants. *Annals of Botany, Old Series*, 11, 29–64.

Roblin, G. and Monmort, J. (1984) Effects of prostaglandins E, precursors and some inhibitors of prostaglandin biosynthesis on dark- and light-induced leaflet movements in *Cassia fasciculata* Michx. *Planta*, 160, 109–12.

Romero, G.A. and Nelson, C.E. (1986) Sexual dimorphism in *Catasetum* orchids: forcible pollen emplacement and male flower competition. *Science*, 232, 1538–40.

Rothert, W. (1904) Ueber die Wirkung des Aethers und Chlorforms auf die Reizbewegungen der Mikroorganismen. *Jahrbücher für Wissenschaftliche Botanik*, 39, 1–70.

REFERENCES

Russo, V.E.A. (1977) Ethylene-induced growth in *Phycomyces* mutants abnormal for autochemotropism. *Journal of Bacteriology*, 130, 548–51.
von Sachs, J. (1887) *Lectures on the Physiology of Plants*, translated by H. Marshall Ward, Lecture 34, Oxford: Clarendon Press, 587–602.
Saimi, Y and Kung, C. (1982) Are ions involved in the gating of calcium channels? *Science*, 218, 153–6.
Sakai, H., Mabuchi, I., Shimoda, S., Kuriyama, R., Ogawa, K. and Mohri, H. (1976) Induction of chromosome motion in the glycerol-isolated mitotic apparatus: nucleotide specificity and effects of anti-dynein and myosin sera in the motion. *Development, Growth, and Differentiation*, 18, 211–19.
Salisbury, F.B. (1963) *The Flowering Process*, Oxford: Pergamon, 161.
Salisbury, F.B. and Ross, C.W. (1985) *Plant Physiology*, 3rd edn, Belmont, CA: Wadsworth.
Salisbury, F.B. and Ross, C.W. (1978) *Plant Physiology*, 2nd edn, Belmont, CA: Wadsworth, 355.
van Sambeek, J. W. and Pickard, B. G. (1976a) Mediation of rapid electrical, metabolic, transpirational and photosynthetic changes by factors released from wounds. I: Variation potentials and putative action potentials in intact plants. *Canadian Journal of Botany*, 54, 2642–50.
van Sambeek, J.W. and Pickard, B.G. (1976b) Mediation of rapid electrical, metabolic, transpirational and photosynthetic changes by factors released from wounds. II: Mediation of the variation potential by Ricca's factor. *Canadian Journal of Botany*, 54, 2651–61.
van Sambeek, J.W. and Pickard, B.G. (1976c) Mediation of rapid electrical, metabolic, transpirational and photosynthetic changes by factors released from wounds. III: Measurements of CO_2 and H_2O flux. *Canadian Journal of Botany*, 54, 2662–71.
Saniewski, M., Banasik, L. and Rudnicki, R.M. (1979) A note on the activity of prostaglandins in a few plant bioassays. *Bulletin de l'Académie polonaise des Sciences. Classe II. Série des Sciences Biologiques*, 27, 775–80.
Satir, P. and Ojakian, G.K. (1979) Plant cilia. In W. Haupt and M.E. Feinleib (eds) *Encyclopedia of Plant Physiology, New Series*, vol. 7, Berlin: Springer-Verlag, 229–49.
Satter, R.L. (1979) Leaf movements and tendril coiling. In W. Haupt and M.E. Feinleib (eds), *Encyclopedia of Plant Physiology, New Series*, vol. 7, Berlin: Springer-Verlag, 442–84.
Satter, R.L. and Galston, A.W. (1981) Mechanism of control of leaf movements. *Annual Review of Plant Physiology*, 32, 83–110.
Saunders, M.J. (1986) Correlation of electrical current influx with nuclear position and division in *Funaria* caulonema tip cells. *Protoplasma*, 132, 32–7.
Schauf, C.L. and Wilson, K.J. (1987) Effects of abscisic acid on K^+ channels in *Vicia faba* guard cell protoplasts. *Biochemical and Biophysical Research Communications*, 145, 284–90.
Schildknecht, H. (1978) Über die chemie der Sinnpflanze *Mimosa pudica* L. *Sitzungsberichte der Heidelberger Akademie der Wissenschaften Mathematisch-naturwissenschaftlichte*, 6, 377–402.

REFERENCES

Schildknecht, H. (1984) Turgorins – new chemical messengers for plant behaviour. *Endeavour, New Series*, 8, 113–17.
Scholz, A. (1976) Lichtorientierte Chloroplastenbewegung bei *Hormidium flaccidum*: Perception der Lichtrichtung mittels Sammellinseneffekt. *Zeitschrifte für Pflanzenphysiologie*, 77, 406–21.
Schroeder, J.I. and Hagiwara, S. (1989) Cytosolic calcium regulates ion channels in the plasma membrane of *Vicia faba* guard cells. *Nature*, 338, 427–30.
Schroeder, J.I. and Hedrich, R. (1989) Involvement of ion channels and active transport in osmoregulation and signalling of higher plant cells. *Trends in Biological Sciences*, 14, 187–92.
Schroeder, J.I., Hedrich, R. and Fernandez, J.M. (1984) Potassium-selective single channels in guard cell protoplasts of *Vicia faba*. *Nature*, 312, 361–2.
Schroeder, J.I., Raschke, K. and Neher, E. (1987) Voltage dependence of K^+ channels in guard cell protoplasts. *Proceedings of the National Academy of Sciences, USA*, 84, 4108–12.
Schwintzer, C.R. (1971) Energy budgets and temperatures of nyctinastic leaves on freezing nights. *Plant Physiology*, 48, 203–7.
Seitz, K. (1979) Cytoplasmic streaming and cyclosis of chloroplasts. In W. Haupt and M.E. Feinleib (eds), *Physiology of Movements, Encyclopedia of Plant Physiology, New Series*, vol. 7, Berlin: Springer-Verlag, 150–69.
Serlin, B.S. and Roux, S.J. (1984) Modulation of chloroplast movement in the green alga *Mougeotia* by the Ca^{2+} ionophore A23187 and by calmodulin antagonists. *Proceedings of the National Academy of Sciences, USA*, 81, 6368–72.
Shen-Miller, J. (1970) Reciprocity in the activation of geotropism of oat coleoptiles grown on clinostats. *Planta*, 92, 152–63.
Sheriff, D.W. and Ludlow, M.M. (1985) Diaheliotropic responses of leaves of *Macroptilium atropurpureum* cv. Sirato. *Australian Journal of Plant Physiology*, 12, 151–71.
Shropshire, W., Jr (1979) Stimulus perception. In W. Haupt and M.E. Feinleib (eds) *Physiology of Movements, Encyclopedia of Plant Physiology, New Series*, vol. 7, Berlin: Springer-Verlag, ch. 1.2, pp. 14–15.
Sibaoka, T. (1954) Conduction mechanism of excitation in the petiole of *Mimosa pudica*. *Science Reports of the Tôhoku University, 4th Series (Biology)*, 20, 362–9.
Sibaoka, T. (1966) Action potentials in plant organs. *Symposium of the Society for Experimental Biology*, 20, 165–84.
Sibaoka, T. (1973) Transmission of action potentials in *Biophytum*. *Botanical Magazine (Tokyo)*, 86, 51–62.
Simons, P.J. (1982) Why plants need aspirin. *New Scientist*, 95, 847.
Simons, P.J. (1991) Aspirin helps the garden grow. *New Scientist*, 5 January, 20.
Sinyukhin, A.M. and Britikov, E.A. (1967a) Generation of potentials in the pistils of *Incarvillea* and lily in connection with the movement of the stigma and pollination. *Soviet Plant Physiology*, 14, 393–403.
Sinyukhin, A.M. and Britikov, E.A. (1967b) Action potentials in the reproductive system of plants. *Nature*, 215, 1278–80.

REFERENCES

Sláma, K. and Williams, C.M. (1965) Juvenile hormone activity for the bug *Pyrrhocoris apterus*. *Proceedings of the National Academy of Sciences, USA*, 54, 411–14.

Small, J. (1917) Irritability of the pollen-presentation mechanism in the Compositae. *Annals of Botany, Old Series*, 31, 261–8.

Smith, A.P. (1974) Bud temperature in relation to nyctinastic leaf movement in an Andean giant rosette plant. *Biotropica*, 6, 263–6.

Smith, J.A.C. and Milburn, J.A. (1980) Osmoregulation and the control of phloem-sap composition in *Ricinus communis* L. *Planta*, 148, 23–34.

Smith, J.E. (1788) Some observations on the irritability of vegetables. *Philosophical Transactions of the Royal Society*, 78, 158–65.

Smith, P.T. (1984) Electrical evidence from perfused and intact cells for voltage-dependent K^+ channels in the plasmalemma of *Chara australis*. *Australian Journal of Plant Physiology*, 11, 303–18.

Sommer, U. (1988) Some size relationships in phytoflagellate motility. *Hydrobiologie*, 161, 125–31.

Southwick, F.S. and Hartwig, J.H. (1982) Acumentin, a protein in macrophages which caps the 'pointed' end of actin filaments. *Nature*, 297, 303–7.

Spanjers, A.W. (1981) Bioelectric potential changes in the style of *Lilium longiflorum* after self- and cross-pollination of the stigma. *Planta*, 153, 1–5.

Stamp, N.E. (1989) Efficacy of explosive vs. hygroscopic seed dispersal by an annual grassland species. *American Journal of Botany*, 76, 555–61.

Stebbins, G.L. (1974) *Flowering Plants. Evolution Above the Species Level*, London.

Steinhardt, R.A. and Alderton, J. (1988) Intracellular free calcium rise triggers nuclear envelope breakdown in the sea urchin embryo. *Nature*, 332, 364–6.

Steuer, E.R., Wordeman, L., Schroer, T.A. and Sheetz, M.P. (1990) Localization of cytoplasmic dynein to mitotric spindles and kinetochores. *Nature*, 345, 266–8.

Stong, C.L. (1961) A young amateur experiments with a plant that collapses its leaves when it is touched. *Scientific American*, 204 (March), 181–92.

Stong, C.L. (1962) How to make an electrocardiogram of a water flea and investigate other bioelectric effects. *Scientific American*, 206 (January), 145–54.

Stong, C.L. (1964) Experiments in phototaxis: the response of organisms to changes in illumination. *Scientific American*, 128–34.

Stong, C.L. (1966) How to cultivate the slime molds and perform experiments on them. *Scientific American*, 214 (January), 116–21.

Suzuki, A., Mori, M., Sakagami, Y., Isogai, A., Fujino, M., Kitada, C., Crag, R.A. and Clewell, D.B. (1984) Isolation and structure of bacterial sex pheromone, cPD1. *Science*, 226, 849–50.

Sydenham, P.H. and Findlay, G.P. (1973) The rapid movement of the bladder of *Utricularia* sp. *Australian Journal of Biological Sciences*, 26, 1115–26.

Sydenham, P.H. and Findlay, G.P. (1975) Transport of solutes and water by resetting bladders of *Utricularia*. *Australian Journal of Plant Physiology*, 2, 335–51.

Szmelcman, S. and Adler, J. (1976) Change in membrane potential during bacterial chemotaxis. *Proceedings of the National Academy of Sciences, USA*, 73, 4387–91.

REFERENCES

Tahvanainen, J., Julkunen-Tiitto, R. and Kettunen, J. (1985) Phenolic glycosides govern the food selection pattern of willow feeding leaf beetles. *Oecologia*, 67, 52–6.

Takahashi, H. and Jaffe, M.J. (1984) Thigmomorphogenesis: the relationship of mechanical perturbation to elicitor-like activity and ethylene activity and ethylene production. *Physiologia Plantarum*, 61, 405–11.

Tamiya, T., Miyazaki, T., Ishikawa, H., Iriguchi, N., Maki, T., Matsumoto, J.J. and Tsuchiya, T. (1988) Movement of water in conjunction with plant movement visualized by NMR imaging. *Journal of Biochemistry*, 104, 5–8.

Tanada, T. (1968) Substances essential for a red, far-red light reversible attachment of mung bean root tips to glass. *Plant Physiology*, 43, 2070–1.

Taylorson, R.B. and Hendricks, S.B. (1979) Overcoming dormancy in seeds with ethanol and other anesthetics. *Planta*, 145, 507–10.

Taylorson, R.B. and Hendricks, S.B. (1980) Anesthetic release of seed dormancy – an overview. *Israel Journal of Botany*, 29, 273–80.

Theophrastus (1946) *Enquiry into Plants*, vol. 1, part 4.7.2, translated by A. Hort, London: Heinemann.

Thompson, C.M. and Wolpert, L. (1963) The isolation of motile cytoplasm from *Amoeba proteus*. *Experimental Cell Research*, 32, 156–60.

Tokufumi, H. and Suda, S. (1975) Thigmotropism in the growth of pollen tubes of *Lilium longiflorum*. *Plant and Cell Physiology*, 16, 377–81.

Tomati, U., Grappelli, A. and Galli, E. (1988) The hormone-like effect of earthworm casts on plant growth. *Biology and Fertility of Soils*, 5, 288–94.

Tompkins, P. and Bird, C. (1974) *The Secret Life of Plants*, London: Allen Lane.

Toong, Y.C., Schooley, D.A. and Baker, F.C. (1988) Isolation of insect juvenile hormone III from a plant. *Nature*, 333, 170–1.

Toriyama, H. (1953) Observational and experimental studies of sensitive plants. I: The structure of parenchymatous cells of pulvinus. *Cytologia*, 18, 283–91.

Toriyama, H. (1955) Observational and experimental studies of sensitive plants. VI: Migration of potassium in the primary pulvinus. *Cytologia*, 20, 367–77.

Toriyama, H. (1962) Observational and experimental studies of sensitive plants. XV: The migration of potassium in the petiole of *Mimosa pudica*. *Cytologia*, 27, 431–49.

Toriyama, H. (1966) The behaviour of the Sensitive Plant in a typhoon. *The Botanical Magazine*, 79, 427–8.

Toriyama, H. and Jaffe, M.J. (1972) Migration of calcium and its role in the regulation of seismonasty in the motor cell of *Mimosa pudica* L. *Plant Physiology*, 49, 72–81.

Toriyama, H. and Satô, S. (1968) Electron microscope observation of the motor cell of *Mimosa pudica* L. II: On the contents of the central vacuole of the motor cell. *Proceedings of the Japanese Academy*, 45, 175–9.

Tronchet, A. (1962) Quelques aspects de la sensibilité de vrilles. *Annales Scientifiques de l'Université de Besançon, 2nd series, Botanique*, 18, 3–17.

Tronchet, A. (1977) *La Sensibilité des Plantes*, Paris: Masson.

Tretyn, A., Kendrick, R.E., Bossen, M.E. and Vredenberg, W.J. (1990) Influence of

REFERENCES

acetylcholine agonists and antagonists on the swelling of etiolated wheat (*Triticum aestivum* L.) mesophyll protoplasts. *Planta*, 182, 473–9.

Tsoupras, G., Luu, B. and Hoffman, J.A. (1983) A cytokinin (isopentenyl-adenosylmononucleotide) linked to ecdysone in newly laid eggs of *Locusta migratoria*. *Science*, 220, 507–9.

Tsurumi, S. and Asahi, Y. (1985) Identification of jasmonic acid in *Mimosa pudica* and its inhibitory effect on auxin- and light-induced opening of the pulvinules. *Physiologia Plantarum*, 64, 207–11.

Ueda, T., Hasumoto, M., Akitaya, T. and Kobatake, Y. (1985) Two steady states in membrane potential deflection in relation to chemoreception and chemotaxis by *Physarum polycephalum*. *Protoplasma*, 124, 219–23.

Umrath, K. (1934) Über die elektrischen Erscheinungen bei thigmischer Reizung der Ranken von *Cucumis melo*. *Planta*, 23, 47–50.

Umrath, K. (1943) Über die Erregungssubstanz von *Berberis*. *Protoplasma*, 37, 346–9.

Umrath, K. (1966) Durch Erregungssubtsanz ausgeloste Spaltoffnungsschliebewegung. *Zeitschrift für Pflanzenphysiologie*, 55, 217–23.

Umrath, K. and Kastberger, G. (1983) Action potentials of the high-speed conduction in *Mimosa pudica* and *Neptunia plena*. *Phyton*, 23, 65–78.

Umrath, K. and Thaler, I. (1980) Auslösung von blattbewegungen bei *Mimosa* und von Krümmungen von *Lupinus*-hypokotylen, gedeutet durch freisetzung von errungssubstanz und auxin. *Phyton*, 20, 333–48.

Van der Pijl, L. (1940) On the sensitive flowers of *Gentiana quadrifaria*. *Annales de Jardin botanique de Buitenzorg*, 49, 89–98.

Vick, B.A. and Zimmerman, D.C. (1984) Biosynthesis of jasmonic acid by several plant species. *Plant Physiology*, 75, 458–61.

Visscher, S.N. (1980) Regulation of grasshopper fecundity, longevity and egg viability by plant growth hormones. *Experientia*, 363, 130–1.

Visscher, S.N. (1982) Plant growth hormones affect grasshopper growth and reproduction. *Proceedings of the 5th International Symposium of Insect–Plant Relationships*, Wageningen: Pudoc, 57–62.

Vogelmann, T.C. (1984) Site of light perception and motor cells in a sun-tracking lupin (*Lupinus succulentus*). *Phyiologia Plantarum*, 62, 335–40.

Vogelmann, T.C. and Björn, L.O. (1983) Response to directional light by leaves of a sun-tracking lupin (*Lupinus succulentus*). *Physiologia Plantarum*, 59, 533–8.

Walker-Simmons, M., Hollander-Czytko, H., Andersen, J.K. and Ryan, C.A. (1984) Wound signals in plants: a systemic plant wound signal alters plasma membrane integrity. *Proceedings of the National Academy of Sciences, USA*, 81, 3737–41.

Waller, J.C. (1925) Plant electricity. I: Photo-electric currents associated with the activity of chlorophyll in plants. *Annals of Botany, Old Series*, 39, 515–38.

Watanabe, S. and Sibaoka, T. (1973) Site of photo-reception to opening response in *Mimosa* leaflets. *Plant and Cell Physiology*, 14, 1221–4.

Watts, F.Z., Miller, D.M. and Orr, E. (1985) Identification of myosin heavychain in *Saccharomyces cerevisiae*. *Nature*, 316, 83–5.

Webster, J., Davey, R.A., Duller, G.A. and Ingold, C.T. (1984) Ballistospore discharge in *Intersonilia perplexans*. *Transactions of the British Mycological Society*, 82, 13–29.

REFERENCES

Weinberger, P. and Measures, M. (1968) The effect of two audible sound frequencies on the germination and growth of a spring and winter wheat. *Canadian Journal of Botany*, 46, 1151–8.

Weintraub, M. (1951) Leaf movements in *Mimosa pudica* L. *New Phytologist*, 50, 357–82.

Werk, K.S. and Ehleringer, J. (1986) Effect of nonrandom leaf orientation on reproduction in *Lactuca serriola* L. *Evolution*, 40, 1334–7.

Whitehead, C.S., Halevy, A.H. and Reid, M.S. (1984) Roles of ethylene and 1-aminocyclopropane-1-carboxylic acid in pollination and wound-induced senescence of *Petunia hybrida* flowers. *Physiologia Plantarum*, 61, 643–8.

Widder, E.A. and Case, J.F. (1982) Distribution of subcellular bioluminescent sources in a dinoflagellate *Pyrocystis fusiformis*. *Biological Bulletin*, 162, 423–48.

Wildon, D.C., Doherty, H.M., Eagles, G., Bowles, D.J. and Thain, J.F. (1989) Systemic responses arising from localized heat stimuli in tomato plants. *Annals of Botany*, 64, 691–5.

Wilkins, M.B. (1984) *Advanced Plant Physiology*, London: Pitman.

Williams, S.E. (1976) Comparative sensory physiology of the Droseraceae – the evolution of a plant sensory system. *Proceedings of the American Philosophical Society*, 120, 187–204.

Williams, S.E. and Bennett, A.B. (1982) Leaf closure in the Venus flytrap: an acid growth response. *Science*, 218, 1120–2.

Williams, S.E. and Pickard, B.G. (1972) Receptor potentials and action potentials in *Drosera* tentacles. *Planta*, 103, 193–221.

Wilson, E.B. (1925) *The Cell in Development and Heredity*, 3rd edn, New York: Macmillan.

Wolken, J.J. (1975) *Photoprocesses, Photoreceptors and Evolution*, New York: Academic Press.

Wolniak, S.M., Hepler, P.K. and Jackson, W.T. (1983) Ionic changes in the mitotic apparatus at the metaphase/anaphase transition. *Journal of Cell Biology*, 96, 598–605.

Yin, H.L., Hartwig, J.H., Mauyana, K. and Stoessel, T.P. (1981) Ca^{2+} control of actin filament length. Effects of macrophage gelsolin on actin polymerisation. *Journal of Biological Chemistry*, 256, 9693–7.

Young, S. (1975) *Electronics in the Life Sciences*, London: Macmillan.

Zeiger, E., Moody, W., Hepler, P. and Vaela, F. (1977) Light sensitive membrane potentials in onion guard cells. *Nature*, 270, 270–1.

Zimmerman, U. (1980) Pressure mediated osmoregulatory processes and pressure sensing mechanism. In R. Gilles (ed.), *Animals and Environmental Fitness*, Oxford: Pergamon.

Zimmerman, U. and Beckers, F. (1978) Generation of action potentials in *Chara corallina* by turgor pressure changes. *Planta*, 138, 173–9.

Zimmerman, U. and Steudle, E. (1978) Physical aspects of water relations of plant cells. *Advances in Botanical Research*, 6, 45–117.

Zurzycki, J. (1959) *Encyclopedia of Plant Physiology*, vol. XVII(2), Berlin: Springer-Verlag, 940–78.

FURTHER READING

Bilderbeck, D.E. (1985) Regulators of plant reproduction, growth and differentiation in the environment. In *Encyclopedia of Plant Physiology, New Series. Hormonal Regulation of Development III*, Berlin: Springer-Verlag, 653–706.
Haupt, W. and Feinleib, M.E. (1979) Physiology of movements. In A. Pirson and M.H. Zimmermann (eds) *Encyclopedia of Plant Physiology, New Series*, vol. 7, Berlin: Springer-Verlag.
Pickard, B.G. (1973) Action potentials in higher plants. *The Botanical Review*, 39, 172–201.
Roblin, G. (1979) *Mimosa pudica*: a model for the study of the excitability in plants. *Biological Reviews*, 54, 135–53.
Sibaoka, T. (1969) Physiology of rapid movements in higher plants. *Annual Review of Plant Physiology*, 20, 165–84.
Simons, P.J. (1981) The role of electricity in plant movements. *New Phytologist*, 87, 11–37.

Addendum to Proofs

Touchiness, Cold Shock and Calcium

Further evidence that calcium is involved in the touch responses of plants has come from plants genetically fitted with a special calcium probe which luminesces. Whenever calcium is released by the plant cell the luminance increases. So when the plant was touched the calcium rose dramatically. Interestingly, the calcium also rose whenever the plant was rapidly chilled or attacked by a fungus. Clearly the calcium response is an important step in warning the plant of outside threat from its environment (Knight, M.R., Campbell, A.K., Smith, S.M. and Trewavas, A.J. (1991) Transgenic plant aequorin reports the effect of touch and cold-shock and elicitors on cytoplasmic calcium. *Nature*, 352, 524–6).

The effects of cold shocks on plants has also been recently reviewed. Whilst many plants trigger an action potential when suddenly chilled, not all of them do. However, they all respond with a drop in electrical voltage, which seems to correspond to a rise in calcium (Minorsky, P.V. (1989) Temperature sensing by plants: a review and hypothesis. *Plant, Cell and Environment*, 12, 119–35; Minorsky, P.V. and Spanswick, R.M. (1989) Electrophysiological evidence for a role for calcium in temperature sensing by roots of cucumber seedlings. *Plant, Cell and Environment*, 12, 137–43).

The presence and importance of calcium ion channels in plants has also been recently reviewed. There are a variety of different calcium ion channels in plant cell plasma membranes, including those activated by voltage, abscisic acid (plant hormone) and stretch. Organelles can act as stores for calcium and can have their own types of calcium channels, one activated by calcium ions and the other triggered by 1,4,5-triphosphate ($InsP_3$). In animal muscles, $InsP_3$ passes messages inside the endoplasmic reticulum, releasing calcium for muscle contraction (Schroeder, J.I. and Thuleau, P. (1991) Ca^{2+} channels in higher plant cells. *The Plant Cell*, 3, 555–9).

Plant Hormones

The role of acetylcholine has recently been reviewed (Tretyn, A. and Kendrick, R.E. (1991) Acetylcholine in plants: presence, metabolism and mechanism in action. *Botanical Review*, 57, 33–73).

The nerve transmitter norepinephrine has been reported to occur in duckweeds, and is involved in the flowering of the plant (Takimoto, A., Kaihara, S., Shinozaki, M. and Miura, J. (1991) Involvement of norepinephrine in the production of a flower-inducing substance in the water extract of *Lemna*. *Plant Cell Physiology*, 32, 283–9).

Plants and animals have been found to contain allene oxides, precursors of a variety of prostaglandin-like molecules such as the hormone jasmonic acid. This supports the reports of prostaglandin effects in plants (Song, W.-C. and Brash, A.R. (1991) Purification of allene oxide synthase and identification of the enzymes as a cytochrome P-450. *Science*, 253, 781–3).

Light Sensors

The rhodopsin light sensor – similar to the rhodopsin in our own eyes – has been reported in *Chlamydomonas* to trigger minute electric currents which have all the hallmarks of an action potential. When the current is triggered the flagella motor changes its beating pattern and the alga swims in a different direction. These electrical signals were driven by calcium ions, and triggered in a thousandth of a second – a hundred times faster than in our retinas. This is further evidence that apparently simple creatures like *Chlamydomonas, Paramecium* and *Haematococcus* show a surprising similarity to an animal neuromotor system: a sensor picks up changes in light and turns on an electric signal which then redirects a motor (Harz, H. and Hegemann, P. (1991) *Nature*, 351, 489).

Plant Movements in General

An excellent book on plant movements has recently been published: *Plant Tropisms and Other Growth Movements* by James W. Hart, published by Unwin Hyman, London, 1990. Although as the title suggests it mainly deals with growth movements, it covers some of the reversible movements described in this book. It is also heartening to see that Hart includes a chapter on touch stimulation and the role of electrical phenomena, although how this ties in with the traditional textbook view of hormone physiology is still a grey area.

Index

ABA, see abscisic acid
Abe, S., 154
abscisic acid (ABA), 137, 188–91, 193, 194, 216
Acacia lopantha, 141, 279, 282
Acetabularia, 211
acetylcholine, 182–4, 190, 226, 230
Achenbach, F., 156
Achlya ambisexualis, 193, 198, 275
Acokanthera oppositifolia, 196
Acostaea, 53
actin, 4, 5, 149, 226
action potentials, 49, 69, 201, 208
   associated with movement, 44, 66, 100, 211–13, 224–5
   associated with nerve-like activity, 80–92, 97, 153, 233
   definition of, 268
   method of production, 87–8
   propagation in animal nerves, 185–7
   due to temperature changes, 215–16
   in wounded plants, 213–15
actomyosin, 149, 150–1, 154, 158, 160, 229–30
   in amoebae, 156–9
   associated with cytoplasmic streaming, 169–70
   mode of action, 152–3
adenosine triphosphate, see ATP

Adler, J., 62
Aeschynomene, 96, 279
   growing conditions for, 253
African hemp, 39
African marigold, 127
Albizzia, 5, 7, 138, 142, 145
Aldrovanda vesiculosa, 83–6, 90, 274, 281, 282
   growing conditions for, 253
Alexander the Great, 129
alfalfa, see lucerne
algae, 154
   anaesthesis of, 174
   cytoplasmic streaming in, 159–60
   electrical activity in, 108–9, 113–17, 156
   movement of, 108–20, 154
   muscles in, 154, 160
   swimming speed of, 277
alkali bee, 29
Allamand, Jean, 234
Allen, R.D., 170
Allogromia, 171
Allomyces, 65
amoebae
   actomyosin in, 156–9
   animal and fungal, 4
   movements of, 4, 155, 156–9
   muscles in, 156–9

INDEX

Amoeba proteus, 105, 157
Amphora coffeaeformis, 154
Anabaena, 109
anaesthesia, 173–7
Anagallis arvensis, 127
Androsthenes, 129
Angiospermae, 231
animals
  action potentials of, 278
  classification of, 9
Anodonta, action potential of, 278
anther, definition of, 269
Anthocerotae, 231
apical dominance, 200
Apocynum androsemifolium, 41, 279, 283
  growing conditions for, 254
Applewhite, P.B., 146, 184, 202, 230
Arabidopsis, 208
Arabis arenosa, 277
Arceuthobium, 25
Archaebacteria, 112
Arctotis, 5, 51, 280
  growing conditions for, 254
Aristotle, 13, 235
Arthrobtrys, 71, 274
Ascelpias, 196
ascomycetes, 16
Aspergillus nidulans, 172
aspirin, 177–81, 214, 246, 247
Asteraceae, 42, 51
ATP
  energy storage in, 65–6, 149–53, 169, 178, 183
  in fuelling plant movement, 45, 65, 149–53, 157, 171, 220, 222
Australian quinine bush, 27
auxins, 194, 207, 220
  involvement of in wounding, 216
Avena sativa, 274–5

Backster, C., 203
bacteria
  anaesthesis of, 174
  electrical activity in, 62–5
  photosynthesis in, 112
  sense of taste in, 61–2

swimming speed of, 277
  voltage-sensitive ion channels in, 185
barberry, see *Berberis*
Barnard, C., 176
Barron, G.L., 19, 20
bastard toadflax, 127
Bates, G.W., 190
Bauhinia, growing conditions for, 254
Bean, W., 52
bean rust fungus, 209–10, 274
Bennet, A.B., 92
Berberis, 35–9, 44, 142, 240, 279, 282
  anaesthesis of, 174
  effect of Ricca's factor on, 180
  electrical activity in, 175, 278
  growing conditions for, 255
Bert, Paul, 173
Biddington, N.L., 210
Biddulphia, 119
Biedermann, W., 98
Bignonia, 280, 284
  growing conditions for, 255
Bignoniaceae, 49, 284
bindweed, 127
bioluminescence, 231
Biophytum, 1, 5, 6, 94, 96, 97, 99, 274, 279, 282
  growing conditions for, 255
Bird, C., 203
bird's nest fungus, 16
bladderwort, 1, 49, 72–3
  trap evolution of, 225
  see also *Utricularia*
Blatt, M.R., 162, 187
blue-green algae, 9–10, 117, 228
  light-sensitive movement of, 108–10
  phototaxis in, 109
Bonner, J.T., 68
Bopp, M., 75
Bose, J.C., 98–9, 141–2, 203, 220
Braam, J., 208
Brachythecium, spore dispersion of, 22
Brassica napus, brassinolide in, 194
brassinosteroids, 193
Bremekamp, C.E.B., 60
Briggs, W.R., 134–5, 163
Britikov, E.A., 49

INDEX

Britz, S.J., 118, 134–5, 163
Brown, R., 35, 93
Brucke, Ernest, 104
Bryonia dioica, 274
Bryophyta, 231
Bryopsis, chloroplast movements in, 163
Buchen, B., 86–7
bullrush common cat-tail, 194
Bunning, E., 35, 44, 45, 133
Burdon-Sanderson, J., 43–4, 79, 80–1, 98–9, 199, 221, 233, 235
Burdwood, William, 171
bur-marigold, memory in, 201
bushman's poison, 196
butterworts, 49, 74

calcium, 91–2, 102
  as a calmodulin property changer, 164
  deposits of in tannin vacuoles, 104
  role in cell division, 169
calcium ions, 136–7, 162
  as electrical charge carriers, 64–5, 86–9
  in controlling muscle contraction, 152–3, 154–5
Caleana, 54–5, 58, 280
Calendula pluvialis, 127
Callitriche heterophylla, 211
calmodulin, 162, 164, 169, 208, 215, 226
  production by plant wounding, 214–15
calyx, 37
cAMP, 68–9, 189, 226
Canadian pondweed, 162, 248
Cande, W.Z., 167
Candida, 194
de Candolle, A.P., 135, 201
Cape sundew, 77
cardiac glycosides, 195–6
carotenoids, 111–12
carpel, definition of, 269
cassava, 7, 126
Cassia, 96, 255–6, 279, 282
castasterone, 194

Catalpa bignoniodes, 51, 256
Catasetum, 5, 31, 32, 263, 281
  growing conditions for, 256
  cell components, functions of, 148
  cell division, 163–72, 211, 249
Centaurea, 4, 5, 42–4, 51, 279
  growing conditions for, 257
  sense of touch in, 219, 240
Ceratium, 277
Cereus, 279, 282
  growing conditions for, 257
Chaos chaos, electrical activity in, 157
Chara, 149, 150, 208, 211, 220
  cytoplasmic streaming in, 159, 232
  electrical activity in, 154, 159, 185–7
  chemotropism, 250
Chlamydomonas, 1, 4, 64, 119
  light sensor in, 229, 236
  phototaxis in, 112–14, 249
  sense of taste in, 67
chloride ions
  function of in driving trigger movement, 46–7
  as a membrane-voltage balance, 136
chlorophyll, 12
chloroplast movement, 249
  different motors of, 161–3
  driven by actomyosin, 151, 163
  light-sensitive, 117–20
  rates of, 277
chloroplasts, 2, 3, 4, 12, 148
  DNA in, 228
Chodat, R., 35, 218
Christalle, W., 49
chromosomes, movement of, 4, 165–9
cilia, 4, 62–5
cilia movement, 164
  processes leading to, 66
Clematis, 5
climbing plants, sense of touch in, 219–21
Closterium, 110
Commelina communis, 247
Compositae, 42, 284
Conidiobolus coronatus, 16
Coniferinae, 231
Convolvulus arvensis, 127

Cooksey, B., 154
Cooksey, K., 154
cornflower, see *Centaurea*
Correns, Carl, 44
Corti, B., 148
Crocus, temperature sensitivity of, 276
cross-pollination, 2, 15–34, 50, 128
   by insects, 35–42, 53–60
Cucumis sativum, reaction time of, 274
cyanobacteria, see blue-green algae
Cycadinae, 231
Cyclanthera, 219, 259, 281
   growing conditions for, 257
Cycnoches, 32
Cynara scolymus, 42, 279
cytokinin, 193, 194
cytoplasm, 2, 3, 37, 148
   division of, 169–72
   in plants, 159–61
cytoplasmic streaming, 4, 149, 159–61, 162, 164, 176, 216
   effect of temperature on, 216
   observation of, 248–9, 252

Dactylella, 5, 70–1
Daphnia, 72
Darwin, C., 13, 14, 31, 32–3, 35, 45, 53–4, 58, 74, 75–92, 77, 124, 131, 142, 174, 205, 218, 241, 242
Darwin, F., 124, 131
Dasyobolus immersus, 16
Davidson, J.G.N., 19
Davies, E., 181, 213, 214
Davis, R.W., 208
Dean, J.M., 95, 127
Dearnman, A., 210
De Mairan, J., 133–4
deoxyribonucleic acid, see DNA
Desbiez, M.-O., 200
desmids, phototaxis in, 109–10
Desmodium motorium, 134–5
   growing conditions for, 257–8
diatoms, movement of, 110, 154
Dicotyledoneae, 231, 282
dictyosomes, 3
Dictyostelium discoideum
   movement of, 110, 157–8

temperature sensitivity of, 276
Digitalis purpurea, see foxglove
digitoxigenin, 195
dinoflagellates, types of, 231
Dionaea muscipula, see Venus's flytrap
Dixon, R., 129
DNA, 2, 3, 12, 66, 163, 175, 228
dolicholide, 194
Dolichos lablab, 194
downy mildew fungus, 18
Drakaea
   labellum movement of, 56–7
   orchid corolla movement in, 280
   wasp imitation of, 58
Drechslera, 17
Drosera, see sundew
Droseraceae, 204
Dryas octopetala, 5
   growing conditions for, 258
   solar energy collection in, 128
Dryopteris, 21
   seed dispersal of, 6, 22
Du Bois Reymond, Emil, 79, 98
duckweed, see *Lemna*
Duddington, C.L., 71
Dutrochet, Henri, 97
dwarf mistletoe, seed dispersion of, 25
dynein, 4, 65, 167

Earnest, D.J., 183
Ecballium elaterium
   fruit dehiscence of, 281
   growing conditions for, 259
   see also squirting cucumber
Eccremocarpus scaber, 219
ecdysone, 191–2, 193
ecdysteroids, 193
Echinocereus, 279, 282
   growing conditions for, 259
Eckert, R., 63, 64, 116, 232
Edgar, L.A., 154
eelworms, 69, 70, 71, 252
Ehleringer, J., 124
Eisner, T., 94
elaters, 21
electrical current, definition of, 268
electronarcosis, 174

INDEX

Eledone moschata, 278
Elodea canadensis, see Canadian pondweed
Emmelin, N., 182
endoplasmic reticulum, 3, 86–9, 172, 224
Enright, J.T., 133
epidermis, definition of, 269
Erickson, R., 48
Erodium, see storksbill
Escherichia coli, 62
   pressure-sensitive ion gates in, 229
   production of insulin in, 197
   sense of taste in, 275
   sense of touch in, 212
   swimming speed of, 277
Eschrich, W., 102, 212
Espeletia shultzeii, 132
ethylene, 210, 213, 215
   associated with pollination, 217–18
   effect of on plants, 207
   relation to sense of touch, 220
   in wounded plants, 216, 235
Euglena, 5
   light-sensitive movements of, 114–16
   photosynthesis in, 114
   phototaxis in, 249
   sense of taste in, 67
   sense of vision in, 115–17
eukaryotes, 12, 64, 227, 228
   classification of, 9
excitation, definition of, 268

Fall panicum grass, 175
feather grass, 247
Feldberg, W., 182
Fensom, D.S., 160
Filicophytina, 231
Findlay, G.P., 46, 74
Fitzgerald, R.D., 58
flagella, 20, 62–5
   movements of, 164
   whip action of, 4
flagella cilia, definition of, 269
flavenoids, 111
Fleurat-Lessard, P.M., 142
flowers, exploding, 27–33

flying duck orchid, see *Caleana*
Forseth, I., 124
Foster, K.W., 112, 113
foxglove, 49, 195
Franks, N.P., 174
Fromm, J., 102
Funaria hygrometrica, 117, 168
   rate of chloroplast movement in, 249, 277
fungi
   anaesthesis of, 174
   carnivorous, 70, 252
   cilia and flagella in, 62–7
   classification of, 9
   explosions of, 15–20
   movement of, 67–8, 151–3
   muscles in, 151–3
   sense of taste in, 68–70
   touchiness in, 209–14
Fusicoccum amygdali, 137

Galston, A.W., 99, 139, 142, 203
Galvani, Luigi, 235
gamete, definition of, 269
Gamow, R.I., 202
gap junctions, 233
Gardner, F.T., 146–7
Gazania nigens, 246
gelsolin, properties of, 158
Gentiana quadrifaria, 283
   corolla movement in, 281
   touch sensitivity of, 60
gentisic acid, 180
Geranium, 5, 7, 22, 27
Gesneriaceae, 96
gibberellic acid, 141
globe artichoke, see *Cynara scolymus*
glutamic acid, 180
Gnetinae, 231
Goebel, K., 60
Goldfussia anisophylla, 51
   style movement in, 280
Goldsmith, M.H.M., 190
Gonyalax polyedra, speed of, 277
grasshopper's cyperus, 192–3
gravitropism, 206
greenhoods, see *Pterostylis*

Gromia, cytoplasmic streaming in, 170–1
Gruenberg, G., 202
Guha, S.C., 218
Gyrodinium, swimming speed of, 277

Haberlandt, G., 35, 37, 44, 82, 97, 129
Hader, D.-P., 108–10
Haematococcus, 113
Hafner, M., 169
Halobacterium halobium, 112, 229
hammer orchid, see *Drakaea*
Haptoglossa mirabilis, 5, 7, 19, 20
Hartmann, Max, 44
Hatano, S., 151
Haupt, W., 119
Hazum, E., 196–7
Heath, O.V.S., 138
Helianthus, 276
Hendricks, S.B., 175, 176
Hepaticeae, 231
heronbill, 22
histamines, 182–4
Hoch, H.C., 209
Hodick, D., 92
Holmes, E., 202
honeybee, 29
Hooke, Robert, 247
Hooker, J., 35, 54–8
Hormidium, 117, 118, 129
hormones
  animal, 188–98
  definition of, 268
  plant, 188–98
  role in plant movement, 141
  working of, 189–90
Horowitz, A.L., 203
Houwink, A.L., 100, 101, 215
hummingbird, 29, 177
Hydra, 8
  migration of vesicles in, 163
5-hydroxytryptamine, 182–4, 188–9, 226
Hyptis pauliana, 29

IAA, see indoleacetic acid
Iijima, T., 89, 90
Impatiens noli-me-tangere, 26, 281

Incarvillea delavayi, 49, 280
  growing conditions for, 260
indoleacetic acid (IAA) 75, 141, 188–9, 191, 220
inositol phospholipids, 209, 226
inositol 1,4,5-triphosphate, 136–40
insulin, 226, 230
  production of by the brain, 197
ion, definition of, 268
Itersonilia perplexans, 19

Jacobson, S.L., 82, 89
Jaffe, M.J., 39–41, 45, 106, 183, 199, 208, 210, 221, 274
jasmonic acid, 141, 179, 188, 214, 215
jellyfish, 8
Jones, C., 94
Junker, S., 219, 220
juvabione, 192
juvenile hormone, 192, 193

Kamiya, N., 154–6
Kerner von Marilaun, A., 25–7, 39, 95
Kersey, Y.M., 159
kinesin, 4
Kjellberg, B., 128
Knapp, J.L., 41–2
Kunkel, A., 98–9

lablab bean, 194
Lactuca sativa, sense of vision in, 275
Ladzunski, M., 189
Laminaria, anaesthesis of, 175
Layia elegans, sensitivity of, 42
Leach, C.M., 18–19
leafcutter bee, 29
leaf movement
  evolution of, 145–7
  processes leading to, 135–45
van Leeuwenhoek, Antonie, 48
Leidy, J., 170
Lemna, 118
  chloroplast movement in, 249, 277
Lentibulariaceae, 49, 225, 284
Levenhookia pusilla, 48
von Linnaeus, Carl, 130, 134
Liquidambar, 206

Listerata ovata, 32
Litchner, F.T., 92
lithium, 201
Lithops, 93
liverworts, 21, 213
Lloyd, C., 170
Loewy, A.G., 149–51
Loose, D.S., 194, 195
lotus plant, sleep movements in, 130–1
lucerne, 28–9, 30
Lupinus, 5, 128
　sun-tracking movements of, 124–6
Lycopersicon esculentum, 274
Lycophytina, 231

Ma, Y.-Z., 221
Machaerium arboreum, 95, 279
MacRobbie, E.A.C., 196
Mahonia, 37, 38, 282
　reaction time of, 39
　sense of touch in, 240
　stamen movement of, 35, 279
Malamy, J., 179
Malvastrum rotundifolium
　growing conditions for, 260
　sun-tracking movements of, 124–6
Marchantia, spore dispersion of, 22
Margulis, L., 228
marigold, 127
marram grass, 127
Martynia, 49, 284
　growing conditions for, 260–1
　reaction time of, 51
　stigma movement in, 280
Martyniaceae, 49, 284
Masdevallia muscosa, 52–3, 280
Mazus, growing conditions for, 261
McDonald, K.L., 167
McGee-Russell, S., 171
Medicago sativa, see lucerne
Mediterranean electric fish, 234–5
Megachile rotundata, 29
meiosis, definition of, 163
Mesembryanthemum, 282
　growing conditions for, 261–2
　stamen movement in, 280
Mesocarpus, 119

methyl jasmonate, 214
Metraux, J.P., 179
Meyerhof, Otto, 44
Micrasterias, 163
microtubules
　function of, 229
　movement of, 167
milkweeds, 196
Mimosa, 95–6, 129, 234, 282
　action potentials in the sieve tubes of, 212
　anaesthesis of, 173–4
　animal-like behaviour of, 233
　effect of Ricca's factor on, 180–1
　electrical activity in, 86, 96–106, 142, 175, 244
　evolution of, 227
　growing conditions for, 262
　learning behaviour in, 239
　movement of, 102–6, 135, 145, 180, 224, 279
　movement of, processes leading to, 45, 96–8
　muscular contraction in, 105–6, 146, 221, 224
　nerve-like behaviour in, 12, 100–2
　norepinephrine in, 184
　reaction time of light-sensitive movement, 146–7
　sense of touch in, 221, 223
　serotonin in, 184
　sleep movements in, 95, 130, 133–4, 141
　temperature sensitivity of, 215–16
　see also *Mimosa pudica*
Mimosa pudica
　action potential of, 278
　anaesthesis of, 239
　animal-like behaviour of, 2
　application of glycosides to, 196
　camouflage of, 93–4
　habituation of, 201–2, 238–9
　5-hydroxytryptamine in, 189
　leaf movement of, 6, 7
　loss of sensitivity in, 201–2
　motor of, 103, 105

317

Mimosa pudica (cont.)
  movement for protection in, 279
  movement of, effect of calcium ions on, 87
  reaction time of, 274
  touch sensitivity of, 146
  wounding of, 101, 239–40
  see also *Mimosa*
Mimulus, 5, 49–51
  growing conditions for, 262–3
  sense of touch in, 219, 240
  stigma movement in, 280
mitochondria, 3, 12, 148
  amoeboid movement of, 4
  DNA in, 228
  evolution of, 228
  movement of, 163
mitosis, definition of, 163
Mohl, H. von, 135
molecules, light-sensitive, see photoreceptors
Molinia coerula, growing conditions for, 263
Molotok, G.P., 218
monists, classification of, 9
monkey flower, see *Mimulus*
Monocotyledoneae, 231, 284
Mormodes, 32
Mormodes buccinator
  growing conditions for, 263
  pollinia disposal in, 281
Morren, C., 51
Morris, D.A., 191
Morton, William, 173
Moser, I., 133
motor tissue, definition of, 269
Mougeotia, 89, 118–20, 162
movement, definition of, 268
Muller, H., 42–3
musci, 231
myosin, 4, 5, 149, 226

Naitoh, Y., 63, 116
Naravelia, 215
Navisula cuspidata, movement of, 154
Neljubow, Dimitry, 207
Nelson, C.E., 31

nematodes, 16, 19, 69, 70, 71, 252
Neptunia, 96, 100, 102, 279, 282
  growing conditions for, 263
  reaction time of, 274
nerve, electrical signal propagation in, 186–8
Newcombe, F.C., 49, 51
nifedipine, 189
Nitella, 150, 208
  cytoplasmic streaming in, 159, 232
  electrical activity in, 159, 185–7, 278
  nerve-like impulses in, 154
  temperature sensitivity of, 216
NMR, see nuclear magnetic resonance
Noctiluca miliaris, 5, 231
Nomia melanderi, 29
noradrenaline, 184
Nordbring-Hertz, B., 252
norepinephrine, 184, 226
Notocactus, growing conditions for, 264
nuclear magnetic resonance, 104, 236
nuclei, movement of, 168

Ochromonas mimima, speed of, 277
Octopus, action potential of, 278
Oesterheit, D., 112
oestrogen, 194, 226, 230
Ohm's law, 187
Oosawa, F., 151
Opuntia, 39
  growing conditions for, 264
  sense of touch in, 219
  stamen movement in, 280
orchid flower, structure of, 52
organelles, 2, 3, 12, 148
  division of, 161
  in eukaryote cells, 228
  evolution of, 228
  movement of, 4, 161, 171
Ortega, J.K.E., 202
osmosis, 74, 90, 102, 104, 135–7, 139
  plant movement caused by, 45, 46
Osterhout, W.J.V., 175, 213
ouabain, 196
ovary, definition of, 269
ovule, definition of, 269

Oxalidaceae, 96, 282
Oxalis, 96, 118, 146, 279, 283
  growing conditions for, 264

Panicum dichotomoflorum, 175
Paracoccidoides brasiliensis, 195
Paramecium, 4, 5, 62, 114
  cytoplasmic streaming in, 159
  movement of, 116
  nerve-like activity in, 63–5
  voltage-sensitive ion channels in, 185
parenchyma, definition of, 269
Parietaria, 37
Passiflora, sense of touch in, 219
Paton, Sir William, 174
Peart, M.H., 24
Penium, 110
Perezia multiflora, 42
petal movement, 52–60
Petalostigma pubescens, 27
Pfeffer, W.F.P., 35, 44–5, 46, 49, 61, 70, 104, 134, 173, 201
Phaseolus vulgaris, reaction time of, 274
pheromones, 65
phloem
  blocking of by ethylene, 207, 213
  definition of, 160, 269
  movements of, 160
  sieve tubes of, 212
Phormidium uncinatum, 108–10
photophobia, definition of, 108
photoreceptors, 110–11
photosynthesis, 2, 3, 9, 12, 136, 138, 164
  in bacteria, 108–9, 111–14
  in plants, 121–8
  water and carbon dioxide requirements of, 121–3
phototaxis, 112–14, 249–50
  definition of, 108
phototropism, 107, 206
  sense of vision in, 275
Phycomyces, 111
  habituation of, 202
  sense of smell in, 275
  sense of touch in, 276
  sense of vision in, 275

Physarum, 149, 151
  electrical activity in, 69, 154–7
  movement of, 152
  muscles in, 154
phytochrome, 119–20, 138, 139, 140, 141
Pickard, B.G., 85, 86, 180, 181, 208, 213, 215, 218, 220, 235
Pickett-Heaps, J.D., 154
Pilea microphylla, 7, 29
Pilobolus, 16–18, 111
Pinguicula, see butterworts
Pisum sativum, reaction time of, 274
Plantae, 231
Plantago, 7, 127
plantain, 7, 127
plant cells, actomyosin in, 160
plant electrophysiology, 234–5
plant hormones, 188–91
plant movement, 7, 281
  classes of, 10
  evolution of, 145–7
  irreversible, 10, 107, 123–4, 281
  light-sensitive, 107–8, 117, 146–7, 223
  measurement of, 244–5
  moisture sensitive, 21–7, 127
  by osmosis, 45, 46
  processes leading to, 10–11, 43–5, 96–8, 102–6, 135–45, 225
  reversible, 10, 124–9, 279
  rhythmic autonomic, 246–7
  role of hormones in, 141
  sun-tracking, 123–9
  touch-sensitive, 35–42, 146–7, 223
  trigger stimuli of, 5–7
plants
  anaesthesis of, 173–7
  animal-like behaviour of, 34, 202–4, 226–37
  camouflage of, 93–4
  carnivorous, 49, 72–92
  classification of, 9
  coiling in, 219–21
  communication between, 160, 214, 231

cross-pollination of, 2, 15–42, 50, 53–60, 128, 182
ecdysteroids in, 193
effect of ethylene on, 207
electrical dispersion of spores of, 18–19
electrical movement of hormones in, 191
electric self-defence in, 93–106
emotions in, 203–4
evolution of, 1–14
exploding, 15–33
heat generation in, 177–9
hormones in, 234, 236
insulin-like substances in, 197
learning-type behaviour of, 201–2
light requirements of, 222
light sensitivity of, 16–17, 246
main groups of, 231
memory in, 199–201
movement of, see plant movement
muscular contraction in, 105–6, 146, 148–73, 220
nerve-like behaviour of, 1–14, 79–92, 182–4, 226–37
perception in, 128–9
photosynthesis in, 117, 121–8
recording electrical activity in, 242–4
sense of hearing in, 275
sense of smell in, 275–6
sense of taste in, 275
sense of touch in, 93–102, 205–37, 276
sense of vision in, 107–10, 275
sensitivity to gravity of, 276
sleep movements of, 129–33
solar energy collection in, 128–9
sugar transport in, 211–3
temperature control in, 122–3, 131–2
temperature sensitivity of, 215–16, 276
time-keeping in, 183–4
touch-sensitive reaction time of, 274
voltage-sensitive channels in, 188
wounding of, 101, 213–15
plant species, touch-sensitive, relationships of, 282–4

plasmodesmata, 3, 14, 37, 103, 142, 145, 146, 220, 222, 233
Plasmodiophora brassicae, 19, 20
Podocarpus, ecdysteroids in, 193
Podospora fimicola, 16
pollination, 216–18
  electrical signals of, 218
  ethylene, reliance on, 217–18
Polymyxa betae, 6, 19
polysomes, 214
Poroglossum, 53, 280
Portulaca, 39, 44, 45, 282
  growing conditions for, 264–5
  reaction time of, 41, 274
  sense of touch in, 219
  stamen movement in, 280
potassium, 89–92, 102
  deposits of in tannin vacuoles, 104
potassium ions
  driving plant movement, 136–8
  as electrical charge carriers, 64–5, 87–9, 145, 185–7, 222
  function of in driving trigger movement, 46–7
prokaryotes, 9, 12
proplastids, 4, 117
Prorocentrum mariae-lebouriae, 277
prostaglandins, 141, 214, 226
proteinase inhibitors, 214, 215
proteins, actin-binding, families of, 152, 154
protists, classification of, 9
protoplast, 3
Pseudomonas aeruginosa, 277
pseudopodia, definition of, 156
Psilophytina, 231
Pterostylis longifolia, 5, 54, 284
  growing conditions for, 265
  orchid corolla movement in, 280
puffballs, spore dispersion of, 17
pulvini
  definition of, 269
  electrical signal response of, 102
  as light sensors, 129
  movement of, 102–6, 136, 139–45, 145, 222
Pyrocystis fusiformis, 231

Racusen, R.H., 138, 142
Rana esculenta, action potential of, 278
Ray, J., 141
receptor potential, definition of, 268–9
Rehmannia, 280
  growing conditions for, 265
Rhizina inflata, 16
rhodopsin, 111, 229, 230
  in bacteria, 112–14
ribosomes, 3
Ricca, U., 97–8
Ricca's factor, 98, 100–1, 180–1, 213, 215, 218, 235
Ritter, Johann, 98
Roberts, H.M., 213
Romero, G.A., 31
Roth, J., 197
rotifer, 19, 20
Roux, S.J., 162
rumslind tree, 39
Rushton, William A.H., 110
Russell, J.A.G., 138

Saccharomyces cerevisae, 195
von Sachs, J., 14, 35, 58, 60, 81
salicylic acid, 177–81, 188, 214
Salisbury, F.B., 210
Salix, 177
Samanea saman, 130, 142–3, 188
van Sambeek, J.W., 181, 213, 235
sarcoplasmic reticulum, 87, 152–3, 164
Satter, R.L., 138, 142, 145
Sauromatum guttatum, 177–9
scarlet pimpernel, 127
Schildknecht, H., 141, 180
Schlauchzellen, 101, 102
Schrankia, 94–5, 279
  growing conditions for, 265–6
Schurman, David, 194
Schuster, A., 181, 214
sclerotium, cultivation of, 252
Scouleria, 21–2
Scrophulariaceae, 49, 284
Scurainia urae, swimming speed of, 277
sea anemones, 8
seed dormancy, 175–7
seed germination, 175, 176

Seifriz, William, 151
Selaginella, 118
sense of vision, definition of, 8
sensitive briar, see *Schrankia*
Sensitive Plant, see *Mimosa*; *Mimosa pudica*
Serlin, B.S., 162
serotonin, see 5-hydroxytryptamine
Shropshire, W., Jr, 275
Sibaoka, T., 89, 90, 96–7, 101, 129, 232
Siberian sow-thistle, 127
Sicyos, 8, 219
Sievers, A., 92
Silphium lacinatum, 123–4
Sinykhin, A.M., 49
sirenin, 65
Slama, K., 192
Slayman, C.L., 99, 203
slime moulds, 67–8, 157
  cultivation of, 250–2
  electrical activity in, 69–70, 154
  movement of, 67–8, 155, 157–9
  sense of taste in, 67–70
slug, formation of, 68
Small, J., 43
Smith, A.P., 95, 127, 132
Smith, J.E., 37
Smithia, 96, 279
  growing conditions for, 266
Smyth, R.D., 112
Sonchus siberius, 127
Spanjers, A.W., 181, 218
Sparmannia africana, 4, 39, 40, 44, 278
  growing conditions for, 266
  sense of touch in, 221, 240
  stamen movement of, 40, 280
Spathodea, growing conditions for, 266
Spemann, Hans, 44
Spermatophytina, 231
Sphaerobolus, 16
Sphenophytina, 231
spiderplant, 168, 248
spirochaetes, 228
Spirogyra, 170, 249
sponges, 8

squirting cucumber, 5
  fruit ejection of, 7, 10, 26
stamen
  definition of, 170
  movement of, 4, 5, 42–3
stamen filaments
  definition of, 270
  sensitivity of, 38
Stamp, N.E., 25
steroid glycosides, definition of, 195–6
stigma, definition of, 270
stigmasterol, 194
stinging nettle, 182
  explosion of, 29
  stamen movement of, 281
Stipa, 247
St John's wort, 132
Stoeckenius, W., 112
stomata, 121–3, 126
  definition of, 270
  movement of, 136–9, 144–5, 221–2
stomata guard cells, stretch-sensitive ion channels in, 211
storksbill, 22, 24–5
  common, 25
  as a hygrometer, 247
Strobilanthes anisophylla, 51
style, definition of, 270
styleworts, 48
Stylidium, 5, 45–8, 90, 92, 280, 284
  growing conditions for, 266–7
Suda, S., 217–18
sundew, 5, 74–5, 78, 85–6, 281, 282
  growing conditions for, 258
  reaction time of, 274
  sense of taste in, 75, 241–3
  sense of touch in, 241–3, 276
Szent-Gyorgyi, Albert, 149, 151

Tagetes erectus, 127
tamarind tree, 129
Tamarindus indica, 129
Tanada, T., 138, 183
tannin, 87–9
tannin vacuoles, 104, 106, 142, 145, 146, 223
taste, definition of, 8, 269

Taylorson, R.B., 175, 176
Tecoma, 51, 280
  growing conditions for, 267
Theophrastus, 35, 129–30
Thesium alpinum, 127
Thimann, Kenneth, 85
Thiospirillium jenemie, speed of, 277
Tillandsia, 7, 127
Tokufumi, H., 217–18
Tompkins, P., 203
Toong, Y.C., 192
Toriyama, H., 102, 104, 105, 106, 201
Torpedo electricus, 234–5
Torrenia fournieri, 51, 280
touch, definition of, 8, 269
touch-me-not, 26, 281
Tracheophyta, 231
Tradescantia, see spider plant
transpiration stream, definition of, 269
Tremella, 148
triffid, 1, 232
Tronchet, A., 219
tropomyosin, 152–3
troponin, 152–3
truffles, 196
turgorins, 141
tutsan-leaved dogbane, 41
twayblade, 32
Typha latifolia, 194
typhasterol, 194

Ueda, T., 69
Ulva, chloroplast movements in, 134, 163
Umrath, K., 35, 101–2, 180, 220
Urtica dioica, see stinging nettle
Utricularia, 49, 280, 281, 284
  growing conditions for, 267
  see also bladderwort

vacuoles, 3, 4
Vallisneria spiralis, 162, 248–9
Van der Pijl, L., 60
variation potential, 100
vascular tissue, definition of, 270
Vaucheria, 118

Venus's flytrap, 10, 75–92, 100, 234, 281, 282
  anaesthesis of, 174
  animal-like behaviour of, 2, 204, 233
  ATP in, 45
  communication in, 220
  digestive glands of, 218, 237
  electrical activity in, 14, 37, 79–92, 98–100, 174, 233, 235, 244, 278
  events leading to the trap closure, 83
  evolution of, 224, 227
  growing conditions for, 259
  growth movement of, 221
  memory in, 199–201, 236, 240
  nerve-like behaviour of, 1, 12, 14, 79–80
  reaction time of, 274
  sense of touch in, 211, 240–1
  sensor cells of, 90
vesicles, 3
vision
  definition of, 269
  evolution of, 113–14
Visscher, S.N., 193
Vogelmann, T.C., 128–9
Volta, Alessandro, 235
Volvox, 10, 233
  phototaxis in, 249
voodoo lily, heat generation in, 177–9

Waller, J.C., 138
Warburg, Otto, 44
Wardia, 21
Warren, John Collins, 173
Watanabe, S., 129
waterwheel plant, see *Aldrovanda vesiculosa*
Weber, W., 75
Webster, J., 19
Weintraub, M., 104
Weisenseel, M.H., 156
white blood cell, 230
  action potentials in, 69
  actomyosin in, 158
Williams, S.E., 80, 85, 86, 90, 92, 204
Wilson, E.B., 228
Wilson, J.M., 94
Wolken, J.J., 115–16
wood sorrel, see *Oxalis*

xylem, definition of, 97, 270

Yen, L.-F., 221

Zebrina pendula, 248
Zucchini, 212
Zurzycki, J., 277